Oxford Figures

Francis Bacon's *Advancement and proficience of learning* was translated into English by a Fellow of Lincoln College and published in Oxford in 1640, with a richly symbolic title page in which Oxford's pillar (on the left) is in sunlight while Cambridge's (on the right) is in shadow. Between them the ship of human intellect sets sail above a quotation from the Book of Daniel, 'many shall run to and fro and knowledge shall be increased.'

Oxford Figures

800 Years of the Mathematical Sciences

Edited by

John Fauvel
Raymond Flood
and
Robin Wilson

OXFORD

UNIVERSITY PRESS

OXFORD
UNIVERSITY PRESS

Great Clarendon Street, Oxford, OX2 6DP

Oxford University Press is a department of the University of Oxford.
It furthers the University's objective of excellence in research, scholarship,
and education by publishing worldwide in

Oxford New York

Athens Auckland Bangkok Bogotá Buenos Aires Calcutta
Cape Town Chennai Dar es Salaam Delhi Florence Hong Kong Istanbul
Karachi Kuala Lumpur Madrid Melbourne Mexico City Mumbai
Nairobi Paris São Paulo Singapore Taipei Tokyo Toronto Warsaw

and associated companies in Berlin Ibadan

Oxford is a registered trade mark of Oxford University Press
in the UK and in certain other countries

Published in the United States
by Oxford University Press Inc., New York

A catalogue record for this book is available from the British Library

Library of Congress Cataloging in Publication Data
Oxford Figures: 800 Years of the Mathematical Sciences / edited
by John Fauvel, Raymond Flood and Robin Wilson.
ISBN 0 19 852309 2
1. Mathematics–Study and teaching
(Higher)–England–Oxford–History.
2. Mathematics–England–Oxford–History.
3. Mathematicians–England–Oxford–Intellectual life.
4. Mathematicians–England–Oxford–Social life and customs.
5. University of Oxford–History. I. Fauvel, John. II. Flood, Raymond.
III. Wilson, Robin J.
OA14.G73 0947 1999   510'.71'142574–dc21   99–27797

Typeset by EXPO Holdings, Malaysia
Printed in Great Britain
on acid-free paper by
The Bath Press.

Foreword

Today mathematics is flourishing at Oxford as never before. The Savilian professorship of geometry, of which I have the honour to be a recent holder, was one of a number of Chairs established in the Renaissance, more than half-way through the story of Oxford mathematics.

But this is not just a story about the pursuit of mathematical research which happens to be taking place in an old town in southern England. It is a story of the intellectual and social life of a community—indeed, of the interaction of several communities: the mathematical community, the students and townsfolk of Oxford, and the wider scientific community in Britain and throughout the world. For 800 years, young men (and latterly young women) have come to Oxford from far and wide to study mathematics. Some have then moved on, with their new knowledge and training, while others have stayed at Oxford to further their studies and to contribute to the education of the next generation.

As this book reveals, the subject matter covered by the term 'mathematics' has changed profoundly down the ages. I doubt whether any of today's Oxford mathematicians can understand the mathematics taught and researched at Merton College in the fourteenth century—and I think that even Sir Henry Savile, the founder of the Savilian Chairs, would be impressed at the sophistication and power of the mathematics that today's undergraduates are expected to have studied before gaining their degrees. Some areas of their study he would not recognize—and I fear that he would disapprove of the way that some areas of the curriculum have disappeared. His beloved Euclid is no longer taught at Oxford—and who is to say that students today actually have a deeper geometrical insight than those whom Sir Henry Savile taught? Today's students can do many more things and can find many more answers, but that is not the same thing.

Of course, change through history is not a smooth upwardly progressive development. The study of mathematics at Oxford has had many ups and downs. There have been periods of widespread intellectual endeavour. There have been times when many scholars have joined efforts to move subjects forward, and times when only one or two isolated scholars and dedicated teachers have kept the flag flying. But characteristic of many periods have been the mathematicians at the vanguard of those seeking to find and preserve manuscripts and papers from earlier times, and have prosecuted their studies in full awareness of participating in a long historical tradition, a story of vigour, energy, and the pursuit of truth.

Research into the history of mathematics, and of science generally, is leading to a much better picture of the stages that have led up to the achievements of today. Through the different aspects of the history of Oxford mathematics described by the contributors to this book, we can see how the subject developed, century by century, and may be led to speculate as to what the next eight centuries may bring.

I. M. James
Mathematical Institute, Oxford

CONTENTS

CHAPTER I

800 years of mathematical traditions

John Fauvel

If the Elizabethan antiquary William Camden is to be believed, the teaching of mathematics at Oxford, along with the foundation of the University itself, dates from the time of Alfred the Great in the ninth century:

> In the year of our Lord 886, the second year of the arrival of St. Grimbald in England, the University of Oxford was begun ... John, monk of the church of St. David, giving lectures in logic, music and arithmetic; and John, the monk, colleague of St. Grimbald, a man of great parts and a universal scholar, teaching geometry and astronomy before the most glorious and invincible King Alfred.

By such a reckoning, we could now be celebrating over 1100 years of mathematical traditions at Oxford. Alas, the founding of the University by King Alfred is but one of the stories of Oxford's mythical past for which there is now considered to be insufficient historical evidence. The University of Oxford can, however, safely trace its existence back over at least 800 years, which confirms its respectable antiquity while recognizing it as younger than the universities of Bologna and Paris.

The University seems to have come into being gradually during the twelfth century, as groups of scholars gathered to learn and study together at 'that catarrhal point of the English Midlands where the rivers Thames and Cherwell soggily conjoin', in the evocative words of Jan Morris. In 1188, the historian Gerald of Wales visited Oxford and recorded the existence of several faculties of study. This date may stand for the start of the University as well as any, not least because it happily matches that of a major fund-raising appeal by the modern University launched in 1988. Throughout the succeeding eight centuries, the mathematical arts and sciences have been pursued at Oxford, taught, used, and explored in a variety of ways.

The medieval heritage

From its early years, Oxford has drawn upon mathematical influences on its ways of thinking. By comparison with the University of Paris,

The earliest mathematical book to be published in Oxford was *Compotus manualis ad usum Oxoniensium*, printed by Charles Kyrforth in February 1520. Slight and unoriginal in mathematical content, its opening image records astronomical instruments as well as bearing evidence of the scribal culture upon which Oxford built its traditions of scholarship.

Portrait of Robert Grosseteste, Oxford's first Chancellor and a strong influence on the early intellectual life of the University, from a fourteenth-century manuscript.

Oxford in the Middle Ages was more strongly influenced by a tradition of Christian Neoplatonism. In this tradition (as for Plato himself) mathematics played an important role, as revealing God's design for the universe. An early Oxford teacher, and Chancellor of the University, was Robert Grosseteste. In what amounts to a research programme for investigating the world through mathematical, specifically geometrical, modelling, he wrote:

The usefulness of considering lines, angles and figures is the greatest, because it is impossible to understand natural philosophy without these ... For all causes of natural effects have to be expressed by means of lines, angles and figures, for otherwise it would be impossible to have knowledge of the reason concerning them.

Oxford scholars of succeeding generations would, in effect, attempt to put this programme into practice. Grosseteste's most famous admirer, the Franciscan friar Roger Bacon, went even further in proclaiming the importance of mathematics:

He who knows not mathematics cannot know the other sciences nor the things of this world ... And, what is worse, those who have no knowledge of mathematics do not perceive their own ignorance and so do not look for a cure. Conversely a knowledge of this science prepares the mind and raises it up to a well-authenticated knowledge of all things.

It was at Oxford in the next century that some of the most sophisticated mathematical discussions of the Middle Ages took place, in which mathematics was used and developed along the lines of the programme that Grosseteste and Bacon had advocated. By this time there was also a strong Aristotelian influence upon Oxford's mathematical studies.

Roger Bacon's study on Folly Bridge later became a place of pilgrimage for scientists. Samuel Pepys visited it in 1669, remarking: 'So to Friar Bacon's study: I up and saw it, and gave the man 1s ... Oxford mighty fine place.' It was pulled down in the eighteenth century for road widening.

Most of the scholars concerned with this development were Fellows of Merton College, founded in 1264. Prominent in this group was Thomas Bradwardine, who showed, in effect, how the programmatic utterances of Grosseteste and Bacon could bear real fruit in the mathematical analysis of nature: 'it is mathematics which reveals every genuine truth, for it knows every hidden secret'. In the hands of Bradwardine and his fellow scholars, the mathematization of magnitudes went beyond anything thought of in Greek geometry. Members of the Merton School attempted to quantify intensities of light, heat, sound, hardness, and density, as well as magnitudes that mathematicians have still not managed to quantify effectively: knowledge, certitude, charity, and grace.

How far such developments would have gone we cannot know. Bradwardine died of bubonic plague (the 'Black Death') in 1349. Many other Oxford scholars and at least a third of the population of England died of the same cause over those terrible years, and the University rapidly fell into a decline mirroring that of the country. Despite the ravages and devastation of the Black Death and other social and political miseries, teaching and learning did continue, albeit in a rather

Merton College's fourteenth-century Mob Quad has seen six centuries of mathematical activity.

depressed state, but not for three centuries did Oxford even begin to regain the intellectual vigour of its glorious early days.

The tradition of the mathematical sciences

If one had to nominate the single most influential event of Oxford's long mathematical history, it would be the founding by Sir Henry Savile of two endowed Chairs in 1619. Through the Savilian Chair of Geometry and the Savilian Chair of Astronomy a conception of the breadth of the mathematical sciences, as well as a tradition of how to teach and study and further them, was passed down from generation to generation. Although in an institutional sense the Chairs began to drift apart in the nineteenth century, under the pressure of examinations and shifting faculty alliances, research into pure mathematics and applied and cosmological mathematics has remained a notable dimension of Oxford activity up to the present day.

Throughout much of the period after the founding of the Savilian Chairs, the two roles seem to have been more or less interchangeable: a leading Oxford mathematician would generally be appointed to the next Chair to fall vacant and could tilt his interests accordingly, or indeed carry on regardless. It can be surprising now to notice the range of many of the Savilian professors, able to write as knowledgeably on music theory as on architecture, and observe an eclipse as attentively as preparing scholarly mathematical editions for the press. To some extent this breadth, shown also in the interchangeability of Chairs, reflects the breadth achieved by any well-trained professional mathematician of the time. One would find the same set of interests among leading math-

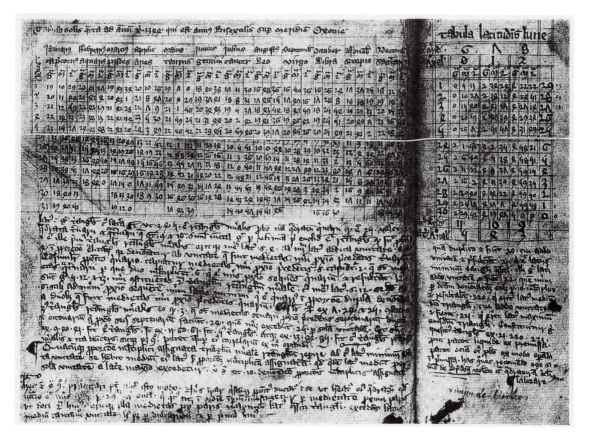

Oxford almanac for 1344, with notes on square numbers written by the Mertonian Simon Bredon. Owners of the manuscript have included John Dee and Thomas Allen.

ematicians throughout Europe, so that is not in itself exceptional—except that the size and continuity of the Oxford tradition has enabled the tradition of the mathematical sciences to flourish, with more or less vigour, over such a long period.

The early seventeenth century saw also the founding by Sir William Sedley of an endowed Chair in Natural Philosophy; this is, in effect, what we would now call physics, or perhaps applied mathematics. Holders of the Sedleian Chair—the first, Edward Lapworth, was appointed in 1621—contributed further to the range of mathematical sciences taught in Oxford, notably in the eighteenth century when they shared in promoting Newtonian science alongside the Savilian professors: indeed, for the latter part of that century Thomas Hornsby managed to fill both the Sedleian Chair and the Savilian Chair of Astronomy, as well as various other positions, for a period of nearly 30 years. Over the past century the Sedleian professor has been a distinguished applied mathematician, to give leadership to that side of Oxford's mathematical teaching and research activity.

Besides the Sedleian Chair, scientific principles were taught from the early eighteenth century onwards by a reader in experimental philosophy, a function discharged for over a hundred years, from 1730 to

1838, by one or other of the Savilian professors who tended to accumulate offices so long as professorial income was derived largely from student fees. It was while David Gregory was Savilian Professor of Astronomy in the 1690s that the style was developed in Oxford of teaching Newtonian science demonstratively through experiments, with lectures given in a room on the first floor of the Ashmolean Museum, now the Museum of the History of Science.

Further subjects, too, have entered the purview from the latter part of the nineteenth century. Following an abortive attempt in the mid-Victorian period to introduce statistics into Oxford studies, through the unlikely alliance of Florence Nightingale and Benjamin Jowett, mathematical statistics finally reached Oxford in the 1890s, in effect, in the researches of the professor of political economy, Francis Edgeworth. But only in the 1930s, with the founding of the Institute of Statistics, was the subject officially integrated within the University, and it was not until 1948 that a professor of statistics was appointed, David Champernowne. During the latter twentieth century another subject too, computer science, grew in significance and came to play an important role in Oxford's academic provision. Without claiming computer studies as a part of mathematics, the Oxford style of doing computer science has had a philosophical and logical cast which seems a continuation under another name of the long tradition of Oxford mathematical sciences.

The logical tradition

The history of logic has long been intertwined with that of mathematics, while never quite forming the same area of inquiry. In the great medieval pedagogical classification of subjects into the trivium and the quadrivium, logic, or 'dialectic', was part of the trivium, alongside grammar and rhetoric, to be studied before and more prominently than the four mathematical subjects of the quadrivium—arithmetic, geometry, music, and astronomy. At Oxford, logic has been taught throughout the history of the University, sometimes by the same people as taught mathematical sciences. At the research level, the great Mertonians of the fourteenth century, such as Thomas Bradwardine, Richard Swyneshead, William of Heytesbury, John Dumbleton, and Richard Billingham, reached unparalleled heights of sophistication and insight into logical matters, arguably not to be reached again for another five centuries, and by the end of the century their influence had spread across Europe.

In the succeeding centuries logic continued to be taught as a fundamental part of the general education which Oxford hoped to instil in all undergraduates. In the words of John Wallis, the Savilian Professor of Geometry who himself wrote a textbook on logic in 1687, the purpose of teaching logic was to lay 'the foundations of that learning, which they are to exercise and improve all their life after', explaining its merits as being

INSTITUTIO
LOGICÆ,
Ad communes ufus accommodata.

Per *JOHANNEM WALLIS*, S.T.D.
Geometriæ Profefforem Savilianum, Oxoniæ;
& Societatis Regalis Londini Sodalem.

Editio Quarta, auctior & emendatior.

OXONII,
Typis *Leon. Lichfield*, Impenfis *Henr. Clements*,
*Ant. Peifley, J. & S. Wilmot, Steph. Kiblewhite,
Steph. Fletcher,* & *Edw. Whiftler.* 1715.

John Wallis's *Institutio logicae*, first published in 1687, was intended to provide a foundation of undergraduate learning.

to manage our reason to the best advantage, with strength of argument and in good order, and to apprehend distinctly the strength or weakness of another's discourse, and discover the fallacies or disorder whereby some other may endeavour to impose upon us, by plausible but empty words, instead of cogent arguments and strength of reason.

For almost as long as logic has been taught at Oxford, there have been complaints about it from students, both during their studies and in retrospect. Besides perennial charges that it was too dry and boring, a long-debated point of substance was whether logic and mathematics were both necessary for mind-training purposes, or which was preferable. An anonymous writer in 1701 (possibly the 1690s undergraduate John Arbuthnot) clearly differentiated what he saw as the advantages of each, with the moral victory to mathematics:

Logical Precepts are more useful, nay, they are absolutely necessary, for a Rule of formal Arguing in public Disputations, and confounding an obstinate and perverse Adversary, and exposing him to the Audience or Readers. But, in the Search of Truth, an Imitation of the Method of the Geometers will carry a Man farther than all the Dialectical Rules.

While the pedagogical arguments between logic and mathematics continued for another two centuries, mathematics gradually superseded logic, in effect, as an instrument for training the mind (and strong views were expressed, too, on behalf of classics as the only effective mind-sharpening study). By the late nineteenth century, Oxford students who were in line for a pass degree had, besides offering Latin and Greek, to choose between mathematics (quadratic equations and some propositions from Euclid) and logic. A guide for undergraduates advised caution in making the choice:

Logic will, most probably, be untried ground to the student, and we advise him to consider well before he chooses this in preference to Mathematics … he must be prepared for hard, and, perhaps, uninteresting reading, demanding not only close attention, but fair powers of memory.

The late nineteenth century also saw a resurgence of interest in logical research, to which the Christ Church mathematics lecturer, Charles Dodgson (Lewis Carroll), made notable and influential contributions, sometimes underrated on account of their apparent whimsicality. In exploring new logical ideas as well as how to teach logic more effectively, Dodgson's work can be seen to lie within a long tradition of Oxford logical activity. Throughout the twentieth century logic continued to form a part of Oxford concerns, moving more into the area of mathematical philosophy.

The instrumental tradition

Oxford is fortunate in having a most impressive collection of mathematical and scientific instruments, situated in one of its finest seventeenth-century buildings. The Old Ashmolean building was opened in

Lewis Carroll's game of *"Logic"* (1886) was an attempt to teach the principles of the syllogism. It proved most effective when he himself was doing the teaching, which he did at women's colleges and girls' schools in Oxford.

Alfred Swinburne's *Picture logic* of 1875 employed vivid imagery to help Oxford undergraduates acquire a rudimentary logical understanding: 'Large or small, young or old, flesh or stone, you must all pass through, for Professor Logic isn't particular about the matter, all he concerns himself with is the form.'

THE LOGICAL SAUSAGE MACHINE.

Cista mathematica, the great triple-locked
mathematical instrument chest, was part of
Sir Henry Savile's provision for the study
of mathematical sciences at Oxford. The
historian R. T. Gunther speculated 'whether
at the time of its making all the mathematical
learning within the University was to be
compressed within its eleven feet of capacity'.

1683, as a museum and centre of scientific activity, and returned to housing what became the Museum of the History of Science in 1925. Although few of the instruments were made locally, and not all of them were used in the University, the presence of the collection is a strong reminder that an instrumental dimension has always accompanied mathematical teaching and research in Oxford.

Many instruments associated with Oxford have been associated with astronomical concerns in a calculational or observational way, from Richard of Wallingford's remarkable albion in the fourteenth century, to the instruments constructed by John Bird for the Radcliffe Observatory four and a half centuries later. There are a large number of sundials around Oxford, some designed by early Savilian professors. Other instruments, extant or known about from a variety of records, were used for mathematical calculation or in connection with the range of applied mathematical sciences taught at Oxford as part of a liberal education.

Mathematical models too, in the sense of physical objects showing geometrical shapes, have long accompanied teaching. An early such model, perhaps used by the Savilian Professors of Geometry in the seventeenth century, takes the form of a five-sided alabaster column, each face representing one of the classical orders of architecture, with a display of models of the five Platonic solids. This association between the five classical orders and mathematical learning is found again in the 'Schools quadrangle', where the early Savilian professors taught, and was referred to in a course syllabus issued by David Gregory in 1700 proposing a practical geometry lecture on 'the five orders of pillars and pilasters'. A visitor to the Bodleian Library in 1710 described the alabaster model as one of the sights of Oxford. The plaster mathematical models bought by the University in the late nineteenth century for geometry teaching are a more recent example of this tradition.

The antiquarian tradition

Later ages would look back and see the early fourteenth century as a pinnacle of scholarly achievement. The predominance of Oxford thought in medieval scholarship is particularly evident from the studies of the sixteenth-century textbook writer Robert Record, himself an Oxford graduate and a Fellow of All Souls College. Record's interest in, and careful reading of, the manuscripts of his predecessors exemplify another tradition of Oxford scholarship. A passage from *The castle of knowledge* (1556) illustrates Record's understanding of the nation's mathematical past. In the course of a survey of earlier cosmographical writings, he remarks:

Dyvers Englyshe menne have written right well in that argument: as
Grostehed, Michell Scotte, Batecombe, Baconthorpe and other dyvers,
but fewe of their bookes are printed as yet ...

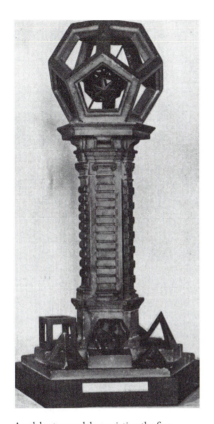

An alabaster model associating the five classical orders of architecture with the five Platonic solids, probably used in the seventeenth century for geometry teaching.

The pillars decorating the central tower of the Schools quadrangle (1613–24) correspond to the five orders of architecture. At this end of the quadrangle the Savilian professors lectured, the professor of astronomy setting up his instruments in the tower.

At least three of these names are of Oxford figures. We have already met Grostehed—that is, Grosseteste (Old French for 'big head'). The Carmelite scholar John Baconthorpe, who is supposed to have been the great-nephew of Roger Bacon, was educated at Oxford and Paris, and taught at Oxford in the early fourteenth century. Although his teachings were mostly theological, he wrote a treatise on astronomy which Record seems to have known something about. The third name mentioned by Record, William Batecombe, is believed to have studied at Oxford and taught mathematics, but little is known about him other than that a manuscript ascribed to him was in Robert Record's library. There is no good evidence that the fourth name mentioned by Record, the early thirteenth-century translator Michael Scot, was associated with Oxford, although some manuscripts of his work are in the

Robert Record's cosmology textbook *The castle of knowledge* demonstrated a humanist scholar's respect for and knowledge of the past.

Bodleian Library. Nevertheless, the fact that three of the four names cited by Record and stretched over three centuries were of Oxford mathematical writers is a small pointer to the supremacy of Oxford in English mathematical life in the medieval period.

When in the sixteenth and seventeenth centuries mathematicians and other scholars looked back towards a distant golden age of British learning, it was the reputation of Oxford scholars that came to mind, particularly those working in the century from Grosseteste to Bradwardine. The very activity of seeking to preserve and reinterpret the records of the past is itself a feature of academic life in which

Oxford has played a major role, both in respect of its own past and that of the various disciplines which have been cultivated at the University; this was particularly evident during the seventeenth century. The national cataclysms of those times—notably the dissolution of the monasteries a hundred years before, and widespread unrest during the Civil War in which Oxford took the losing, Royalist, side—produced among scholars an acute sense of the vulnerability of documents and the need to preserve and treasure them.

Thomas Allen was such a scholar. One of the most influential of Oxford figures from the 1560s until his death in 1632, he began collecting manuscripts at a time when the monasteries had but recently been dispersed, and when some Oxford colleges were replacing their older manuscript collections with printed books. It is in great measure due to Allen that the Bodleian Library, itself a product of these years, has one of the largest collections of manuscripts reflecting English monastic learning, including an exceptional record of medieval mathematics and science. His successors, such as the antiquaries John Aubrey and Anthony à Wood, continued to show a deep and vigilant concern for the fate of manuscripts as they preserved records and testimony of all kinds (and levels of reliability) to ensure that knowledge about their contemporaries, as well as scholars of earlier generations, was not lost. It is thanks to Aubrey and Wood that we have such detailed accounts of many mathematicians of the sixteenth and seventeenth centuries.

The historical research tradition

During the seventeenth century the Oxford antiquarian interest in preserving the records of the past became subsumed in an interest in the history of mathematics itself. A number of influences came together to create a historical approach to mathematics which has characterized the Oxford tradition. Sir Henry Savile was crucial: in laying down statutes for the Chairs of geometry and astronomy, Savile gave instructions for the subjects to be taught by exegesis of the great texts of the past, principally Euclid's *Elements* and Ptolemy's *Almagest*. Another influence on the style of mathematics was the creation at this time of the Bodleian Library, by Savile's friend Sir Thomas Bodley. This, together with a smaller library attached to the Savilian Chairs, ensured the availability of many books and manuscripts for scholarly purposes.

The influence of the time is also seen in the spirit of investigation evident in the work of the Royal Society, founded in 1660 by a group of people many of whom had Oxford connections. In the activities of the Royal Society the history of a subject of concern was an essential part of how it presented itself for understanding, categorization, or analysis. Thus, when John Wallis wrote his great *Treatise of algebra* (1685) as a study of algebra seen through its development from earliest times, it

Throughout the history of Oxford, libraries have played a central role in mathematical teaching and scholarship: left, Merton College Library, 1370; right, the Whitehead Library in the Mathematical Institute, 1960.

was at once the first algebra text on this scale and the first history of mathematics written in English. The way in which Wallis presented mathematics as something intrinsically bound up with its history can be seen as a development particularly characteristic of Oxford, in which knowledge from the past is explored and built upon in order to make future progress, as much in mathematics as in any other subject.

Another way in which the Oxford historical approach to mathematics showed itself was in the production of a number of major editions of ancient mathematical texts, edited by senior University mathematicians under the auspices of the University Press. This began in the 1640s, with an edition of Arabic astronomical tables worked on by the first two Savilian Professors of astronomy. This kind of production was another consequence of Savile's insistence that his professors know about the past in order to make future progress. He recognized the extent to which astronomy, in particular, is an intrinsically historical discipline, in which records made by earlier observers are a crucial dimension of current analysis and understanding.

The tradition of major editions persisted for some time after the age of Wallis, but it was not until the nineteenth century that another Oxford mathematician showed comparable historical interest and enthusiasm. Stephen Rigaud, Savilian Professor, successively, of Geometry and Astronomy from 1810 to 1839, played a major part in the renaissance of the history of mathematics—of Newtonian studies, in particular—that was taking place in the early nineteenth century. Besides more general themes, he took pains to explore Oxford's own contributions to historical scholarship. In a paper to the Ashmolean Society in 1836, for example, Rigaud described the textual history of

ΑΡΧΙΜΗΔΟΥΣ
ΤΟΥ
ΣΥΡΑΚΟΥΣΙΟΥ
Ψαμμίτης,
καὶ Κύκλε Μέτρησις.
ΕΥΤΟΚΙΟΥ ΑΣΚΑΛΩΝΙΤΟΥ,
εἰς αὐτὴν ὑπόμνημα.

ARCHIMEDIS
SYRACUSANI
Arenarius,
Et Dimensio Circuli.
EUTOCII ASCALONITÆ,
in hanc Commentarius.

Cum Versione & Notis *Joh. Wallis*, SS.Th.D.
Geometriæ Professoris Saviliani.

OXONII.
E Theatro SHELDONIANO. *1676.*

ΑΡΧΙΜΗΔΟΥΣ
ΤΑ ΣΩΖΟΜΕΝΑ
ΜΕΤΑ ΤΩΝ
ΕΥΤΟΚΙΟΥ ΑΣΚΑΛΩΝΙΤΟΥ
ΥΠΟΜΝΗΜΑΤΩΝ.

ARCHIMEDIS
QUÆ SUPERSUNT OMNIA
CUM
EUTOCII ASCALONITÆ COMMENTARIIS.

EX RECENSIONE
JOSEPHI TORELLI, VERONENSIS,
CUM NOVA VERSIONE LATINA.

ACCEDUNT
LECTIONES VARIANTES EX CODD. MEDICEO ET PARISIENSIBUS.

O X O N I I,
E TYPOGRAPHEO CLARENDONIANO.
MDCCXCII.

ON

THE ARENARIUS

OF

ARCHIMEDES.

BY

STEPHEN PETER RIGAUD, M. A.
SAVILIAN PROFESSOR OF ASTRONOMY.

OXFORD,
PRINTED BY S. COLLINGWOOD, PRINTER TO THE UNIVERSITY, FOR
THE ASHMOLEAN SOCIETY.
MDCCCXXXVII.

Treatments of a work of Archimedes over two centuries from 1676, 1792, and 1837 illustrate the Oxford tradition of historical scholarship.

Archimedes' *Arenarius* or *Sand-reckoner*: this work has borne an Oxford imprint since Wallis's critical edition of 1676, incorporated by Torelli in his 1792 Oxford *Archimedes*, using manuscripts found in Oxford libraries from the early sixteenth century to the late eighteenth century. He pointed out, too, that one of the few English editions of this work was made by a Wadham College undergraduate, George Anderson, in 1784.

Rigaud's interest in history had been aroused by a failed attempt at the end of the eighteenth century to sustain the Oxford tradition of publishing important historical mathematical editions. In 1784 manuscripts of Thomas Harriot—perhaps the most highly regarded Oxford-educated mathematician since the Middle Ages—were rediscovered in a country house in Sussex, having been thought lost since the 1630s. An edition of the more significant papers was proposed, to be published by the University Press, and the papers were examined by Abraham Robertson, shortly to become the Savilian Professor of Geometry. He reported back that the papers were in no state for publication and would not contribute to the advancement of science. But when Rigaud took over the Savilian Chair of Astronomy and re-examined them 30 years later, he realized that Harriot's papers were indeed rich in content and full of interest; he worked on them to some effect, but his death prevented a major opus.

To bring the story up to date, interest in the Harriot papers eventually revived in the 1960s and a Thomas Harriot Seminar was founded,

Robert Gunther, the founder of the Museum of the History of Science, pictured in about 1930 with a wood and pasteboard astrolabe by the seventeenth-century Oxford instrument maker John Prujean; this astrolabe is illustrated on page 72.

meeting initially in Oxford, now in Durham and Cambridge. An article published in 1969 by some of those involved described the Harriot papers as 'one of the most important bodies of unpublished English scientific manuscripts, and probably the most complete collection of working observations and calculations revealing the scientific art, in one of its most original practitioners'. But the problem still remains of reducing them to order and putting them in a fit state for publication.

The history of science, as something distinct from but closely related to the history of mathematics, was pursued in various ways in twentieth-century Oxford. A relatively small area of interest tends to be buffeted in the storms of institutional politics, and it was only with considerable struggle that Robert Gunther was able to establish the Museum of the History of Science in the Old Ashmolean building, based on the scientific instrument collection of his friend Lewis Evans. Gunther's 14 volumes on early science in Oxford (1923–45) contain a considerable amount of material on which subsequent scholarship could build. In 1972 a Chair in the history of science was established at Oxford, associated with the modern history faculty.

Oxford's tradition of historical research into the history of the mathematical sciences has been extended by the award of doctoral degrees for research into the subject. The 1990s saw Julia Nicholson's thesis on the development of group theory, Eileen Magnello's on the pioneering statistician Karl Pearson, and Eleanor Robson's on mathematics in ancient Mesopotamia. Besides observing that all these degrees were gained by women, it is worth noting that each was done within a different area of the University: in the Faculty of Mathematics, the

CJmpzeſſum eſt pzeſens opuſculũ in ce=
leberrima vniuerſitate Ozonienſi p
me Carolum Kyzfoth. Jn vico
diui Joãnis baptiſtemozã
trahẽte Innodñi.M.D.rir.Meſis
vero februarij,die,v.

Compotus manualis.

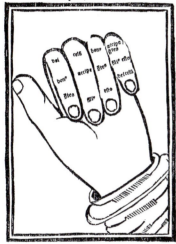

Si forſan neſcis que littera tu vſil ſit
Et quotus eſt ſolis annus 2gnoſcere queris
Annos a vſii:denas ãnis przius octo
Perqʒ quater ſeptrin vſii diuiſeris annos
Pat quotus iſt ſolis ãnus quirquid remanebit

Compotus manualis ad usum Oxoniensium, 1520.

Wellcome Unit for the History of Medicine, and the Oriental Institute, respectively. Thus scholarly concerns for the history of mathematics are found throughout the University.

The publishing tradition

The earliest book with some slight mathematical content to be printed at Oxford seems to have been *Compotus manualis ad usum Oxoniensum*, printed by Charles Kyrforth in 1520. This 16-page booklet, intended explicitly for Oxford students, explained how to make calculations for the date of Easter and other movable Christian feasts by counting on the hand. It took a further century for another mathematical book to be published in Oxford, Sir Henry Savile's lectures on Euclid's *Elements*, printed by John Lichfield and James Short in 1621.

Mathematical works only really began to be published in Oxford some time after Archbishop Laud's statutes of the 1630s laid the framework for a University Press. The years following the Civil War, from 1647 to the Restoration in 1660, marked one of the most productive periods for the printing of mathematical texts in Oxford. In 1648 *Canicularia*, an edition of Arabic astronomical tables by the first Savilian Professor of Astronomy, John Bainbridge, and completed by his successor John Greaves, was printed using Arabic type purchased from Leyden at Laud's instigation. John Wallis, the newly appointed Savilian Professor of Geometry, was thereafter the leading figure in Oxford's mathematical publishing. Under his guidance an edition of William Oughtred's innovative algebra text *Clavis mathematicae* was published in 1652, as well as his own *Arithmetica infinitorum* and other research works in the late 1650s.

The latter half of the century saw the University Press develop its printing activities, making use initially of the newly built Sheldonian Theatre, designed by the Savilian Professor of Astronomy, Christopher Wren. In the 1690s Wallis's mathematical works were published in three massive volumes, including his editions of some classical mathematics texts. In the next decade the Press embarked on a series of major mathematical editions: David Gregory's edition of Euclid's *Elements* (1703), Edmund Scarburgh's *English Euclide* (1705), and Edmond Halley's edition of Apollonius's *Conics* (1710), as well as a number of smaller works. Little of this kind was published for much of the eighteenth century, but after a long gap an edition of the third important classical Greek mathematician was published in 1792—Torelli's edition of the collected works of Archimedes.

The major nineteenth-century influence upon the Oxford University Press was a mathematician. Bartholomew Price, Sedleian Professor of Natural Philosophy, was Secretary to the Delegates—in effect, the chief executive of the Press—from 1868 to 1884. Described by one of OUP's historians as 'the architect and founder of the modern Press', Price

Cl. V. Iohannis Bainbrigii,
Aftronomiæ,
In celeberrimâ Academiâ Oxonienfi,
Profeffioris Saviliani,
CANICVLARIA.

Unà cum demonftratione
Ortus *Sirii* heliaci,
Pro parallelo inferioris Ægypti.

Auctore Iohanne Gravio.

Quibus accefferunt,
Infigniorum aliquot Stellarum Lon-
gitudines, *&* Latitudines,

Ex Aftronomicis Obfervationibus
Vlug Beigi,
Tamerlani Magni nepotis.

OXONIÆ,
Excudebat Henricus Hall, Impenfis
Thomæ Robinson, 1648.

Canicularia, by the first two Savilian professors of astronomy, was one of the first Oxford books to use Arabic type.

exemplified to a remarkable degree the Oxford 'amateur' tradition whereby an academic of the right calibre with no special training successfully assumes a leading executive function. Under Price's guidance and inspiration, the Press blossomed into a major modern publishing house; one of the early titles in his newly created Clarendon Press series was the *Treatise on electricity and magnetism* by the Cambridge physicist James Clerk Maxwell. Price was not the only mathematician contributing to the management of the Press. Henry Gerrans, a Fellow and tutor of Worcester College, was a Delegate for a long period into the twentieth century, and took particular interest in supporting educational books.

In 1930 the University Press took on the publishing of a major mathematical journal, the *Quarterly Journal of Mathematics*. This had been published in Cambridge hitherto, under the increasingly eccentric editorship of James Glaisher, its editor for half a century from 1878 until his death in 1928. On Glaisher's death, G. H. Hardy—at this time Savilian Professor of Geometry at Oxford—successfully urged that both editorship and publication of the journal be transferred to Oxford. An editorial board was formed of three young Oxford mathematicians, Theodore Chaundy, William Ferrar, and Edgar Poole, and the journal went from strength to strength, bolstering along the way Oxford's research reputation within the international mathematics community.

At the same time, the University Press was seeking to develop its mathematics and science publishing side, and took the opportunity to make a thorough overhaul of its mathematical printing. The University Printer, John Johnson, went to considerable pains to study the work of foreign printers, buy new type, adapt and recut type faces, and set specimen pages for comment by a range of leading mathematicians. At that time few printers had experience of setting mathematics by machine, and the result was a major contribution to knowledge and expertise in the subject. The *Quarterly Journal* editor most closely involved with this activity was Theodore Chaundy, who was later to collaborate on a classic text, *The printing of mathematics*. In the words of Johnson's colleague Kenneth Sisam, the eventual result was to reach a standard of mathematical printing worthy of 'the recent renascence of mathematical studies in Oxford'.

The popularizing tradition

Although not formally written into the Savilian or other statutes, several Oxford mathematicians down the centuries have interpreted their role as having a dimension of explaining the purpose and conclusions of their subject to a wider public. Even before the time of Sir Henry Savile, an exposition of the advantages and purpose of studying arithmetic was incorporated in Robert Record's influential textbook *The ground of artes*, in which he described how arithmetic was needed for the study of music, medicine, law, grammar, philosophy, divinity, and for military uses, as well as the more obvious cases of astronomy and geometry.

The refounding in Oxford of the *Quarterly Journal of Mathematics*, in 1930, signalled and promoted the increased research activity of Oxford mathematics led by G. H. Hardy.

An early example of professorial outreach may be seen in the work of Edmond Halley, who explained in 1715 the implications of a forthcoming solar eclipse in a widely distributed handbill, to reassure people who might otherwise 'be apt to look upon it as Ominous, and to interpret it as portending evil to our Sovereign Lord, King George and his Government'. He also used the event to organize careful observations of the precise timings of the eclipse in different parts of the country, perhaps the first systematic scientific observation on such a large scale, whose results are still of value for estimating the size of the Sun.

THE PRINTING OF
MATHEMATICS

AIDS FOR
AUTHORS AND EDITORS
AND RULES FOR
COMPOSITORS AND READERS AT THE
UNIVERSITY PRESS, OXFORD

By

T. W. CHAUNDY
University Reader in Mathematics

P. R. BARRETT
Mathematical Reader at the University Press, Oxford

and

CHARLES BATEY
Printer to the University

OXFORD UNIVERSITY PRESS
LONDON: GEOFFREY CUMBERLEGE
1954

Title page and image from Chaundy *et al.*'s *The printing of mathematics*.

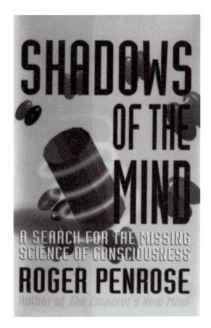

Writing books of popular scientific exposition is a tradition long associated with Oxford mathematicians.

One of Halley's successors as Savilian Professor of Geometry, Baden Powell, was also concerned with the public understanding of science, and through various publications strove to draw attention to

the advantages attending the prosecution of science, of the evils sometimes supposed to be involved in it, of … its connection with physical and still more with moral civilisation; … the state of public opinion respecting its claims; the recognition of scientific instruction as a branch of education.

In the late twentieth century a notable example in the same vein is the popular work of the Rouse Ball Professor of Mathematics, Roger Penrose: indeed, the benefaction which set up this professorship, from the will of the Cambridge mathematician and historian of mathematics Walter William Rouse Ball in 1925, spoke of the hope that the new mathematics Chair 'would not neglect its historical and philosophical aspects'. In his subsequent appointment as Gresham Professor of Geometry Roger Penrose continued another tradition too, of links between Oxford mathematicians and the public lectures set up in the late sixteenth century by the City of London financier Sir Thomas Gresham (see page 80).

The literary tradition

A characteristic of Oxford academic life over almost the whole eight centuries of its existence has been a literary self-awareness, so that we have the benefit of centuries of descriptions of Oxford activities, mathematical and otherwise. An early account of Oxford student life is found in *The Canterbury tales*, written in the late fourteenth century, where Geoffrey Chaucer told a salacious tale involving Nicholas, a poor

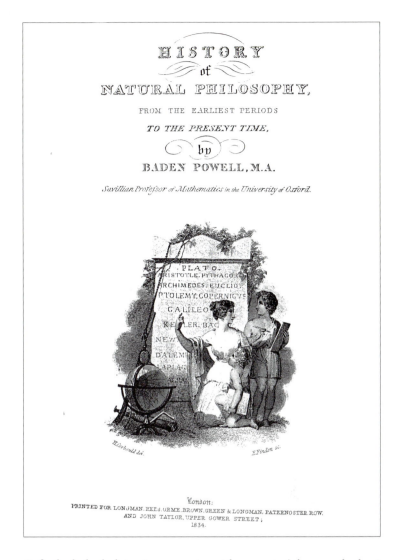

HISTORY
of
NATURAL PHILOSOPHY,
FROM THE EARLIEST PERIODS
TO THE PRESENT TIME,
by
BADEN POWELL, M.A.

Savilian Professor of Mathematics in the University of Oxford.

PLATO.
ARISTOTLE. PYTHAGOR
ARCHIMEDES. EUCLID
PTOLEMY. COPERNICVS
GALILEO
KEPLER. BAC
NE
D'ALEM
LAPLAC

London:
PRINTED FOR LONGMAN, REES, ORME, BROWN, GREEN & LONGMAN, PATERNOSTER ROW,
AND JOHN TAYLOR, UPPER GOWER STREET;
1834.

The nineteenth-century Savilian professor Baden Powell wrote his history of the physical and mathematical sciences as a way of promoting scientific awareness and removing public prejudices and misconceptions about the pursuit of science.

Oxford scholar lodging in a room in a rich carpenter's house, who kept by his bed a copy of Ptolemy's *Almagest*, an astrolabe, and counters for making calculations ('augrim-stones'):

> His Almageste and bokes grete and smale,
> His astrelabie, longinge for his art,
> His augrim-stones layen faire a-part
> On shelves couched at his beddes heed.

Chaucer's knowledge of astronomical instruments and practices was deep. Besides poetry he wrote, in the 1390s, two astronomical texts, *A treatise on the astrolabe* and *Equatorie of the planetis*. These texts are notable, not least for being written in the vernacular (rather than, as hitherto, in Latin), fostering a tradition of writing on mathematical and scientific matters in the English language. The awareness of Oxford science and instruments shown in his writings suggests that Chaucer was educated at Oxford, although there is no extant archival evidence to this effect.

Chaucer was not alone in representing the typical Oxford student as used to using an astrolabe, the slide-rule or electronic calculator of its day. A disputation of 1420 recounted the nocturnal adventures of one Robert Dobbys of Merton, attempting to find his way back to college—a distance of barely a couple of hundred yards if traversed in a direct line:

It is related of him that one night after a deep carouse, when on his way from Carfax to Merton, he found it advisable to take his bearings. Whipping out his astrolabe he observed the altitude of the stars, but, on getting the view of the firmament through the sights, he fancied that sky and stars were rushing down upon him. Stepping quickly aside he quietly fell into a large pond. 'Ah, ah', says he, 'now I'm in a nice soft bed I will rest in the Lord.' Recalled to his senses when the cold struck through, he rose from the watery couch and proceeded to his room where he retired to bed fully clothed. On the morrow, in answer to kind inquiries, he denied all knowledge of the pond.

It was Chaucer, though, who provided the definitive image of an Oxford academic, in his portrayal of the clerk of Oxenford. The final line of his description here stands as a noble image to which many Oxford figures have aspired down the centuries.

Of studie took he most cure and most hede.
Noght o word spak he more than was nede,
And that was seyd in forme and reverence,
And short and quik, and ful of hy sentence.
Souninge in moral vertu was his speche,
And gladly wolde he lerne, and gladly teche.

The mathematical research tradition

While the early days of medieval Oxford represent a golden age of mathematical research on an international level, it was only in the seventeenth century that research began to be developed as an explicit activity in which dons and the better students might be expected to join. A research dimension was implicit in Henry Savile's statutes for the Savilian Chairs, since the professors were expected to develop their subjects alongside their teaching duties, and the early Savilian professors set a high standard in this respect.

John Wallis, in particular, Savilian Professor of Geometry throughout the second half of the seventeenth century, created in effect the first mathematical research school at the University since the Middle Ages, in the sense that he actively encouraged his younger colleagues and former students to solve problems that arose from work he was doing, and found means to publish and publicize their solutions. Younger mathematicians whose work he promoted in this way included Christopher Wren, William Neile, and William Brouncker. Throughout his life Wallis was forthright in ensuring that the work of

John Wallis.

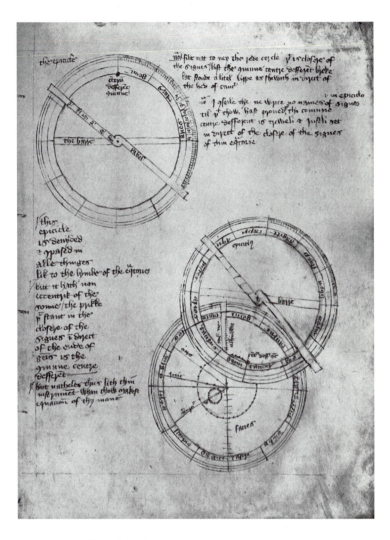

A page from a 1392 manuscript, probably in the hand of Geoffrey Chaucer, discusses in English an astronomical instrument, the equatorium (a planetary computer); the work may be based on an earlier Latin version by Simon Bredon, a Fellow of Merton College.

others was published for the good of the wider mathematical community, and even reprimanded Isaac Newton for not making his work more readily available. Indeed, the contrast with Newton points up the great achievement of Wallis in creating his research school. Newton did not, on the whole, gather round him younger scholars whose work he assisted and promoted—although he was effective in later years in supporting the career aspirations of young men whose work he admired—but had a far more complex and anguished relationship with his contemporaries.

Notwithstanding the strong state in which Wallis left Oxford mathematics, and the capable mathematicians who were his immediate successors, research activity seemed to run into the ground from 1720 or so, and little more is heard in this respect until well into the next century.

Although nineteenth-century Oxford had some distinguished mathematical researchers—notably, two Savilian professors, Henry Smith and James Sylvester—there was never a widespread research school,

Merton College's astrolabe is engraved 'Lat.52.6/m Oxonia' and was made around 1360 for use in Oxford. It was probably bequeathed to the College by Simon Bredon.

and such an ethos did not move far below the professorial level. There were two societies that fulfilled some of the functions of a research seminar, the Ashmolean Society from 1828, where scientific researchers could discuss and share their work, and the Oxford Mathematical Society from 1888. But new creative mathematical activity was fairly restricted. To some extent, sheer overloading of the college tutors and lecturers might be thought responsible: their teaching and examining duties would preclude time for research, particularly if there was not seen to be any point or advantage in it. Research, including explicit training in research techniques, did not seem particularly relevant. In a 1925 talk to the Oxford Mathematical and Physical Society (as the Oxford Mathematical Society became), the retired Waynflete professor Edwin Elliott recalled the frame of mind in which he taught as a young don in the 1870s and 1880s, teaching an effective syllabus of which he was proud:

But how about research and original work under this famous system of yours, I can fancy someone saying. You do not seem to have promoted it much. Perhaps not! It had not yet occurred to people that systematic training for it was possible.

At the time that Elliott gave this talk the Savilian professor was G. H. Hardy, who had come over from Cambridge after the First World

J. J. Sylvester, founder of the Oxford Mathematical Society.

War. Hardy succeeded in stimulating the research activity that Sylvester had hoped to encourage, and also urged the formation of a Mathematical Institute to embody such aspirations. By the time that a new purpose-built Mathematical Institute finally came about, in 1966, the research activity to fill it was already strongly under way.

The examination tradition

Over the past two centuries, examinations have come to occupy a prominent place and now form a key role in the Oxford undergraduate experience. Indeed, in the words of Baden Powell, the examination process 'will always be the first moving principle of the whole machinery'. The situation in the late eighteenth century is sometimes felt to be symbolized in the reminiscences of John Scott (later, Lord Chancellor Eldon), a student at University College, who called the examination 'a farce in my time', recounting his experience with the examiners on 20 February 1770 in these terms:

"What is the Hebrew for the place of a skull?" I replied "Golgotha."
"Who founded University College?" I stated (though, by the way, the point is sometimes doubted) "that King Alfred founded it." "Very well, Sir," said the Examiner, "you are competent for your degree."

Perhaps Scott exaggerated his memories for the sake of an after-dinner joke; even so, that it could be told at all indicates that things were different then.

From the beginning of the nineteenth century, the examination process was refined, with several changes being progressively introduced in the final examinations. Fresh examinations, too, were introduced at different stages of the undergraduate career, alongside changes in the structure of the Schools of the University. By the early twentieth century, examinations had begun to weigh heavily, not only on students but also on their teachers. J. J. Sylvester had already in the previous century described the setting of examination papers as 'this horrible incubus', and William Ferrar, a Fellow of Hertford College, wrote in his memoirs about the toll the examination process exacted from the small number of mathematics dons:

In 1926 or so I began my chore of examining for Oxford University; at first easy tasks like Pass Mods or Science Prelims, but later a steady round of Mods, Finals & the Junior left me with a major examination every year. Further, every year required a paper to be set for the scholarship (entrance) examination in the Hertford group of colleges. Altogether the work of examining within the University was, on looking back on it, unduly heavy. There were only 10 or 12 full time mathematical dons & professors & with so few qualified examiners and so many annual examinations the pressure on each individual don was considerable. I wonder how any of us found either the time or the energy to pursue our own researches; in fact many Oxford dons gave up & confined their mathematics to their tutoring, lecturing & examining.

W. L. Ferrar.

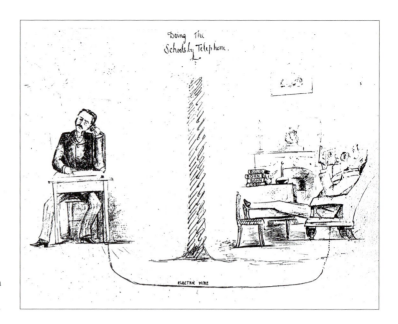

Oxford has always been at the forefront of new technology: in this cartoon of the 1880s a student telephones answers to the geometry examination to his friend in the exam Schools.

The numbers given by Ferrar for the 1920s may be compared with the early 1990s, by which time there were twelve professors, five readers, and over forty university lecturers, a five-fold increase. Since the student numbers have more than kept pace with this rise, and examinations have certainly not diminished in quantity or importance, it is to be hoped that there are economies of scale for the continued academic effort involved.

The tradition of rivalry with Cambridge

Oxford has always maintained a close relationship with the University of Cambridge, which came into existence at around the same time; their histories have many features in common. The very similarity has from the start provoked attempts to create or characterize differences between the two universities, although the most measured judgement must be that of J. I. Catto and Ralph Evans that 'Together, in tandem or in counterpoint, the two collegiate universities would eventually form a single, empirical tradition of ideas.' The free flow and interchange of scholars between the universities over the centuries are evidence of that. Nonetheless, attempts have long persisted to contrast Oxford and Cambridge. As recently as 1952, the music critic of *The Times* perceived a difference in how mathematics was viewed:

At Redbrick they treat mathematics as an instrument of technology; at Cambridge they regard it as an ally of physics and an approach to philosophy; at Oxford they think of it as an art in itself having affinities with music and dancing.

It is certainly interesting to compare the development of mathematical studies at the two universities, since the overall historical record here is somewhat at variance with popular impression and belief. It has been long and widely felt that there is a marked difference of quality: that mathematics at Cambridge is a much more serious study than at Oxford; that Cambridge is the place to study mathematics, while Oxford's genius lies more in the humanities. Thus when foundation appointments were to be made in the 1820s to the new London University, the first English university to be founded since the Middle Ages, *The Times* reported that 'it is known to be the intention to choose classical professors at Oxford, and mathematical at Cambridge', although in the event the classical appointees were from Cambridge too. This judgement has not been without some justification, but is hardly a timeless truth.

Over the past eight centuries, taken as a whole, the alleged mathematical superiority of Cambridge seems mainly to be an artefact of nineteenth-century circumstance. During that century, it happened that the purpose of studying mathematics at the two universities was rather different, as were the conditions and framework within which it was done. While being attentive to the role of mathematics in a liberal education, Cambridge developed, from the mid-eighteenth century onwards, a highly competitive examination culture geared towards ranking students on a mathematical examination, after prolonged coaching. Furthermore, until mid-century no student could go on to take the classical tripos who had not taken, and done well in, the mathematics tripos. This system produced, for more than a century, many young men whose mathematical abilities made a real contribution in, and for, later life: the ranks of the Senior Wrangler and others high in the listings include many of the best mathematicians and scientists of the nineteenth century, as well as some of the best lawyers, clergymen, and other professionals. Oxford, by contrast, did not place such a high premium on competitiveness in the mathematical sphere—for example, honours classifications were never individually ranked—nor indeed were all students expected to study mathematics. Since 1831 Oxford has had University scholarships in mathematics, the senior one of which corresponded in effect to Cambridge's 'Smith's prize', but these were not at first awarded for original work and did not play such an important role in the world outside Oxford.

Although the Cambridge system in its most competitive form was dropped in the early twentieth century, the impression lived on for some time that Oxford mathematical training, and the calibre of people involved, was markedly inferior. Such a belief had a self-fulfilling quality. The effect of this can be seen in the rueful reflections in 1912 of Arthur Joliffe, Fellow and tutor at Corpus Christi College from 1891 to 1920, upon the evidence presented by candidates for Oxford entrance scholarships:

By the 1980s, three of the professors at the Mathematical Institute had received the Fields Medal, the prestigious international award for younger mathematicians sometimes regarded as the mathematical equivalent of a Nobel prize: Michael Atiyah (1966), Simon Donaldson (1986), and Daniel Quillen (1978).

it is undeniable that the average candidate is not as good as the average candidate at Cambridge. The genius from the small grammar school, the promising student from a provincial university, the ablest boy at the large public school, all are sent to Cambridge in preference to Oxford as a rule. Some of the candidates sent to Oxford from large public schools are occasionally so bad that one can only suppose that their masters think that a willingness to come to Oxford is a sufficient qualification for a Mathematical Scholarship there.

In developing his analysis of how Oxford mathematics suffered in comparison with that at Cambridge, Joliffe put his finger on the way low expectations, both of people and of career prospects for mathematicians at Oxford, dampened down the whole process. Nor was this comparison with Cambridge purely abstract. Oxford colleges themselves believed in Oxford's relative mathematical weakness, to the extent of sending their more promising students over to Cambridge to receive special coaching for Oxford examinations.

In the course of the twentieth century, Oxford mathematics made huge strides away from the melancholy picture painted by Joliffe, to such an extent that at one stage in the 1980s three holders of the internationally prestigious Fields Medal were teaching at Oxford. As early as the 1920s, Mary Cartwright, one of the country's leading mathematicians of the twentieth century and the first woman mathematician to be elected a Fellow of the Royal Society, was an undergraduate at Oxford. Again, one of the most famous mathematicians of the 1990s, Andrew Wiles (the Princeton professor who proved Fermat's last theorem) studied at Oxford as an undergraduate, later returning as a Royal Society Research Professor. These individual examples provide evidence both for the quality of the undergraduate intake and for the quality of mathematical

The success of a former Merton College student, Andrew Wiles, in proving Fermat's Last Theorem in the 1990s recalls the international position last held by the University of Oxford when the Merton School was in its prime, six and a half centuries before.

training provided. Perhaps Oxford need no longer feel an automatic sense of mathematical inferiority in comparison with Cambridge.

The traditions live on

In regaining its mathematical strength during the twentieth century, Oxford was but rediscovering its historical traditions. The memory lives on in the Oxford of today not only of the medieval Oxford Schools, but also of the great flourishing of seventeenth-century mathematical science and, above all, John Wallis whose name has, since 1969, been commemorated in a named Chair of mathematics. Indeed, Oxford's knowledge of itself and memory of its past is one of its most enduring traditions, a past that is forever being added to.

We can see this sense of the immanence of history in the way that those elected to professorial Chairs have introduced themselves to their Oxford audience. The Savilian astronomy professor David Gregory opened his inaugural lecture on 21 April 1692 with tributes to Sir Henry Savile, to his new colleague John Wallis, 'the prince of geometers', and to Edward Bernard whom he succeeded, and highly praised his predecessors Seth Ward and Christopher Wren. A hundred and forty years later the list had grown longer: when Baden Powell delivered an introductory lecture in 1832 as Savilian Professor of Geometry, he rehearsed the names of Oxford mathematicians of the distant and recent past:

We are justly proud of the names of Wallis and Briggs; at a subsequent period we claim those of Boyle, Wren, and Gregory, of Halley, Stirling, and Bradley; while, in still later times, we boast a Horsley and a Robertson.

James Sylvester, delivering his inaugural lecture in the same Chair half a century later, mentioned some of the same names, singling out Briggs, Wallis, and Halley, and added a moving tribute to his predecessor Henry Smith. G. H. Hardy's inaugural lecture, 35 years on, explicitly drew attention to the contributions of Briggs, Smith, and Sylvester, characterizing the study of the theory of numbers in England as particularly associated with Oxford and with the Savilian Chair. And so we might expect it to continue, with each generation remembering, celebrating, and building upon the traditions of the past.

CHAPTER 2

Medieval Oxford

John North

It is customary for historians to try to astonish their readers with touches of mathematical genius from the past, if only to prove that while they are only historians, they are in good intellectual company; or, in an Oxford context, they might simply be aiming to show that Oxford was one of the first European universities in point of learning, as well as in point of foundation. There are certainly some mathematical surprises to be found in medieval Oxford—whether or not they are to the modern taste is another matter—but the very idea of singling them out for praise would have surprised the typical medieval scholar. For him, mathematics was an essential part of the curriculum, to be sure, but as a measure of intellectual excellence it was a hurdle, one to be crossed by the student in the Faculty of Arts who wished to pass on to the study of higher things.

'What higher things are there?', one might ask. Arithmetic and geometry were certainly not at the bottom of the heap. Together they were two of the four non-trivial liberal arts, the quadrivium. The others were astronomy and music—but music studied arithmetically, as harmony. The three liberal arts that comprised the so-called trivium were grammar, rhetoric, and dialectic, which for the weaker students might mean little more than reading and writing—in Latin, of course. The *higher* faculties, however, were those of law and medicine, and the highest of all was the Faculty of Divinity. You have only to walk into the Bodleian quadrangle from the Catte Street entrance to see this pattern in stone. The schools for the liberal arts are humble rooms to left and right as you enter, while law and medicine flank the central entrance ahead of you. Straight ahead, at the apex of this array, are the Divinity Schools, a magnificent example of architecture in the perpendicular style—the perpendiculars having been put there by stonemasons, not by mathematicians. Mathematics in the Middle Ages was kept strictly in its place.

This is not to say that the subject was without honour. On the contrary, as a result of the great respect in which first Plato and later Aristotle were held in the Middle Ages, the praises of the subject were very often sung, even by people who had very little idea of what they

The skill, sophistication, and complexity of medieval Oxford learning is illustrated in the great astronomical clock designed for St Albans Abbey by Richard of Wallingford, who studied at Oxford in the 1320s. This is a reconstruction: the original, within a cubic iron framework whose sides were over six feet long, was lost with the dissolution of the monasteries in the sixteenth century.

The Divinity School, built between 1423 and 1483, embodies, in a triumph of the stonemason's craft, the respect due to the most prestigious study of the medieval university curriculum.

were singing about. For Plato, the realm of geometrical forms had a greater reality than the world of experience, of which it was independent. Modern mathematicians are often drawn to this doctrine, perhaps having misconstrued it as a piece of flattery, and overlooking Plato's belief that the realm of geometrical forms had less reality than the realm of ideas. In the early Middle Ages, when Plato's reputation was considerable, they would not have been allowed to get away with such a misconception.

Aristotle is not a philosopher to the modern mathematical taste, but he did much more to form the distinctive Oxford mathematical style than any other writer. Most of his works were available in Oxford in Latin by the end of the thirteenth century, and they provided the starting point for the characteristic Mertonian mathematics of the fourteenth century. Aristotle had sophisticated mathematical arguments and mathematical constructions in his physics and in his astronomy, but perhaps most important of all was his clear enunciation of what is involved in a demonstrative science. To put the matter in somewhat simplistic terms, he taught us (inspired by the proto-axiomatics of earlier Greek mathematics) the nature and value of the axiomatic method. He had insisted that any mathematics must be directly applicable to the world, and he had seen that this gives rise to numerous difficulties; we will come back to some of these later. For the moment, let us stress the point that, in the Christian universities of the Middle Ages, it was by dint of their potential application to the world that the various sorts of mathematics were most valued.

Mathematics and the world

According to the statutes of the early fourteenth century, to incept in the Arts Faculty one must have heard lectures on the *Computus* (a type of work with rules for reckoning Easter and other calendar matters), the *Sphere* (a work on spherical astronomy, but entirely descriptive), the *Algorismus* (with its rudiments of algebra), and six books of Euclid's geometry. Such texts were to be had in various versions, and most involved learning by rote. In the thirteenth century, astronomy and the art of the computus helped to introduce the Hindu-Arabic numerals into Europe, to take their place side by side with Roman numerals and the abacus. Latin translators of the work by al-Khwarizmi on the art of calculating with these numerals wrongly attributed to him the techniques as well as the system of numeration, and a distorted version of his name appears in our word *algorithm*. The essential textbooks were available in over a hundred versions; among the authors of the more famous we find the names of Robert Grosseteste, Oxford's first Chancellor, and John Sacrobosco, apparently an Englishman teaching in Paris who might have spent some time at Oxford.

The *Arithmetic* of Boethius was the standard work in that subject, and is commonly listed as having been bequeathed from one scholar to another. It had been based on an *Introduction to arithmetic*, composed by Nicomachus of Gerasa around 100 AD, and reminds one of nothing so much as Euclid with the proof structure removed. Number was introduced as something to be discussed through metaphorical means, and books on mystical arithmetic proliferated. There was one, for instance, by John Pecham, a late thirteenth-century Oxford Archbishop of Canterbury who wrote on optics. Oxford men generally were poor calculators, despite a Boethian emphasis on simple examples, well illustrated through the game of *rithmomachia*, which enjoyed much popularity throughout the Middle Ages.

Rithmomachia had been devised in the eleventh century to give life to Boethius's *Arithmetic*. A metrical work on the game is ascribed to Thomas Bradwardine. John Shirwood, Bishop of Durham, on his way to the Roman Curia around 1475, visited his exiled fellow Oxonian George Neville, Archbishop of York, and tried to teach him the game. It was never too late to learn! The game was played on what amounted to a double chessboard (8 × 16 squares), with pieces representing certain numerical values. The rules for movement and capture were highly involved, the aim being to arrange four pieces in each of the progressions—arithmetic, geometric, and harmonic. We do not wish to trivialize this splendid game, but it well illustrates a failing of medieval arithmetic: it was a subject taught through simple and limited examples, using only a few numbers and a good memory.

Despite the didactic merits of the elementary texts in use among students, there was hardly an original theoretical point between them.

The game of rithmomachia (from the Greek for 'number battle') was popular among medieval scholars with a taste for complicated number games.

Geometry was often treated as a relatively unstructured empirical subject, in much the same way as happens in schools today, and this obscuring of the axiomatic character of Euclid tended to undo one of the most important of the lessons taught by Aristotle.

The mathematics of astronomy

For the real achievements in Oxford mathematics one must look elsewhere. One should look in the first place to astronomy, where the best scholars received their most advanced mathematical training. The scholar knew a variety of astronomical instruments, ranging from simple quadrants and sundials to highly sophisticated astrolabes and *equatoria*, devices for calculating a planet's position with a minimal use of astronomical tables. Leaving for a moment the astronomers who designed these things, the very act of using them, even by rote, gave the ordinary scholar a feeling for geometrical entities and procedures, for coordinates, and even for transformations between different systems of coordinates. We cannot talk yet of coordinate geometry, in the sense of the handling of loci by means of algebraic equations, but we can most certainly speak of the use of geometrical curves generated by algorithms to provide numerical information, rather as do nomograms.

To give a trivial astronomical example, divide a circle into 365 parts—let us forget the odd quarter—and take a suitably chosen point near to, but away from, its centre. The vectors from that eccentric point to the divisions of the circle can be made to fit the apparent positions of the Sun on successive days of the year. There is nothing new in this: on the contrary, it had been known in Ancient Greece, and when the medieval astronomer used the whole panoply of astronomical models involving such eccentric circles, with epicycles and variable velocities achieved by other types of eccentrics (such as those known as *equants*), he was primarily using theories set out by Ptolemy in the second century of our era, although the parameters were not Ptolemy's. What the medieval astronomer added to tradition was a series of new ways of looking at the old geometrical models, and this was through the instrument known as the equatorium.

As it is often represented, the equatorium was nothing more than a physical simulation of the old Ptolemaic models, and so added nothing much to them. After all, if one person draws circles in a book and tells you about points moving uniformly round them, and another person chooses rather to make brass discs and pivot them around one another so that they can be moved in the same way, what is the difference? There are problems to be solved in the second case—sometimes quite difficult problems—but they are essentially mechanical, not mathematical.

There is a class of equatoria, however, that does not proceed by an obvious simulation of circular movements. Needless to say, the astro-

Richard of Wallingford depicted at work constructing an instrument (dividing a brass disc), in the Abbot's study at St Albans.

nomer calculating planetary positions did not use Ptolemy's geometrical models and work everything out from first principles on each occasion. There were algorithms available, standard procedural rules. Richard of Wallingford, an Oxford astronomer of the early fourteenth century, was perhaps the best English mathematician and astronomer of the Middle Ages. When he designed an equatorium, which he called his *albion*, it was used in a completely non-intuitive manner to emulate the algorithms of such planetary calculation. It contained over 60 scales of various sorts, one of them a spiral of 32 turns, reminiscent of a circular slide-rule—indeed, logarithms apart, that is precisely what it was. His albion carried a universal astrolabe (one with a use not limited to one geographical place) and an eclipse computer. It was revised after his death (1336) by another Oxford scholar, Simon Tunsted, Regent Master of the Franciscans in the mid-century, and was highly influential in southern Europe, even into the seventeenth century. Indeed, the albion taught several important astronomer-mathematicians how to reduce to graphical terms the problem of making one variable a function of any number of independent variables.

Richard of Wallingford wrote one of the first comprehensive trigonometrical texts in Europe, a systematic and rigorous exposition of Menelaus' theorem, and a series of rules that we may regard as trigonometrical identities involving sines, cosines, and chords. The main inspiration for this work came from Ptolemy's astronomy and various texts received through Spain from writers in Arabic. It did not, of course, use our algebraic terminology, and a whole paragraph was often needed to express what we can now express in a single line. The wonder of it is that progress could be made at all using such a difficult medium of expression.

After lecturing at Oxford, Richard left to become Abbot of St Albans, where he built a vast and intricate astronomical clock, the first mechanical clock of which we have any detailed knowledge, and far surpassing most others for the next two centuries. It is always hard to say where mechanical ingenuity ends and mathematical ingenuity begins, but we must at least refer to a highly original section of a treatise on the clock in which Richard set out a theory of gear ratios. He devised a method of achieving a variable velocity solar drive by the use of a carefully calculated oval contrate wheel meshing with a long pinion lying along a radius to the oval. He devised spiral contrate wheels, too, and an ingenious differential gear that could correct the lunar velocity, leaving a theoretical error in the angular velocity of only seven parts in a million. One sometimes reads such ludicrous remarks as that the concept of velocity was not known in the Middle Ages. Those who write such things should be locked in the school of arithmetic and kept there until they have managed to calculate, using any mathematical means whatsoever, the final velocities in the St Albans clock.

Before passing to those who did indeed struggle with some of the deeper aspects of the concepts of velocity and acceleration, let us

The manuscript of Richard of Wallingford's treatise on the theory, construction, and use of the albion.

Front and back faces of the albion designed by
Richard of Wallingford. This instrument, the
only surviving medieval albion, was engraved
in southern Europe.

baculū paftorálem puls
nomen eius.

Richard of Wallingford pointing to his
astronomical clock on the wall of St Albans
Abbey.

mention another Oxford astronomer, a Mertonian. John Killingworth
died in his mid-30s in 1445, a year after writing an *Algorismus*, but his most
notable work was a set of astronomical tables for assistance in drawing up
ephemerides, lists of daily planetary positions. He found mean positions
and adjusted them by adding certain correction terms, so-called 'equations', which themselves were functions of the mean angles. Killingworth
was faced with a situation where the spans of time over which he was
working meant that the correction terms themselves needed correction.
One can think of him as having evaluated a third differential coefficient.
A procedure he gave depended on the proportionality of the differen-
tials of x and y, where x and y are related by the equation

$$\tan x = \sin y \,/\, (k + \cos y).$$

Precisely how he arrived at this result one cannot say. It was not, of
course, by using any general algorithm, but rather by a process of intuition that was helped along by his having worked constantly with tables
of differences. We do not want to play those games that have been
played endlessly in the past, of showing that the calculus was known to
the ancient Egyptians, and space-travel to the Aztecs, but we recall
what was said above in connection with Richard of Wallingford about
those who deny to the Middle Ages any clear conception of velocity
and acceleration. It is simply wrong.

Natural philosophy and sophistical argument

Discussions of the nature of velocity took place in the context of
Aristotelian philosophy, and might be easily brushed aside by those
impatient with mathematical foundations. But since this is where

John Killingworth, the best of the Merton astronomers, lived a century after the main flush of Oxford mathematicians such as Bradwardine and Richard of Wallingford. He died young in 1445 and is remembered on a memorial brass in Merton College Chapel.

Oxford's most original work was done, we will mention four writers on kinematics.

- Thomas Bradwardine left Merton College in 1333 for a succession of ecclesiastical appointments that ended with his appointment as Archbishop of Canterbury, shortly before he succumbed to the Black Death.
- Richard Swyneshead, known as 'the Calculator', outlived Bradwardine by six years and was a shadowy Merton figure; in fact, there were two (or even three) Swynesheads around at the same time.
- William of Heytesbury was at Merton in the 1330s, probably went to The Queen's College at its foundation, and was perhaps Oxford's Chancellor in 1371.
- John Dumbleton was also at Merton and then at Queen's at about the same time.

What characterized the work of these men was a belief that properties—or, more strictly, forms—could be quantified and represented by a geometrical model. Geometry was their mathematical ideal, but their material was always chosen on the strength of its relevance to natural philosophy. In his *Geometria speculativa*, Bradwardine had made a private excursion into the problem of angle sums in star polygons and the problem of isoperimetry, with theorems not to be found in Euclid. (Euclid might have put this down to the fact that Bradwardine had lower standards than he, for the work is not always well argued.) The real starting point for the new Mertonian thinking was not Euclidean, however, but Aristotle's writings on the continuum, and commentaries on them, especially by Averroës. What is the nature of change, of beginning and ending? Are there finite numbers of indivisibles in a continuum, as the atomist Walter Chatton maintained? (Chatton was an Oxford Franciscan; another atomist was Henry Harclay, who had been Oxford's Chancellor in 1312.) When Bradwardine wrote a book on the continuum, he could list five different doctrines held in his day, a fact that speaks volumes for the vitality of the discussion. He is particularly interesting, however, not for such things as his anticipations of Galileo on motion, but for the way he examined different theories with a view to their *mutual consistency* and their *compatibility with Euclidean geometry*.

Bradwardine's best-remembered work is his *Tractatus de proportionibus velocitatum in motibus* ('Treatise on the ratios of speeds in motions') of 1328. It is usual to quote with a flourish the law of motion given there, translate it into a logarithmic equation, and put aside all questions of whether it is empirically acceptable or not. If it does nothing else, this persuades us that perhaps history was right to forget Bradwardine and to remember Galileo. The point to notice, however, is that what Bradwardine had done was to work out a mathematical relation between velocities, forces, and resistances that was compatible

Title page of the first printed edition (1495) of Thomas Bradwardine's *Geometria speculativa*. Bradwardine had written it in Oxford, perhaps in the early 1330s, as an aid to students, although it was not a required text.

with medieval-Aristotelian postulates of motion. His was a piece of conceptual analysis that used logic and mathematics as a tool. In places he made progress, as in applying geometrical methods, where he had a familiar notation, a familiar language; but in places he became bogged down for want of a decent algebraic notation, as we might now put it.

With Aristotle's writings on the continuum and change go the philosopher's writings on infinity and the debate over the eternity of the world. Bradwardine, for the record, was a rare case of a scholar who defended the reality of infinite space, but here there is a long procession of writers actively engaged in discussing the problem of infinity. They would begin with a series of supposedly self-evident axioms, and then go on to show that these were violated by the existence of infinite sets. Thus, in an eternal world, the Sun's revolutions are infinite in number, but the Moon's are more than 12 times as many. (The argument is often ascribed to Galileo, but goes back at least as far as Philoponos.) Does this show that an eternal world is absurd? In retrospect, we value more the attempts made to resolve the paradox—and there were many—than those of men who wanted to use it to prove that the world was created a finite number of years ago.

Thus a clear impression emerges, not of the contents of the various discussions, which are very extensive, but of the fact that one should not expect to find them in books with 'mathematics' in their titles. There was, for example, a logical exercise that students had to undergo called *sophis-*

tical argument, and in the context of that we can find much of mathematical value. Heytesbury's *Sophismata*, which spawned numerous works and was known to Galileo, was at heart a series of exercises in mathematical logic (as we should now call it), taking examples from physics, theology, and indeed from wherever it could find them. Richard Swyneshead, in the same context of sophisms, developed something like a set of techniques for summing infinite series. Again he used geometrical models, such as when he took a rectangle's area as a quantity of heat and its width as the degree of the heat (the temperature). Such geometrical analogies were seized upon by Nicole Oresme and other Parisian philosophers in mid-century, when they extended the Mertonian theories of motion.

In the sixteenth century, Swyneshead's logico-mathematical techniques passed to Italy, where they earned a reputation for British barbarism—a typical humanist response to logical and mathematical style. They were really being attacked, however, for attracting a large following among Italian scholars, and the *Liber calculationum* was printed several times for student use in Padua, Pavia, and elsewhere. A century later we find Leibniz, a man not unknown for his treatment of the continuum, praising Swyneshead for introducing mathematics into philosophy. The Mertonian kinematicists had been dead for three centuries. Richard of Wallingford, not a Merton man but a Gloucester Hall contemporary, had been forgotten for half of this time. Swyneshead was the last of the Merton group to receive any such attention. After that, all was silence, more or less, until our own time. And soon, given the evolution of our own curriculum, silence will perhaps descend once more.

Further reading

A comprehensive history of medieval Oxford learning, including mathematics and the other sciences, will be found in *The history of the University of Oxford*, Vol. I, *The early Oxford Schools* (edited by J. I. Catto) and Vol. II, *Late medieval Oxford* (edited by J. I. Catto and Ralph Evans), Clarendon Press, Oxford, 1984, 1992; other volumes continue the history after 1500.

The work of Richard of Wallingford is edited, translated, and commented upon in J. D. North's *Richard of Wallingford* (3 vols.), Clarendon Press, Oxford, 1976.

In *A source book of mediaeval science* (edited by E. Grant), Harvard University Press, 1974, there are numerous excerpts from the texts actually used in the universities, especially in Paris and Oxford, and much useful introductory material and bibliography.

The standard biographical source for Oxford scholars of the Middle Ages is A. B. Emden's *A biographical register of the University of Oxford to A.D. 1500* (3 vols.), Clarendon Press, Oxford, 1957–9.

TRACTATUS SECUNDUS
DE NATURÆ SIMIA SEU
Technica macrocosmi historia
in partes undecim divisa.

AVTHORE
ROBERTO FLUDD ALIAS DE FLUCTIBUS
armigero et in Medicina
Doctore Oxoniensi.
In Nobili Oppenheimio
Ære Iohan-Theodori de Brÿ Typis Hieronÿ=
mi Galleri. Anno CIↃ IↃCXVIII.

CHAPTER 3

Renaissance Oxford

John Fauvel and Robert Goulding

The period which started with Henry Tudor's victory at Bosworth Field in 1485 and ended with the death of his granddaughter Elizabeth in 1603 was one of the most vigorous periods of English history, and one of the most important for the development of its educational structures. Over this time the country experienced the Reformation, the state religion going back and forth between Catholicism and Protestantism, the dissolution of the monasteries and the redistribution of their wealth, many people being burned for a variety of reasons, wars and threats of wars, the Spanish Armada, and the seeds of the British Empire being sown in a mixture of military might, voyages of discovery, increased trade, and intellectual achievement. Literacy increased—there was a massive expansion in the availability of printed books—and numeracy too: the military, navigational, geographical, and trade expansion over the period was predicated on a growing facility with practical mathematics. In all of this the University of Oxford played its part.

It was admittedly not the most distinguished period for Oxford mathematics. We cannot point to many Oxford figures in the class of Bradwardine and the Merton School of two centuries before, nor indeed of Henry Briggs or John Wallis in the next century. True, the greatest English mathematician of the time, Thomas Harriot, was an Oxford graduate, but he had little officially to do with the University during his subsequent career. It was, overall, a period in which Oxford turned out some graduates with a modest competence in mathematics, a few ready to turn their hand to the wide range of mathematical arts and sciences—land-surveying, navigation, making sundials, casting horoscopes, calculating Easter, aiming guns, building fortifications—which constituted mathematical practice.

This chapter aims to give some impression of Oxford's role in promoting and fostering mathematical activities over this period. In the early years of the Tudor dynasty there is little of mathematical interest to report, and so our story really begins in the 1520s once the influence of King Henry VIII began to make itself felt after his accession in 1509. The early years of the young King's reign saw the rise of humanism

The title page of Robert Fludd's *Ape of nature* (1618) surveys mathematical arts studied at Oxford in the late Renaissance period. Clockwise from the ape's pointer are arithmetic, surveying, perspective, painting, fortification, engineering, sun-dialling, cosmography, astrology, geomancy, and music.

Erasmus Williams was a Fellow of New College, and a skilful surveyor who mapped college lands. The memorial brass erected in Tingewick church after his death in 1608 shows him turning his back on his attainments, including music, astronomy, and geometry, when he decided to devote his life to the church.

and the spread of new learning at court and in the universities, promoted by the King and his advisors. In the late 1510s and early 1520s, Henry VIII was beginning to take interest in overseas enterprise and the structures to support this. He and Cardinal Wolsey both wished to encourage the teaching of mathematical sciences, amongst the other new learning, and Wolsey founded a mathematical lecture at Oxford which was given at Corpus Christi College by Nicolaus Kratzer, a Munich-born scholar who also lectured on astronomy at the King's behest. With Wolsey's fall and subsequent religious and political turbulence, however, the momentum was not sustained, and the story of Oxford mathematics during the rest of the century is a somewhat fitful one, but showing a gradual growth in confidence which culminated in efforts to understand the neglect of mathematical studies and put it right.

Just as the beginning of the period of interest to us is symbolized by Kratzer's mathematical lectures given in the 1520s, so we will continue the story to another set of lectures, delivered in Oxford in 1620. These lectures on the first book of Euclid's *Elements* were given by Sir Henry Savile to mark his foundation of two professorships, the Savilian Chairs of Geometry and Astronomy. These Chairs, whose holders will be discussed in later chapters, were to have an enormous influence on the course of Oxford mathematics.

The student experience

The time-honoured curriculum of trivium and quadrivium, described in the previous chapter, was the formal structure leading to the Master of Arts degree until about the mid-sixteenth century. It is hard to pin down to what extent the formal quadrivium was studied in depth by how many students. The intention was certainly there: the 1549 Edwardian statutes, issued to both Oxford and Cambridge Universities, state:

The Student freshly come from a grammar school, mathematics are first to receive. He is to study them a whole year, that is to say, arithmetic, geometry, and as much as he shall be able of astronomy and cosmography.

It is unclear how far these obligations were sustained in subsequent reigns; throughout the sixteenth century, it seems that students entering Oxford were obliged to attend lectures on mathematics, but the evidence suggests that even apparently compulsory lectures were often rather poorly attended in practice.

Indeed, what reason did the sons of nobles and gentlemen have for studying the mathematical arts? Their training at the University was intended to give them sufficient grace and sophistication to serve in some public office, and their needs were not obviously met by either

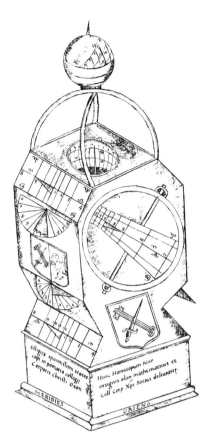

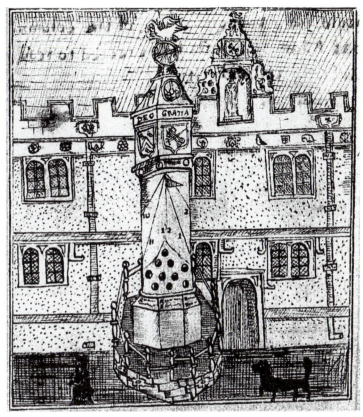

Robert Hegge, a Fellow of Corpus Christi College, drew two college sundials in the 1620s: left, a sundial made by Nicolaus Kratzer in the 1520s which stood in the college orchard up to the early eighteenth century; right, the 'Pelican' sundial in the front quadrangle, designed by Charles Turnbull in 1579.

the old quadrivium type of mathematics, theoretical and divorced from practical application, or the new type of practitioner mathematics which was more generally associated with tradesmen. Nor were there examinations, in the modern sense, on the contents of lecture courses to encourage assiduous study. Furthermore, through its blurring into astrology, mathematics was also considered by some to have dubious associations. In the next century, Francis Osborne commented:

my Memory reacheth the time, when the generality of People thought her [mathematics] most useful *Branches*, *Spels* and her *Professors*, *Limbs of the Devil*; converting the Honour of *Oxford*, due for her (though at that time slender) Proficiency in *this Study*, to her Shame: Not a few of our then foolish *Gentry*, refusing to send their Sons thither, lest they should be smutted with the *Black-Art*.

Even the humanist influence on mathematical study was ambivalent. The leading educational thinker Juan Luis Vives, friend of Erasmus, Wolsey, and Sir Thomas More, spent some time at Oxford in the 1520s, where he was, like Kratzer, a fellow of Corpus Christi College. Vives took the view that mathematics is almost a remedial subject, where not positively unhealthy:

The mathematical sciences are particularly disciplinary to flighty and restless intellects which are inclined to slackness … Often those students

Cuthbert Tunstall, a humanist scholar and friend of Sir Thomas More, studied at Oxford, perhaps at Balliol College, in the 1490s, and rose to become Bishop of Durham. He wrote the first arithmetic book to be published in England, *De arte supputandi*, in 1522.

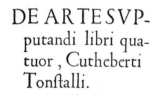

who have no bent for the more agreeable branches of knowledge, are most apt in these severe and crabbed mathematical studies … anxious inquiry into such mathematical problems leads away from the things of life, and estranges men from a perception of what conduces to the common weal.

With friends like Vives—and such views as these echoed down the century—it is not surprising that parents would be chary of entrusting their sons to a curriculum in which mathematics was seen to play too large a part.

What young men actually studied might not correspond exactly to the statutory curriculum. The system seems to have been pretty flexible: some students left after a year or two, some after their Bachelor of Arts degree, some studied other things as took their fancy, some took no degree but treated it as a finishing school, and some came up very early—nearly a fifth of students were aged 14 or under in late Elizabethan Oxford. Students seem to have studied according to their needs and interests, and if the statutes did not quite suit them they could ask for dispensations of one sort or another. In 1580, Thomas Harriot put in a supplication to be let off some statutory obligation *causa quia non sine maxima incommodo* ('because it is very inconvenient').

We have evidence about the textbooks used in the late Elizabethan period from probate inventories of students and recent graduates who had died—sadly, a not uncommon occurrence. The arithmetic of Gemma Frisius, a mixture of number theory and commercial arithmetic first published in 1540, was the most popular textbook—more popular than Cuthbert Tunstall's *De arte supputandi*, for example. Tunstall was at Oxford for some of his studies (possibly at Balliol in around 1491) and pursued his career in the church; he was eventually Bishop of Durham. His book was rather dry and did not contain such up-to-date material as Frisius's did. Most textbooks of the period were in Latin, as that was the language of teaching, but some students had copies of the first popular arithmetic textbook in English, *The ground of artes*, written by the Oxford graduate Robert Record.

At this time the tutorial system was beginning to find its feet, as the medieval academical halls gave way to endowed colleges as the centre of undergraduate life. Students had not only University and college lectures, but also a tutor *in loco parentis* who might give a lot of guidance and help—indeed, a very rich student might sweep his tutor off on a European tour or incorporate him into his household. This was the century in which students became more and more organized into colleges, and in which various new colleges were founded, so there was vigorous renewal in the air.

How was the teaching done? Mathematical lectures were given throughout the period, as far as we can make out, but there is no complete record. Generally speaking, the University provided at least three

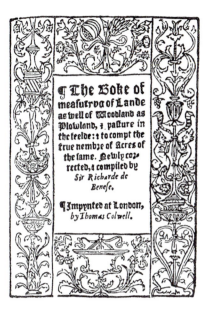

The first surveying book in English, *The boke of measuryng of lande* (c. 1537), was written by Richard Benese, an Augustinian friar who in 1519 had supplicated for an Oxford law degree. The historical preface was written by his friend Thomas Peynell, who was educated at the College of St Mary the Virgin in Oxford in the 1510s. Books such as this provided the quantitative skills needed for opening up the Tudor land market.

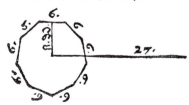

A Fellow of The Queen's College, Thomas Crosfield, penned a verse one day in 1628 on his mixed reactions to Aaron Rathborne's book on surveying:

> Besides a new Surveyor I did see
> Compos'd by Aaron Rathborne, which may be
> A book much useful for a gentleman
> But as for others few or will or can
> delight therein, because they have small land
> for to survey: but only live by hand,
> or by their wits: and such we scholars be
> that have no lands; and there's the misery …

lecturers each year, in geometry, arithmetic, and astronomy, and most colleges seemed to put on lectures on mathematical subjects, often during the long vacation. We know only a few names of those involved. The text of some lectures are extant; in particular, the manuscripts of Henry Savile's lectures on astronomy in the 1570s (discussed later in the chapter) are held in the Bodleian Library.

We gain some insight into how lectures might have been conducted from the instructions concerning two lectureships set up by Sir Thomas Smith at The Queens' College, Cambridge, in 1573. After making a bequest to support lectureships in geometry and arithmetic, he cautioned:

> The which two lectures are not to be redd of the reader as of a preacher out of a pulpit, but 'per radium et eruditum pulverum' as it is said, that is with a penn on paper or tables, or a sticke or compasse in sand or duste to make demonstracon that his schollers maie both understand the reader and also do it themselves and so profit.

Evidently some lecturers continued the medieval tradition of reading lectures like a preacher, as otherwise he would not have needed to tell them not to: and we may notice too that blackboards had not been invented.

How many students were involved in all this? From the 1570s onwards, Oxford began to keep matriculation records. Somewhere between 250 and 400 young men a year started to study at Oxford in the late Tudor period. Even though not all would stay the course, the University experience did affect quite a substantial proportion of the population. Indeed, in 1621, 784 students matriculated, a number not again reached until 1883. The actual statistic here may be misleading, as a number of years' entries may have matriculated in one, but the spirit is not misleading that Oxford grew to be larger under the Tudors and Stuarts in absolute terms than at any time before the late nineteenth century, and of course as a proportion of the population (a tenth of today's population) even more so.

Five Oxford mathematicians

To exemplify different aspects of the Tudor Oxford experience in a more personal way, we highlight five people who between them reflect different facets of Oxford mathematics over the period:

- Robert Record, the first and greatest of mathematics textbook writers in English;
- Henry Billingsley, who combined being a translator of Euclid with being a prosperous London merchant;
- Thomas Allen, representing the Oxford tutorial system at its highest, an enormous personal influence on generations of students for 60 years;

- Henry Savile, the establishment figure who moved smoothly up the academic ladder and whose influence on subsequent Oxford mathematics is unparalleled;
- Thomas Harriot, Oxford's star pupil of the Tudor period—indeed, England's only world-class mathematician in the two or three centuries leading up to Newton.

Robert Record

At the beginning of the Tudor period there were no indigenous mathematics textbooks in Britain. The growth in numeracy and wider mathematical practice of the later sixteenth century was made possible by the fine series of arithmetic, geometry, cosmography, and algebra textbooks written by Robert Record in the 1540s and 1550s. Record was the first person to consider afresh what mathematics should be taught, and how to teach it in the circumstances of the time: increasing literacy, growing trade and prosperity, and developing technological capabilities for peace and war. Record was Welsh, born in Tenby in about 1510, and graduated with a Bachelor of Arts degree from Oxford in 1531. He was elected a Fellow of All Souls College in that year, and studied medicine before removing to Cambridge, where he took an MD (Doctor of Medicine) degree in 1545. He became a civil servant, and was surveyor of the mines and monies in Ireland in the early 1550s, but did not manoeuvre very adroitly in the political see-saw of the times, and died in prison in 1558.

Robert Record's influential series of mathematics textbooks popularized the subject, as well as having a teaching function.

Record's textbooks remain remarkable examples of pedagogy. Most are written in dialogue style, between Scholar and Master, with a sensitive ear for the difficulties in understanding that learners go through, a remarkably clear prose style, a surprisingly radical wit, and a thoughtful humanist perspective. Record emphasized repeatedly that it is the *reason* for something that is important in deciding whether to accept it, not the *authority* of whoever has said it. In his cosmology text, *The castle of knowledge* (1556), he discussed the grounds for relying on the work of Ptolemy:

No man can worthely praise Ptolemye … yet muste you and all men take heed, that both in him and in al mennes workes, you be not abused by their autoritye, but evermore attend to their reasons, and examine them well, ever regarding more what is saide, and how it is proved, than who saieth it: for autorite often times deceaveth many menne.

The teaching decisions that he took in constructing his books seem to have sprung from the practical experience of teaching at Oxford. Although the documentation is missing, there has always been a tradition at Oxford to this effect. In the words of Anthony à Wood:

'Tis said that while he was of All Souls coll. and afterwards when he retired from Cambr. to this university, he publicly taught arithmetic, and the grounds of mathematics, with the art of true accompting. All which

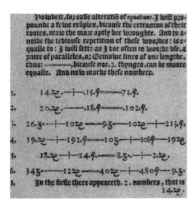

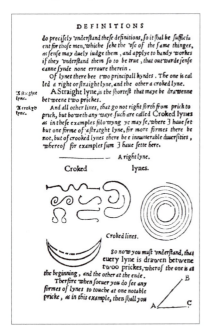

The Oxford scholar Robert Record wrote the earliest set of mathematics textbooks in English. The modern 'equals' sign was first introduced in his algebra text *The whetstone of witte* (1557), while his geometry text *The pathway to knowledg* (1551) contained a lively discussion of lines as an introduction to Euclidean geometry.

he rendred so clear and obvious to capacities, that none ever did the like before him in the memory of man.

As with other figures we are considering in this period, Record was an outstanding scholar, deeply imbued with a sense of the mathematical heritage that had come down from earlier times, whether in manuscript or by tradition. He was interested in British antiquities and was one of the first students of the Anglo-Saxon language. This nexus of interests may indeed characterize what can be seen over the next couple of centuries as an Oxford mode of mathematical scholarship, balancing the heritage of the classical and British past with a sense of the needs of the present and future.

Henry Billingsley

One of the most significant English book productions of the sixteenth century was the great translation of Euclid's *Elements* which appeared in 1570. This was the first time that a full translation had been made into English—Robert Record's 1551 geometry text *The pathway to knowledg* was a textbook based upon elementary Euclidean geometry, rather than a scholarly edition.

The translation was due to Henry Billingsley, a wealthy merchant who later became Lord Mayor of London. According to Wood, Billingsley was a student at Oxford, where he got to know 'an eminent mathematician called Whytehead, then or lately a fryar of the order of St Augustine in Oxon'. After the dissolution of the monasteries, Billingsley took in Friar Whytehead and looked after him, and Whytehead in turn taught his host mathematics to a standard of excellence:

When Whytehead died he gave his scholar all his mathematical observations that he had made and collected, together with his notes on *Euclid's Elements*, which he had with great pains drawn up and digested. Afterwards our author Billingsley translated the said *Elements* into English …

The way Anthony à Wood authenticates the account is a good example of the Oxford antiquarian tradition—that tradition which from the Middle Ages onwards has striven to preserve manuscripts and knowledge of the past in as reliable a way as possible:

what I have said relating to him and Billingsley concerning mathematics, I had from the mathematical observations of our antiquary Brian Twyne, and he from the information of that noted mathematician Mr Th. Allen of Gloc. hall, and he from an eminent physician called Rob. Barnes MD, who was elected fellow of Merton coll 29 Hen 8 Dom 1537, and remembred, and had some acquaintance with Whytehead and Billingsley.

This book is mainly remembered now for its extraordinary preface, due to the Cambridge-trained scholar John Dee, classifying and discussing an enormous range of mathematical arts and sciences, from astronomy and architecture to studies that he called 'trochilike' and

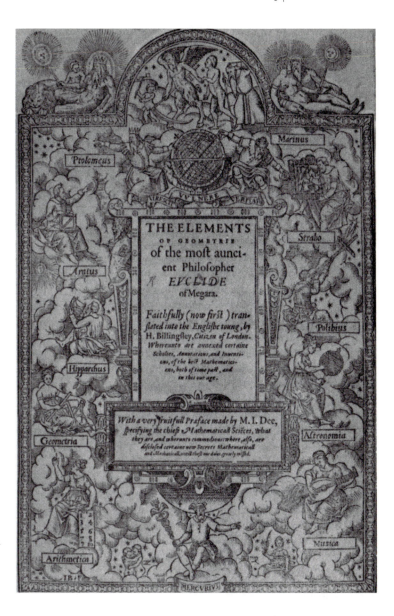

The first English edition of Euclid's *Elements*, published in 1570, was based on the work of a sixteenth-century Oxford friar called Whytehead. The attribution to 'Euclid of Megara' would within a few years be questioned by Henry Savile.

'zographie'. Dee's *Mathematicall praeface* was very influential in supporting the growing belief in the role of mathematics in practical applications, and there are many contemporary testimonies to its importance.

Besides its mathematical content, Dee's *Praeface* considered issues around publishing a classical text in the vernacular. Because university teaching was done in Latin, translating such a major university text into English was a radical act and Dee needed to provide reassurance that no harm was intended:

For, the Honour, and Estimation of the Vniuersities and Graduates, is, hereby, nothing diminished … Neither are their Studies, hereby, any whit hindred.

Indeed, Dee went on to say, it would be all to the good if students were to come up to university better prepared with knowledge of basic arithmetic and geometry. He argued further that wider public appreciation of the role and value of the mathematical sciences could only redound to the universities' credit:

Well may all men coniecture, that farre greater ayde, and better furniture, to winne to the Perfection of all Philosophie, may in the Vniuersities be had: being the Storehouses and Threasory of all Sciences, and all Artes, necessary for the best, and most noble State of Common Wealthes.

These issues of the public perception of mathematics and its neglect at the universities were also of concern to Henry Savile, as we shall see.

The Euclid edition which Dee's preface accompanied was essentially a product of Oxford scholarship, if Wood's account is true, and thus was one of the first of a succession of major classical mathematical editions associated with Oxford produced by different scholars over the next two and a half centuries.

Thomas Allen

Thomas Allen.

Thomas Allen was a Scholar of Trinity College in 1561, and moved in 1570 to Gloucester Hall where he studied and exerted a great influence upon generations of students and contemporaries, until his death at the age of 90 in 1632. Allen was a great collector of mathematics books and manuscripts; he gave a large number of his books to Trinity College Library, and the old manuscripts he collected are now in the Bodleian Library, of which he was one of the most active early supporters. Allen was close friends with John Dee and the Earl of Northumberland, in whose household he spent some time. Amongst Allen's Oxford circle were Thomas Harriot and Nathaniel Torporley, Harriot's friend and mathematical executor, who for a time in the 1580s was secretary to the French mathematician François Viète.

Allen also acted as a focus for some of the fears people had about the propriety and associations of mathematics (something which John Dee attracted also), as can be seen from the account of his life written later in the seventeenth century by John Aubrey:

The great Dudley, Earle of Leicester, made use of him for casting of Nativities, for he was the best Astrologer of his time. Queen Elizabeth sent for him to have his advice about the new Star that appeared in the Swan or Cassiopeia (but I think the Swan) [but actually Cassiopeia] to which he gave his Judgement very learnedly.

In those darke times, Astrologer, Mathematician, and Conjurer were accounted the same things; and the vulgar did verily beleeve him to be a Conjurer. He had a great many Mathematicall Instruments and Glasses in his Chamber, which did also confirme the ignorant in their opinion.

The reputation of mathematics as a dark and dangerous Oxford subject did not start with Allen. In 1551, for example, the Royal

Commissioners who visited the University to consolidate its reformation along Protestant lines are reputed to have taken particular exception to mathematical manuscripts. According to John Aubrey, writing over a century later:

My old cosen Parson Whitney told me that in the Visitation of Oxford in Edward VI's time they burned Mathematical bookes for Conjuring bookes, and, if the Greeke Professor had not accidentally come along, the Greeke Testament had been thrown into the fire for a Conjuring booke too.

This account was confirmed by Aubrey's friend Anthony à Wood, who commented: 'Sure I am that such books wherein appeared Angles or Mathematical Diagrams, were thought sufficient to be destroyed because accounted Popish, or diabolical, or both.' These memories may be somewhat exaggerated. Although medieval manuscripts *were* lost over this period, and the King's Commissioners certainly caused havoc in Duke Humfrey's Library and other libraries, the pattern of loss suggests general neglect as much as ideological factors. Some half-century after these events, Sir Thomas Bodley refounded Duke Humfrey's Library with the aid of Allen and Savile.

Sir Henry Savile. John Aubrey recorded that 'He was an extraordinary handsome and beautiful man; no lady had a finer complexion.'

Henry Savile

Henry Savile was renowned in his lifetime both for his mathematical abilities and for his encyclopedic knowledge of ancient texts. Through the statutes he set up for the Chairs he founded at the end of his life, he did more than anyone to influence the course and style of future mathematics teaching at Oxford. Savile matriculated at Brasenose College in 1561 and was elected Fellow of Merton College in 1565, graduating with a Bachelor of Arts degree in the following year. He took his Master of Arts in 1570, in which year he also began to lecture on Ptolemy's *Almagest*.

Towards the end of the 1570s, Savile began to make arrangements for a European tour, establishing contacts amongst humanists with mathematical interests. In 1578 he left for France, and from there travelled as far east as Wroclaw and as far south as Rome. By the time he arrived back in England, in late 1582, Savile had made lifelong friends among the leading astronomers and mathematicians of Europe, and had studied deeply both contemporary astronomical controversies and ancient mathematics in manuscript.

Upon his return from the continent, Savile became tutor of Greek to Queen Elizabeth, and was elected Warden of Merton in 1585. In 1595 he was appointed Provost of Eton College. He continued to hold the Wardenship of Merton while residing at Eton, and governed both with uncompromising strictness, not to say arrogance; this, and his constant absence from Merton, made him rather unpopular amongst

Ralph Kettell was president of Trinity College from 1599. According to John Aubrey, he kept a close eye on the college teaching and frequently intervened, as during a geometry tutorial: 'As they were reading of inscribing and circumscribing Figures, sayd he, I will shew you how to inscribe a Triangle in a Quadrangle. Bring a pig into the quadrangle, and I will sett the colledge dog at him, and he will take the pig by the eare, then come I and take the Dog by the tayle and the hog by the tayle, and so there you have a Triangle in a quadrangle; quod erat faciendum.

the Fellows. Aubrey records Savile's impatience with superficial brilliance:

> He could not abide Witts: when a young Scholar was recommended to him for a good Witt, "Out upon him, I'le have nothing to doe with him; give me the ploding student. If I would look for witts, I would go to Newgate [prison]: there be the Witts."

Savile's notes for his lectures on the *Almagest*, delivered between 1570 and 1575 and now held in the Bodleian Library, are an invaluable source for attitudes towards mathematics at Oxford. They are also the only extensive lecture notes we have surviving from this period. The lectures certainly made an impression amongst Savile's contemporaries; Wood reports that, as a result of these lectures, Savile grew 'famous for his learning, especially for the Greek tongue and mathematics'.

Reading these lectures and his other manuscripts now confirms his contemporaries' judgement: Savile was evidently deeply imbued in the mathematical literature, ancient, medieval, and contemporary. We have an alphabetical list composed by Savile of mathematical authors he knew, often with a brief description of their works and sometimes a judgement on their value. He includes all the ancient astronomers and mathematicians, often also giving biographies, which he presumably drew on for the biographical section of his astronomy lectures. Medieval and contemporary astronomers are also exhaustively covered, as are some authors who could not properly be called astronomers but who had written on the subject from a philosopher's point of view. Although Savile did not work closely with other English mathematicians, he certainly knew Thomas Digges and John Dee, and read with care their writings on the New Star of 1572.

Peter Ramus also receives an entry, but unfortunately Savile adds no other information. At Oxford, Ramism attracted barely a fraction of the support that it enjoyed at Cambridge, where heated debates broke out between Aristotelians and followers of the French iconoclast. Henry Savile, however, had read with close attention Ramus's popular introduction to mathematics, the *Prooemium mathematicum* of 1567—and even borrowed entire well-turned sentences for his own lectures (which he also called *Prooemium mathematicum* and in which he included a biographical history of the art from its mythical beginnings, just as Ramus had done). Ramus had attempted to reduce each branch of knowledge to a handful of easily memorized facts and rules, directed towards a single practical end. Savile shared both Ramus's concern for clarity and his desire to make the mathematical sciences relevant for university students. He believed, however, that mathematics was a much broader endeavour than Ramus had allowed; his introductory lectures are scattered with attacks on the Ramist programme, sometimes sitting uncomfortably amongst Ramus's own words.

Arisippus Philosophus Socraticus, naufragio cum ejectus ad Rhodiensi litus animadvertisset Geometrica schemata descripta, exclamavisse ad comites ita dicitur, Bene speremus, Hominum enim vestigia video *Vitruv. Architect. lib 6 Præf.*

Henry Savile began his course of lectures on Ptolemy's *Almagest*, on 10 October 1570, with the story of the shipwreck of Aristippus. This classical reference provided an evocative image for Oxford mathematics for the next two centuries, as in David Gregory's 1703 edition of Euclid's *Elements*; see also pages 129 and 161.

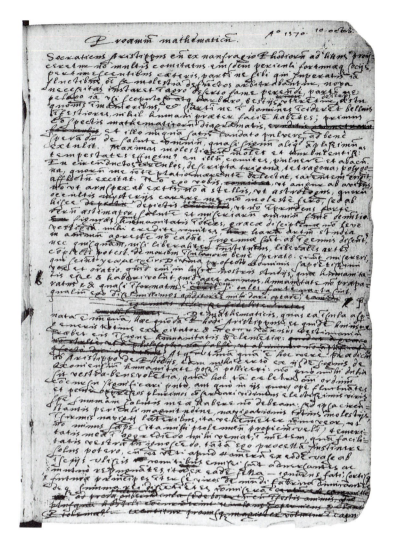

In the introduction to his lectures, Savile provided his own reasons for studying mathematics. The classes who sent their sons to Oxford associated mathematics with two unacceptable kinds of men: magicians and tradesmen. To circumvent such concerns, Savile claimed that a third, completely acceptable, kind of person has a proper claim to mathematical study: the humanist. He recounted the tale of the Socratic philosopher Aristippus who, shipwrecked on the coast of Rhodes, was assured of the inhabitants' civilized nature when he discovered mathematical figures drawn in the sand. This classical image became almost the symbol of Oxford's mathematical aspirations, providing the frontispiece for the great early eighteenth-century editions of the classical Greek mathematicians.

Savile saw many shortcomings in the Oxford of 1570. In a revealing part of his lectures, he pondered upon the reasons for the neglect of mathematics. There were times, he said, when Oxford was famous for its mathematicians and scientists, and he looked forward to the time—following, it is implied, his series of lectures—when the University would again see men of the calibre of Richard Swyneshead, Roger Bacon, and Richard of Wallingford.

Savile identified four reasons for this fall in standards. The first was indifference towards the mathematical sciences. To the average student of Oxford, Savile said, mathematics just does not seem an important component of his education. The three other causes were educational shortcomings that Savile would spend the rest of his life trying to rectify: no one reads the works of the best mathematicians, there are no teachers to explain the difficult points, and no established 'method' for the teaching of astronomy. He himself had been collecting and translating the works of the 'best mathematicians'; in these lectures he displayed an eminently humanist approach to the teaching of mathematics. He began with what his students were most comfortable with, an exposition of the history of astronomy, crammed with investigations that were more truly philological than scientific, such as the dating of ancient figures and the authenticity of their works, and only then introduced them to the technicalities of the subject.

Savile constructed his lectures as a commentary (albeit a very discursive one) on the *Almagest*, but within that familiar structure he managed to combine a detailed exposition of classical astronomy with the refinements of Regiomontanus and the 'new astronomy' of Copernicus. So far as we know, Savile was unique at Oxford in this method of teaching. Elsewhere in these lectures he criticized the traditional methods of teaching astronomy, where students laboriously copied out arrangements of circles from textbooks such as Sacrobosco's *Sphere*: an affront to both the spirit of ancient astronomy and the ideals of modern humanism. For Savile, astronomy could not be learnt in such a second-hand way. He wanted to ground his students in the whole of ancient mathematics so that they could participate directly in Ptolemy's complex arguments and constructions, and this grounding itself was to be built upon the students' linguistic and historical skills. It is not surprising that Savile did not find these high ideals upheld by the University lecturers in mathematics and astronomy, but it was this holistic approach that he eventually enshrined in the statutes of the Savilian Chairs.

On Queen Elizabeth's second visit to Oxford in 1592, Savile delivered an oration in Latin in which we see some of the same preoccupations that were evident in his lectures 20 years before. The Queen's visits to Oxford were very much of the nature of a tour of inspection. All faculties were put through their paces, each day one or more presenting a disputation on several pre-arranged theses. Savile was moderator of the disputation in moral philosophy, and the theses

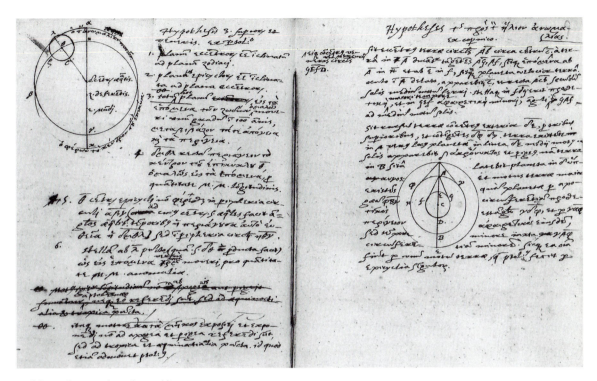

Savile's 1570 lectures, the earliest public teaching in Oxford of the new astronomy for which we have evidence, presented the Ptolemaic (left) and Copernican (right) models for the Sun's apparent motion.

to be disputed were: 'Can military and literary studies thrive together in a state? Should astrologers be banished from the state? Do justice and injustice consist in law or nature?' In his speech summing up the debate, of which we have an eye-witness report by a visiting Cambridge scholar, Savile defended the various arts by demonstrating their use to the state. Savile found the usefulness of astronomy to lie in the regulation of the calendar, citing the authority of Plato, while arithmetic and geometry are both important in military affairs, as was shown by Archimedes. This conception of the use of mathematics is rather different from that of Savile's earlier lectures, but can be explained as a plea to the Queen for continued patronage of the universities, even in those subjects of no immediately obvious benefit to the state.

In 1620 Savile delivered a series of lectures at Oxford to mark his foundation of the Savilian Chairs of Geometry and of Astronomy. His audience included the two men he had chosen to continue his work, Henry Briggs and John Bainbridge. While ostensibly an introduction to the first book of Euclid's *Elements*, a goal which they achieve very successfully, the lectures are even more about the method of teaching mathematics that Savile had developed throughout his life, and whose continuation he intended to promote through his two professors. Savile commended the writings of Euclid and Ptolemy 'alone, or almost alone' as the central texts upon which lectures should be based. The

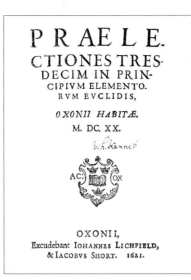

PRAELE.
CTIONES TRES-
DECIM IN PRIN-
CIPIVM ELEMENTO.
RVM EVCLIDIS,

OXONII HABITÆ.

M. DC. XX.

Joh. Kennet

AC: OX

OXONII,
Excudebant IOHANNES LICHFIELD,
& IACOBVS SHORT. 1621.

Sir Henry Savile's lectures on Euclid's *Elements* were published in Oxford in 1621.

Almagest was still to be valued for its quotation of ancient Babylonian observations, for its 'carefully thought out, very ingenious hypotheses' which still corresponded to phenomena, and for its valuable methodology and instruction on instrument making. While discussing the works of Euclid and Ptolemy, he mentioned several times that he had deposited these in the mathematical library *ad usum professorum meorum* ('for the use of my professors'). Savile gave the highest praise to Averroës, Avicenna, and the Arabic commentators and translators of Euclid, saying that in their abilities they equalled the Greeks, and in practice far surpassed them.

There was evidently a wide range of students present at these lectures, since Savile apologized for a digression on methods of proof, but said that it might be useful for the younger members of the audience. Evidence of the inspirational effect of Savile's lecturing—and testimony too, perhaps, to the benefits of enforced attendance at lectures—is found in the career of Henry Gellibrand, a student at Trinity College who showed no great promise until:

At length upon the hearing of one of sir Hen. Savile's mathematic lectures by accident, or rather to save the sconce of a groat, if he had been absent, he was so extreamly taken with it, that he immediately fell to the study of that noble science …

Gellibrand was subsequently elected Gresham Professor of Astronomy in London, developing a particular interest in navigational and naval mathematics.

The 1620 lecture course must have seemed to Savile his last teaching appearance at Oxford. At times he took the opportunity, as one would in a farewell speech, to reflect on the high points of his own career. Savile laid claim to the discovery that Euclid the geometer and Euclid of Megara (the disciple of Plato) were two different figures. They had been identified with each other in the Middle Ages and Renaissance, so that editions of Euclid's *Elements* might appear adorned with a portrait of 'Euclid of Megara'; indeed, the great 1570 Billingsley / Dee edition had attributed the work to him. Savile had separated the two Euclids in his lectures on the *Almagest*, and Commandino did the same in his edition of Euclid two years later. With masterful ambiguity, Savile said that Commandino was moved by the same arguments, 'as it is only fair to believe'.

Overall, we see in these lectures the same placing of mathematics within a humanist framework that had distinguished Savile's teaching a half-century earlier, and which was no doubt meant to be a model for his new professors.

Thomas Harriot

We cannot leave this chapter without a brief account of the greatest mathematician Oxford has produced. Such a judgement may seem provocative, but is defensible, and is intended to draw attention to the

Thomas Harriot. His treatise on algebra *Ars analyticae praxis* (1631), assembled from his papers after his death, contains some remarkable algebraic innovations, even though it is a far from adequate representation of his mathematical work.

strange fact that his name is almost unknown to the wider world. There are understandable reasons for this neglect, though, not least Harriot's own reluctance to share his learning and his discoveries with the world.

'By birth and by education an Oxonian' (in the words of the memorial plaque erected by his employer, Henry Percy, the ninth Earl of Northumberland), Thomas Harriot joined St Mary's Hall, part of Oriel College, in 1577, when he was 17. It was at Oxford that Harriot made lifelong friendships with others interested in the mathematical sciences—notably, Thomas Allen and Richard Hakluyt. During the period when Harriot was an undergraduate, Hakluyt gave lectures on cosmography and geography at Christ Church which he claimed to be the earliest lectures to introduce state-of-the-art geography to Oxford:

in my publike lectures [I] was the first, that produced and shewed both the olde imperfectly composed, and the new lately reformed Mappes, Globes, Spheares, and other instruments of this Art for demonstration in the common schooles, to the singular pleasure and general contentment of my auditory.

After taking his Bachelor of Arts degree in 1580, Harriot was engaged by Walter Ralegh (also a member of Oriel College, earlier in the 1570s), who was looking for someone to assist in the preparations for his colonizing venture to north-east America, to what was soon to be called Virginia. Harriot busied himself with every aspect of navigational theory and practice, and in 1585 he took part in a colonizing expedition on which his role was that of a versatile mathematical practitioner: to advise on navigation, to make astronomical observations, to study the natural and human resources, and to survey and map the new territories.

For most of his subsequent career, Harriot continued in the household of the Earl of Northumberland, as did two other Oxford contemporaries of Harriot with mathematical interests, Robert Hues and Walter Warner. The 'Wizard Earl', as his scientific and mathematical interests had led him to be called, was an enlightened patron and Harriot was for the most part free to pursue his wide interests, not least because from 1605 until after Harriot's death in 1621 the Earl was imprisoned in the Tower of London on suspicion of involvement in the Gunpowder Plot. There are several scattered pieces of evidence showing that he kept in touch with Oxford friends and concerns. For example, some of Henry Savile's continental researches made their way into his papers, probably through the agency of Walter Warner. Again, Thomas Allen solicited support from Harriot (as well as from Northumberland and Ralegh) when Sir Thomas Bodley was refounding the University Library, and one of Harriot's last acts, drafting his will three days before his death from cancer of the nose, was to ask that the 12 or 14 books he had borrowed from Thomas Allen should be returned.

The only book by Thomas Harriot to appear in his lifetime was his report on the 1585–6 expedition to North America in which Sir Richard Grenville attempted to set up a colony on Roanoke Island. One of his responsibilities was to liaise with the inhabitants. The manuscript (right) of his phonetic alphabet for expressing the native language (North Carolina Algonkian) shows the use of some algebraic symbols such as appear in Record's *Whetstone of witte*.

Harriot's extraordinary range of activities, recorded on some 10 000 manuscript pages in the British Library and at Petworth House in Sussex, rather exceed the grasp of any one scholar today to come to terms with. They have been well summarized by D. T. Whiteside:

Harriot in fact possessed a depth and variety of technical expertise which gives him good title to have been England's—Britain's—greatest mathematical scientist before Newton. In mathematics itself he was master equally of the classical synthetic methods of the Greek geometers Euclid, Apollonius, Archimedes and Pappus, and of the recent algebraic analysis of Cardano, Bombelli, Stevin and Viète. In optics he departed from Alhazen, Witelo and Della Porta to make first discovery of the sine-law of refraction at an interface, deriving an exact, quantitative theory of the rainbow, and also came to found his physical explanation of such phenomena upon a sophisticated atomic substratum. In mechanics he went some way to developing a viable notion of rectilinear impact, and adapted the measure of uniform deceleration elaborated by such medieval 'calculators' as Heytesbury and Alvarus Thomas correctly to deduce that the ballistic path of a projectile travelling under gravity and a unidirectional resistance effectively proportional to speed is a tilted parabola—this years before Galileo had begun to examine the simple dynamics of unresisted free fall. In astronomy he was as accurate,

resourceful and assiduous an observer through his telescopic 'trunks'—
even anticipating Galileo in pointing them to the Moon—as he was
knowledgeable in conventional Copernican theory and wise to the
nuances of Kepler's more radical hypotheses of celestial motion in focal
elliptical orbits. He further applied his technical expertise to improving
the theory and practice of maritime navigation; determined the specific
gravities and optical dispersions of a wide variety of liquids and some
solids; and otherwise busied himself with such more conventional
occupations of the Renaissance savant.

Harriot died in 1621, Savile in 1622. With the appointment of the
first Savilian professors in 1619, Henry Briggs and John Bainbridge, a
new chapter in Oxford mathematics opened.

The foundation of the Savilian professorships

Through his arrangements for the Chairs that he endowed, Sir Henry
Savile ensured that his own conception of the mathematical sciences
and the method he had perfected to teach them should continue after
his death. The statutes he composed for the Chairs provided not only
for their maintenance and administration, but also gave very precise
details of how the subjects were to be taught. The Savilian statutes
were subsequently incorporated in their entirety into Archbishop
Laud's statutes of 1634, the defining framework of Oxford studies for
the next two centuries.

The core of the teaching in both subjects as laid down by Savile is
exegesis of ancient authority, although he did not demand slavish
acceptance. The professor of geometry was required to give public
expositions on the whole of Euclid's *Elements*, Apollonius's *Conics*, and
all the works of Archimedes; his notes and observations on these texts
were to be deposited in the University Library. The professor of astronomy had to lecture on Ptolemy's *Almagest*, but with the proviso that
the discoveries of Copernicus, Geber, and other moderns should be
taught within this scheme, and his notes should also be deposited for
common use in the University Library. During his exposition of the
Almagest, he should also teach the students the skills of calculating with
sexagesimal numbers, and introduce them to Proclus's *Sphaera* or
Ptolemy's *Hypotheses planetarum*.

Both professors had other teaching requirements, and had to cover
the necessary groundwork for the understanding of the set texts, and
demonstrate the practical uses of the subject. The professor of geometry had also to give lectures, at times suitable to himself and to the
University, in the subjects of arithmetic, *geodoesia* (practical geometry),
canonics (theory of music), and mechanics. He was also required to
hold hour-long informal gatherings at least once a week to impart the
basic skills of numerical calculation—in English, if he wished. Informal
expeditions should also be made into the countryside 'when weather

and circumstances permit', to demonstrate the *praxis geometriae*. Attendance at both these forms of informal tuition was optional for the students, but we see the necessity of teaching the elements of numeracy to undergraduates fresh from the grammar schools.

The professor of astronomy had similar two-fold additional duties. At suitable times he, like his colleague, had to teach the subjects that related to his own discipline; in his case, these were 'the whole of optics', *gnomica* (sun-dialling), geography, and navigation—or at least the properly mathematical parts of these disciplines. The teaching of astrology was expressly and forcefully forbidden: the professor 'is utterly debarred from professing the doctrine of nativities and all judicial astrology without exception'—thus making Savile's attitude an official element *in perpetuum* of the teaching of mathematics at Oxford. Astronomy was not to become the sterile drawing of circles, never going beyond the word of the standard textbooks, which Savile had also condemned a half-century before. The professor of astronomy was required to build astronomical instruments so as to make his own observations, which had then to be written up and deposited in the mathematical library; by doing this he would be acting *ad imitationem Ptolemaei et Copernici*; this, according to Savile, was the only way to verify or change the astronomy handed down by the ancients. In Savile's foundation, the library had great importance, both in ensuring the advance of knowledge from generation to generation, and in perpetuating his own conception of astronomy, since the library was based around a core of his own manuscripts, covering his entire life's work and including his lecture notes of 1570.

Not only mathematics, but a certain way of doing it, are enshrined in these statutes; philological skills are almost as important as observational and mathematical skills, while never becoming an end in themselves. The first holders of the Savilian Chairs, chosen by Savile himself, exemplified the mixture of sound academic learning and practical experience that he demanded. The closing words of Savile's last lecture in 1620 emphasize how deliberately he planned for the continuation of his influence in passing the torch to Henry Briggs: 'I hand on the lamp to my successor, a most learned man, who will lead you to the inmost mysteries of geometry.'

Further reading

There is no book focusing solely on the Oxford story outlined in this chapter, but a readable and interesting discussion of English mathematical developments over the period, set in the framework of broader historical influences, is found in Antonia McLean, *Humanism and the rise of science in Tudor England*, Heinemann, 1972. More detailed discussion of the situation at Oxford and Cambridge is given in Mordechai Feingold,

The mathematicians' apprenticeship, Cambridge University Press, 1984.

Of the Oxford figures mentioned in this chapter, the only one to have a full modern biography is Thomas Harriot. John W. Shirley's *Thomas Harriot: a biography*, Clarendon Press, Oxford, 1983, is well worth reading, although short on Harriot's mathematics. Aspects of the latter, notably Harriot's navigational work, are dealt with in more detail in *Thomas Harriot: Renaissance scientist* (edited by J. W. Shirley), Clarendon Press, Oxford, 1974. There is a fuller discussion of Robert Record's work, as well as that of Dee and Harriot, in John Fauvel, *The renaissance of mathematical sciences in Britain*, Unit 6 of MA290, *Topics in the history of mathematics*, Open University Press, 1987. Aspects of Savile's work are discussed in Robert Goulding, 'Henry Savile and the Tychonic world-system', *Journal of the Warburg and Courtauld Institutes* **58** (1995), 152–79.

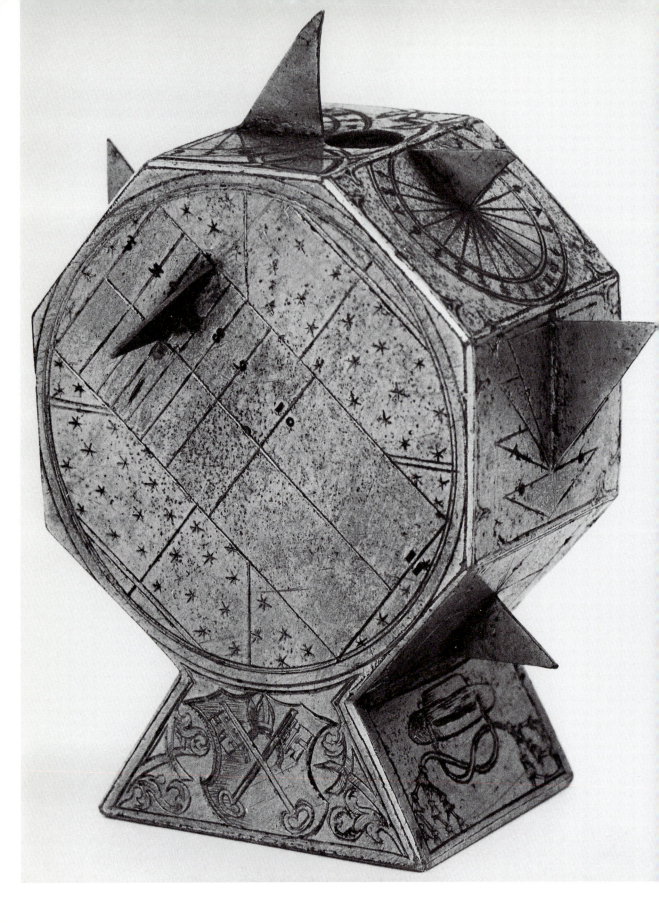

CHAPTER 4

Mathematical instruments

Willem Hackmann

'What more pleasing studies can there be', the Oxford graduate Robert Burton asked in his *Anatomy of melancholy* (1621), 'than the mathematickes, theorick or practick part?' or, as he elaborated in a later edition,

what so intricate or pleasing withall as to peruse Napier's *Logarithmes*, or those tables of artificial sines and tangents not long set out by mine old collegiat, good friend and late fellow student of Christ Church in Oxford, Mr. Edmund Gunter ... or those elaborate conclusions of his *Sector, Quadrant* and *Crosse-Staffe*?

As Nicholas Tyacke has pointed out, Burton was here prescribing mathematics as a cure for melancholy.

Some years later, that scion of seventeenth-century experimental philosophy, Robert Boyle, would recommend the study of mathematics in similar terms. However, the mathematics they had in mind was not the theoretical mathematics of the scholar which had formed the basis of cosmology since the Greeks, but the mathematics that would help with the practical application of knowledge to daily life. It was not the mathematics written in Latin, culminating in Newton's *Philosophiae naturalis principia mathematica* (1687), but the mathematics in the vulgar tongue embodied by such works in the minor key as John Blagrave's *A booke of the familiar staffe* (1594) and John Brown's *The description and use of a joynt-rule* (1661).

Burton's special concerns were with inland navigation, fen drainage, and 'industry [which] is a lodestone to drawe all good things', and Boyle was interested in improving the crafts by combining their trial-and-error methods with the rational techniques of the scholar. He admitted freely his indebtedness to the experience and practices of 'tradesmen'. On the other hand, as he pointed out, natural philosophers in their turn had been responsible for creating new trades, in the manufacture of telescopes, microscopes, quadrants, sectors, globes, and pendulum clocks.

Mathematics became increasingly important, both in the scholarly pursuit of natural philosophy and as the basis for such practical endeavours as navigation, surveying, and gauging. It was also important for the needs of a growing bureaucracy. In 1594, Thomas Blundeville

Cardinal Wolsey's portable polyhedral sundial, probably made by Nicolaus Kratzer in the 1520s, when he was in Oxford as Wolsey's lecturer. This gilt brass instrument is reminiscent of one in Holbein's portrait of Kratzer (page 67).

The frontispiece of John Brown's *The description and use of a joynt-rule* and a rare unsigned example in boxwood from the Museum of the History of Science, Oxford. There are holes for locking rods and a pair of slots so angled that the arms can be set at 60° by inserting a bar.

wrote in the foreword to the reader of *His exercises, containing sixe treatises ... for young gentlemen* interested in these matters:

I Greatly rejoyce to see many of our English Gentlemen, both of the Court and the Country in these dayes, so earnestly given to travell as well by Sea as Land, into strange and unknowne Countries ... And because that to travell by Sea requireth skil in the Art of Navigation, in which it is impossible for any man to be perfect, unless he first have his Arithmetick, and also some knowledge in the principle of Cosmography, and especially to have the use of the Spheare, of the two Globes, of the Astrolabe, and Cross-staffe, and such like instruments belonging to the Art of Navigation, I thought therefor to write the Treatise before mentioned ...

Blundeville's *Treatises* include descriptions of several recent interests: Plancius's world map (1592), John Blagrave's *Mathematical jewell* (1585), Molyneux's globes (1592), Thomas Hood's cross-staff (1590), and Gemma Frisius's instrument called *quadratum nauticum*. Blundeville wrote largely for the instruction of young gentlemen, and not for youths apprenticed to the sea. Among his friends were William Gilbert and the most eminent mathematician and natural philosopher of his age, John Dee. Both were interested in the practical aspects of their studies.

Natural philosophy and empiricism

Blundeville's circle of friends illustrates that it is not possible to make a clear-cut distinction between natural philosophers (those who practise so-called 'high science') and the mechanics or artisans (the practitioners of 'low science'), apart from the fact that for the former it was generally an amateur interest, while for the latter it was their living. The complexity of the origins of natural philosophy is indicated by the divergence in historians' views about it. Both the Hermetic naturalist and the mathematical–scientific traditions have been cited as sources for the 'mechanization' of the world picture during this period, to use Eduard Dijksterhuis's evocative phrase. In his otherwise admirable paper on 'mathematics versus the experimental tradition', Thomas Kuhn produced a rather artificial, and historically sterile, distinction between the mathematical and empirical sciences.

A more fruitful way forward, and one faithful to the period, is Jim Bennett's suggestion to adopt the perspective of the sixteenth and seventeenth centuries for what was considered to constitute the mathematical sciences. The subject was interpreted very broadly, embracing both theory and utility, and practised by men whom we would categorize as scientists or (natural) philosophers and as artisans or mechanics. A key difference between these two groups was one of class and therefore of training and disposition. Economic and social conditions made it possible for practical mathematics to flourish in the sixteenth century. Mathematical practitioners, skilled in devising and using new instruments, became involved in the practical subjects vital to colonial and

John Blagrave, author of *A mathematical jewell and other books on mathematical instruments*, was a student at St John's College. His monument in St Lawrence's Church, Reading, indicates his interests. The mathematical jewell (right) is a form of universal stereographic astrolabe first developed by the Arabs. On the back is an orthographic projection of the sphere.

economic expansion, such as navigation, surveying, horology, cartography, gunnery, and fortification, and a powerful corpus of textbooks in the vernacular was established. In England, the early development was centred on Gresham College in London, founded in 1597, with its Chairs of astronomy and geometry. Important were the college's mercantile setting and its involvement with practical instruction.

Transmission from the continent

According to sixteenth-century authorities such as John Dee and Thomas Hood, England followed the continent of Europe in the dissemination of practical mathematics and associated instruments. Dee, a mathematical genius and founder Fellow of Trinity College, Cambridge, visited Paris, Brussels, and Louvain during the period 1547 to 1550, studying continental mathematical practices, including the design and making of instruments for navigation, surveying, cartography, gunnery, and dialling (the science of the sundial).

While in the Low Countries, Dee worked with Gemma Frisius and Gerard Mercator, and brought back with him a cross-staff and an astronomer's ring designed by Gemma, and two of Mercator's globes. The development of the cross-staff has been traced from Levi ben Gerson to Gemma Frisius, whose seminal book on this instrument was published in 1545. Gemma's version of the cross-staff could be used by the relatively untutored as a hand-held astronomer's staff, as a surveyor's staff, and as a navigator's staff. In surveying, it could measure heights of buildings with no calculation, and serve in a triangulation survey. The cross-staff brought back by Dee from Louvain in 1547 is the earliest example so far discovered of such an instrument in England. An example of this type of staff, made in 1571 by Gemma's nephew Walter Arsenius, is in the British Museum.

In 1563, John Dee wrote that, while the universities had many scholars in divinity, Hebrew, Greek, and Latin, 'our country hath no man (that I ever yet could here of)' skilled in the sciences of number, weight, and measure. In his influential preface to the first English translation of Euclid's *Elements*, published in 1570, Dee outlined the relations between the mathematical sciences and the practical applications of arithmetic and geometry in terms similar to the French educational reformer Pierre de la Ramée (Peter Ramus). Thomas Hood reinforced Dee's view that mathematics was a foreign import. In his published *Speache: made by the Mathematicall Lecturer* of 1588, he traced the relation between mathematics, prosperous civilizations, and expanding empires, and urged the study of mathematical sciences, 'if you thinke it a blessed thing, to compasse the worlde [as Drake did in his circumnavigation of 1577 to 1581 on the *Golden Hind*], and returne again enriched with golde'. His own career is a case in point. He was appointed a mathematical lecturer, following a Privy Council recommendation that citizens should be instructed in military matters.

The mathematical sciences can scarcely be traced back as a continuous tradition in England beyond the mid-sixteenth century. At this time, Arabic numerals began to replace the Roman ones, although they did not entirely disappear from the ultra-conservative Exchequer records until the mid-seventeenth century. Accounts were done using rule boards or cloth, and casting counters known as *jettons*. The first English self-help text on pen-reckoning was the anonymous *An introduction for to learne to recken with pen or work the counters* (1537), but the book that opened the way to self-education was *The ground of artes teaching the worke and practice of arithmetike* of Robert Record, published in London in 1543, and running through 20 editions to 1699 (see pages 44 and 47).

Record may be regarded as the founder of the English school of mathematical practitioners. A generation of craftsmen and instrument makers learned their mathematics from his textbooks. Incidentally, a

The late sixteenth-century Dutch painting *The measurers*, in Oxford's Museum of the History of Science, depicts all manner of measuring techniques from time telling to wine gauging.

Nicolaus Kratzer, King Henry VIII's astronomer and horologer, was painted by Hans Holbein in 1528, surrounded by the instruments of his profession. One of the European scholars who flourished under the young King's patronage of humanist learning, Kratzer was an early Fellow of Corpus Christi College.

further insight into the dependence of England on the skills from the Low Countries at this time is shown by the fact that Reynor Wolfe, the printer of Record's textbook, had not long left his native Drenthe, a province in the north of the Netherlands, to settle in London in 1533. Nearly 50 years later, it was still difficult to find printers who could set up mathematical texts correctly. Indeed, throughout the sixteenth century England imported from the continent, especially from the Low Countries, the skills of book printing and the engraving of plates, maps, and instruments, often by the same craftsman.

Practical mathematics at Oxford

Robert Record studied at Oxford and at Cambridge, and is said to have tutored mathematics privately at both universities before settling in London. He and John Dee had intellectual links with the Oxford school of the thirteenth and fourteenth centuries, through the possession of manuscripts by that early group of brilliant men which included Geoffrey Chaucer, who wrote an early English tract on the astrolabe, and Richard of Wallingford, who became well known for his astronomical instruments—in particular, for a complicated astronomical clock of his devising (see Chapter 2).

Over the next two centuries, practical mathematics in Oxford lay more or less dormant, apart from a few highlights. Nicolaus Kratzer, the German astronomer and instrument maker who became astronomer and 'Deviser of the King's Horloges' to Henry VIII, also spent a brief sojourn in Oxford, attracted by Cardinal Wolsey.

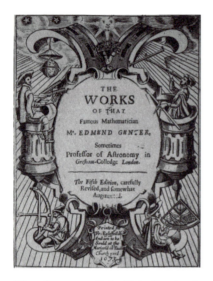

Edmund Gunter studied at Christ Church and was an unsuccessful candidate for the first Savilian Chair of Geometry. The 1673 edition of his *Works* shows various of his instruments, including his quadrant (bottom right).

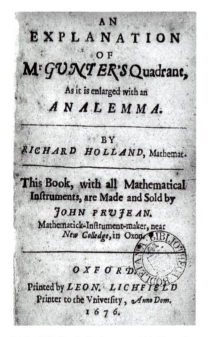

Richard Holland's *Explanation of Mr Gunter's quadrant* (1676) advertised the mathematical instruments of John Prujean.

According to his own evidence, he lectured on elementary astronomy and the construction of the astrolabe. During his brief stay from 1521 (or 1522) until the autumn of 1524, he constructed two famous large polyhedral sundials: the one for the University Church of St Mary survived until 1744, while the other, in the orchard of Corpus Christi College, had disappeared by 1710.

Thomas Harriot, one of the most brilliant mathematicians of his age, was well aware of the utilitarian aspects of his interest. Some time after he came down from Oxford in 1579, Sir Walter Ralegh set Harriot up in his house in the Strand to advise on problems of navigation. This task Harriot tackled both theoretically and practically, by suggesting the best way of making observations with the cross- and back-staffs.

Edmund Gunter devised the famous logarithmic scale named after him. He spent 15 years at Christ Church, first as an undergraduate and then probably as a mathematics tutor, before being elected the third professor of astronomy at Gresham College. Gunter's *Description and use of the sector* had been circulated in Latin manuscript since 1607, before the English translation was published in 1623. One of those who received a presentation copy from his 'old collegiat, good friend and late fellow student' was an admiring Robert Burton. Gunter's sector was derived from similar calculating instruments, devised by Galileo and the London mathematical lecturer Thomas Hood; his 'line of numbers' or logarithmic scale was based on Henry Briggs's table of logarithms (1617). Briggs was appointed from Gresham College to the first Savilian Chair of Geometry, hand-picked by the founder, who had considered Gunter for the post but turned him down. According to Aubrey, Henry Savile sent for Gunter, who came

and brought with him his sector and quadrant, and fell to resolving of triangles and doeing a great many fine things. Said the grave knight, *'Doe you call this reading of geometrie? This is showing of tricks, man!'* and so dismisst him with scorne and sent for Briggs, from Cambridge.

The London and Oxford trades in mathematical instruments

The newly endowed teaching posts, the Savilian Chairs of Astronomy and Geometry (1619) and the Sedleian Chair of Natural Philosophy (1621), were attempts to improve science education at Oxford—in particular, in the mathematical sciences. They also resulted in the need for instruments which the professors had to purchase themselves. The chief instrument makers used by Oxford scholars in the course of the seventeenth century were Elias Allen and his apprentices Christopher Brookes, Ralph Greatorex, and John Prujean; only Prujean was based in Oxford.

Astrolabe made for Queen Elizabeth I, signed and dated by Thomas Gemini (1559), made in gilt brass, 355 mm diameter, with rete for 23 stars. The inscription in the centre records that this instrument was given by Nicholas Greaves in 1659, in memory of the first two Savilian professors of astronomy, John Bainbridge and John Greaves.

Astronomical gilt brass compendium, signed and dated '.HVMFRAY.COOLE.MADE. THIS.BOKE+ANNO.1568'. Among the towns whose latitude is given is Oxford (51′ 50″).

Over this period, London had become established as the pre-eminent instrument-making centre in Europe, a quite remarkable development. The earliest instrument workshop in London was founded by Thomas Gemini. Nothing is known about his early training or when he arrived in England. He may have worked for a time with Gemma Frisius and Mercator at Louvain. Perhaps he started as a engraver—he prepared an edition of Vesalius's anatomy in 1545, for which he was paid £10 by Henry VIII. Gemini may have imported his first rough-cast instruments from Louvain.

Possibly Gemini had as a pupil Humphrey Cole, the best known of the Elizabethan instrument makers by virtue of the relatively large number of Cole's instruments that have survived. These illustrate that he catered for both the connoisseur and the bread-and-butter market. A gilt-brass astronomical compendium in the form of a book with clasps was made for the top end of the market occupied by the wealthy collector. It incorporates a number of instruments for determining time, the height of buildings, a table of latitudes, and an almanac; one wonders how many of the owners could actually use such a compendium. In the collection at the Oxford Museum of the History of Science is one with the canting badge of Richard Jugge, the Queen's printer, engraved on its cover. In the same collection are a brass surveyor's folding rule, signed and dated 1575, a plane table alidade, signed and dated 1582, and, most significant of all, an altazimuth theodolite, signed and dated 1586. This is the earliest surviving theodolite known, based on the 'topographical instrument' described by Thomas Digges in his *Pantometria* (London, 1571) and illustrated in Blaeu's *Atlas* of 1664.

Cole's finest surviving instrument is a large two-foot-diameter astrolabe dated 1575, in the possession of the University of St Andrews. This resembles the astrolabe by Thomas Gemini, made for Queen Elizabeth I, now at the Oxford Museum of the History of Science. Both instruments have on the back a horizontal projection of the sphere derived from the planisphere of Gemma Frisius. The Oxford astrolabe was given to the University by Nicholas Greaves with a batch of other astronomical angle-measuring instruments, made mostly by Elias Allen for the use of the Savilian professors. These instruments had originally been acquired by John Greaves, brother of Nicholas and Savilian Professor of Astronomy, for his scientific journey to the Levant in 1637–40. He may have used this astrolabe for determining the latitude of Rhodes.

Cole came from the north of England, and was employed as an engraver and sinker of stamps at the Mint. Apart from undertaking other engraving commissions, such as for Jugge, he undertook to supply the instruments described in Digges' *Pantometria*. Benjamin Cole, a surveyor, engraver, bookbinder, and map maker living in

The earliest complete theodolite known, signed and dated 'v H v Cole v 1586 v', made of brass and consisting of a horizontal azimuth circle (divided into 360 degrees and containing the points of the compass and a geometrical square) and a vertical (upper) altitude semicircle with a shadow square. Above the compass box hangs a levelling plummet, and there is provision for two fixed and two removable sights.

A universal equinoctial ring dial in the Elias Allen tradition from the late seventeenth century, inscribed 'The Owner Ben Cole Engraver in Oxford, to him that finds it a Reward'. Its inspiration was Gemma Frisius' astronomical ring of the previous century, a simplified version of the armillary sphere equipped with sights to allow measurements to be made of the elevation of stars.

Oxford, may well have been a relative, especially as he practised the same craft. His son and grandson, also called Benjamin, became well-known instrument makers in London, while another son, Maximilian, continued as an engraver in Oxford in the eighteenth century.

Humphrey Cole may have been the father of the English instrument-making trade, but Elias Allen was the first truly professional mathematical instrument maker who lived solely off his products. He was apprenticed in about 1602 to the London engraver and instrument maker Charles Whitwell; he may have specialized initially in surveying instruments, but he soon collaborated with many mathematicians and inventors in putting new forms of instruments into production.

Allen's best-known collaboration was with William Oughtred, a Cambridge-educated mathematician and clergyman who invented several measuring and calculating instruments. The best known of these was their *circles of proportion* (1632), the earliest form of circular slide-rule, which achieved a certain amount of notoriety because of a vehement priority dispute with a London writing-master and mathematics teacher Richard Delamain. John Aubrey, that inveterate seventeenth-century gossip, refers to Delamain's part in unflattering terms, but perhaps it is fairest to describe Delamain as 'the accommodator, not the inventor'. The development led to a host of specialist slide-rules, usually of boxwood, for the different trades.

Allen's workshop also tackled large astronomical quadrants, such as a large mural quadrant of iron and brass, dated 1637. Formerly in the collection of the Savilian Professor of Astronomy, this instrument would have been secured to a meridian wall to observe the altitude of stars as they culminated in the south, after the method of Tycho Brahe. In the seventeenth century, the Savilian professors had their observatory on the tower of the Schools quadrangle, now part of the Bodleian Library. A fine portable astronomical quadrant of about the same date is probably also by Allen, judging by the style of the engraved numerals which resemble those of the previous instrument. It is believed to have belonged to the Savilian Professor of Astronomy, John Greaves.

Allen's shop was the meeting place for a mathematical 'club'. Among its achievements was to agree upon the size of a standard English foot. He trained 14 apprentices, who included Robert Davenport, the earliest recorded instrument maker in Scotland, and John Prujean, who brought the trade to Oxford in the 1660s. Other members of the Allen dynasty of makers were Ralph Greatorex, Walter Hayes, and Edmund Culpeper.

Ralph Greatorex was apprenticed to Elias Allen in 1639. He is known as an important maker from the diaries of Samuel Pepys and John Evelyn and Aubrey's *Brief lives*, and he was a great friend of Oughtred. He may have continued Allen's business. He disappeared from view shortly after the Great Fire of 1666, when he assisted Sir

ELIAS ALLEN.
Apud Anglos Cantianus, iuxta Cranbridge natus Mathematicis Instrumentis ære incidendis sui temporis Artifex ingeniosissimus. Obiit Londini, prope finem Mensis Martij, Anno à Christo nato 1653, senaque ætatis.

A 1666 engraving of Elias Allen by Wenceslaus Hollar.

Holland's Universal Quadrant.

His Altimetrick Quadrant, serving to take Hights by inspection.

Oughtred's Quadrant.

His double Horizontal Dial.

Gunter's Quadrant.

His Analemma.

His Nocturnal.

Collins's Quadrant.

Mr. *Halton's* Universal Quadrat, for all Latitudes, with Mr. *Haley's* Notes.

Orontia's Sinical Universal Quadrant.

Napier's rods.

Mr. *Caswel's* Nocturnal.

Mr. *Haley's* Nocturnal.

Mr. *Thomson's* Pantometron.

Mr. *Pound's* Cylinder-Dial.

Mr. *Edward's* Astrolabe.

Mr. *Hooper's* Dialing Scales.

Scales for Fortification.

Scales for Surveying, Dialing, &c.

And most other mathematical Instruments.

Jonas Moore in surveying the devastated area of the City. All the instruments mentioned in the literature are what we would now term 'philosophical' and are pieces of mechanical engineering: a thermometer, an air pump, a smoke jack, a diving bell, and various water-pumping and fire-extinguishing engines. By the 1650s, experimental apparatus and machinery had become the centre of interest in the circle of Wilkins and Boyle at Oxford. Greatorex assisted Boyle with the development of the first English air pump, before he was replaced by Robert Hooke who moved with Boyle to London. Remarkably, no instruments by him were known until the Oxford Museum of the History of Science purchased an equinoctial ring dial by his hand in 1987, which proves that he continued to produce traditional mathematical instruments.

John Prujean was apprenticed to Allen in 1646. He may have been attracted to establish a workshop in Oxford because of favourable reports from Greatorex about his time in the city, and from Christopher Brookes who had married one of William Oughtred's daughters and had obtained work in Wadham College. On 11 March 1664, Prujean obtained 'privilege' at Oxford (that is, he was given non-academic membership of the University) as a 'mathematical instrument maker'. He published Richard Holland's book on Gunter's quadrant (1676) and several broadsheets describing his instruments.

An interesting document that has survived is an advertisement of 1701, in which Prujean lists over 20 different instruments which he made and sold from his workshop in New College Lane. This is the earliest known catalogue of a provincial British mathematical instrument maker. The named instruments are shown in the margin.

The names have recently been identified, and a high proportion of those listed were Oxford men: Richard Holland was a popular local teacher of mathematics and geography; William Oughtred and Edmund Gunter were well-known mathematicians and designers; John Caswell would be elected Savilian Professor of Astronomy in 1709; Thomas Edwards had recently written *Dialling made easy: or, tables calculated for the latitude of Oxford* (1692); and Edmond Halley would be elected Savilian Professor of Geometry in 1704 and Astronomer Royal in 1720. Of the others, 'Orontia' was Oronce Fine, the sixteenth-century French writer on astronomy, and John Napier was the Scottish mathematician who invented logarithms in 1614 and designed the popular calculating rods named after him. The still living designers were distinguished by the designation 'Mr.', with the exception of Immanuel Halton who had recently died.

Prujean's advertised list was firmly rooted in the seventeenth century, with its emphasis on instruments related to dialling. His *oeuvre* does not appear to have taken account of the rapid strides taking place in surveying, astronomy, and natural philosophy. Perhaps his small workshop did not have the facilities to cope with the manufacturing of

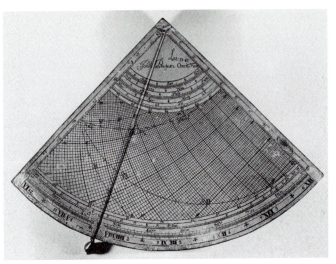

(left) An astrolabe to the design of Thomas Edwards, engraved paper on wood and card, signed 'Joha. Prujean Fecit Oxon' on the latitude plate for 51° 45".

(right) An horary quadrant with plumb-bob, engraved paper on board, signed, 'John Prujean Oxon: Fecit'. On one side (shown) is a projection for latitude 51° 45'; on the reverse side is the Gunter projection with Delamain's version of Oughtred's horizontal instrument in the apex.

the newer, more complicated, instruments. The impression is that he concentrated on traditional instruments made of paper and of brass, used primarily as teaching aids and affordable to undergraduates. This impression is reinforced by the instruments that have survived. A few are of brass, such as a quadrant made for the latitude of Oxford, but he is now better known for paper instruments made from sheets printed from engraved plates and mounted on pasteboard. Two such instruments, an astrolabe and a quadrant, are in the Oxford Museum of the History of Science. There is also a quadrant printed on paper only, while the Bodleian Library holds an unmounted sheet of an astrolabe signed by him.

Prujean spent his entire working life in Oxford, but he did not make a particularly good living, nor did he establish a local instrument-making dynasty. In 1712, the local gossip Thomas Hearne lamented the University's failure to give a place in its charitable hospital to

the Widow of one Mr John Prujean a Mathematical Instrument Maker, who was an ingenious man, and had done a great deal of service to the University for several years, and died very poor, wanting bread. Interest was made for this poor woman by Dr Halley and others but to no purpose.

He may have been succeeded by a Thomas Cookcow, who worked at the same address after Prujean's death. Apart from the publication of a broadsheet on the use of the Gunter's quadrant, he still remains to be wrested from obscurity.

The Orrery collection

The late seventeenth century saw the flowering of experimental philosophy and the development of the instrument-making trade in that

A plane table of wood with brass fittings and magnetic compass, signed and dated 'Iohn Worgan Londini fecit 1696'.

Three rules. From top to bottom: a nine-inch ivory sliding Gunter scale, signed, 'I: Rowley fec'; an ivory gunner's scale with silver slider, signed, 'I. Rowley fecit', for calculating the amount of gunpowder needed for mortars and long guns; an unsigned boxwood 'box scale' with a scale of inches and a proportional parts scale on one side, and plane scales and scales for leagues, rhumbs, chords, sines, tangents, half-tangents, and longitudes on the other.

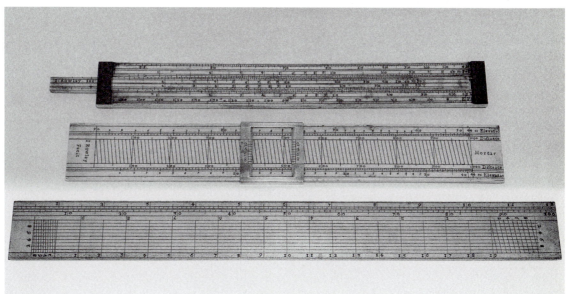

direction. Another area of growth was the development of didactic instruments, both for the connoisseur and the student. One of the finest of such collections was put together by Charles Boyle, fourth Earl of Orrery. At his death in 1731, he bequeathed it and his library of scientific books to his old college, Christ Church, for the use of scholars and students. The remains of the Orrery collection have been in the Oxford Museum of the History of Science for the past 50 years.

A small silver armillary sphere, signed 'I. Rowley fecit', on a circular base with an inset magnetic compass. Within the armillary sphere is the Sun–Earth–Moon system, with the Earth performing its diurnal rotation and the Moon moving round it by means of a train of five wheels. This instrument is stored in a lignum vitae case.

These instruments were not those of a working practitioner of mathematical and physical studies, but those of a wealthy amateur with a keen interest in the practical and theoretical sciences of his day. As such, his instruments were generally commissioned from leading craftsmen, with the work of John Rowley of London predominating.

The collection clearly reflects the three categories of instruments which commanded attention at the end of the seventeenth century:

- didactic instruments which demonstrated the laws of physics, astronomy, and mathematics—these are represented by the armillary sphere in silver and the boxwood geometrical solids;
- everyday instruments used by the surveyor and other craftsmen— these are represented by the boxwood plane table with magnetic compass, the proportional compasses and silver rule, and the three scales (sliding Gunter scale, gunner's scale, and a box scale);
- scientific novelties and optical toys, such as the camera obscura.

But with these, a new theme is opened that goes well beyond this present chapter. What is so striking when studying the development of practical mathematics in this period is the importance of the craft base, the ease with which skills in paper engraving were translated into the skills of making mathematical instruments, and the cohesion of the circle of scholars, mathematical practitioners, and instrument makers involved in this work.

Further reading

Useful sources on the mathematics of Oxford, Cambridge, and London during this period are Mordechai Feingold, *The mathematicians' apprenticeship*, Cambridge University Press, 1984, and J. A. Bennett, *The mathematical science of Christopher Wren*, Cambridge University Press, 1982. The latter scholar, in particular, has highlighted the two strands of mathematics at this time—the University-based 'academic mathematics' concerned with astronomy, and the 'practical mathematics' of the emerging professional class of surveyors, fort builders, navigators, and instrument makers; see, for instance, Bennett's chapter, 'The challenge of practical mathematics', in *Science, culture and popular belief in renaissance Europe* (edited by S. Pumfrey, P. L. Rossi, and M. Slawinski), Manchester University Press, 1991. An important standard work remains Charles Webster's *The great instauration*, London, 1975.

A book which is central to some of the issues dealt with here is Frances Willmoth's *Sir Jonas Moore: practical mathematics and restoration science*, Boydell and Brewer, 1993. The 14 volumes of Robert T. Gunther's *Early science in Oxford*, 1922–45—in particular, Volumes I and II—remain an inexhaustible well of information, even after more recent studies, but because of the frustrating way in which these volumes are organized it is not easy to determine what the bucket may bring up. An important source on practical mathematics and Elizabethan navigation is David Waters, *The art of navigation in England in Elizabethan and early Stuart times*, London, 1958.

Although it is still treated as rather a specialized subject by historians, several more general works on the history of instrumentation have appeared in recent years. A useful introduction is A. J. Turner, *Early scientific instruments, Europe 1400–1800*, Sotheby's, London, 1987. Also, J. A. Bennett's *The divided circle*, Clarendon Press, Oxford, 1987, can be read with profit by those who want to delve deeper into the development of the (mathematical) angle-measuring instruments primarily used in astronomy and surveying, while Maurice Daumas, *Les instruments scientifiques aux XVII et XVIII siècles*, Paris, 1953 (translated into English by M. Holbrook, London, 1972) is a good general source on instruments, including the mathematical variety discussed in this chapter. The suggestions concentrate on mathematical instruments, and no general sources have been cited which deal with natural (or experimental) philosophy instruments of this period, although these are included in the above-mentioned book of A. J. Turner.

CHAPTER 5

The mid-seventeenth century

Allan Chapman

Alongside Oxford's strong contribution to mathematics, the University has long been the seat of a major tradition in observational and theoretical astronomy. Three centuries after the spectacular achievements of the medieval mathematical school which centred upon Merton College, Oxford came to enjoy a reputation in astronomy which lasted from the Jacobean period to the late eighteenth century and beyond. Sir Henry Savile's Chairs of astronomy and geometry, established in 1619, made Oxford the first British university to 'profess' astronomy as an academic discipline, while at its foundation in 1772, the Radcliffe Observatory was said to be better equipped than any other observatory in Europe, including the Royal Observatory at Greenwich.

The pursuit of astronomy and mathematics at both Oxford and Cambridge in the seventeenth and eighteenth centuries was very different from what it is today. Universities did not attempt to train undergraduates in specific disciplines, as is now the case, but to impart certain qualities of mind. The main components of the curriculum were Protestant theology, philosophy, and classical languages and literature, and the degree with which a man left the University was an indication that he was qualified to be a useful and loyal subject, a Christian in the Anglican tradition, and the possessor of a finely honed classical intellect. Unlike modern graduates he was not a specialist, unless he was an older scholar taking one of the three higher doctorates in divinity, law, or medicine, after completing his undergraduate degree. Specialism was regarded as narrow and illiberal.

While this gentlemanly education might appear excessively abstract and linguistic in its basis, it also aspired to teach the intellectual virtues of temperance, balance, and proportion, which possessed conspicuously mathematical components. Astronomy, geometry, and arithmetic derived (along with music) from the medieval quadrivium, the four disciplines that were rooted in proportion, harmony, and spatial relationships. There were grounds, therefore, for the inclusion of certain aspects of mathematics in formal university education, provided that they were related to the business of sharpening the intellect rather than overtly 'conjuring' with numbers, as was more fitting to

Oxford mathematics from the 1620s onwards was dominated by the legacy of Sir Henry Savile. The professorial Chairs he founded had an unparalleled influence on the institutional structure of Oxford mathematics. His monument in the antechapel of Merton College shows him flanked by Ptolemy and Euclid, two authors whose work he expounded in lectures at Oxford.

The Schools quadrangle was built under the inspiration of Sir Thomas Bodley between 1613 and 1624. The Savilian Professor of Astronomy lectured in the first floor to the north (on the left) of the central tower, while the Savilian Professor of Geometry lectured in the first floor to the south. The tower itself was fitted out by John Bainbridge as an astronomical observatory.

accountancy or handicraft. Astronomy and geometry, in particular, were deemed worthy of serious academic cultivation, relating as they did to the glories of God's celestial creation and man's perception of the three-dimensional world of nature.

Astronomy and geometry were undergoing exciting new developments in the 1600s, for they showed ways through which practical subjects could cast light upon the most profound philosophical issues. The great voyages of discovery in the wake of Columbus had shown how practical men had effectively undermined classical geographical theories about the land and sea ratios of the globe. The maps and astronomical coordinates by which their discoveries were presented and argued, moreover, gave an unaccustomed philosophical cogency to the once simple handicrafts of measuring and reckoning. Likewise, the physical construction of the universe and the Earth's possible motion around the Sun had emerged as matters of the greatest intellectual importance that could be settled not by debate, but by precise observation and calculation.

In these respects, therefore, astronomy and mathematics became subjects fit for inclusion in a gentleman's education, in so far as they refined his intelligence and demonstrated the glory of God. And, as astronomy emerged as a 'new' science in the seventeenth century—with the heliocentric universe replacing the geocentric one of antiquity—it brought new challenges into the realm of higher learning, not least of which was the status of empirical evidence in the definition of philosophical truth.

A simple stone with only his name engraved on it marks the grave of the first Savilian Professor of Geometry, Henry Briggs, set into the floor of the antechapel of Merton College. This contrasts with the elaborate monument on the wall nearby, commemorating the first astronomy professor, John Bainbridge.

Sir Henry Savile's professors

In the same way that Merton College had provided the foundation for Oxford's rich medieval mathematical tradition, so it was to provide an invaluable service to the seventeenth-century astronomical revival. In the early seventeenth century it was the Warden of Merton, Sir Henry Savile, who in 1619 founded the Chairs of astronomy and geometry which still bear his name. Emphasizing, as he did, the *intellectual* nature of these disciplines within a liberal education, he outrightly dismissed the illustrious Edmund Gunter at the professorial job interview, because Gunter's demonstration of his mathematical inventions was deemed by Savile to be no more than 'tricks' (see Chapter 4).

John Bainbridge was the first man to hold Savile's astronomical Chair, and Henry Briggs the geometrical one. Bainbridge was an active teacher who represented, along with Briggs, a tradition which was to hold for several future Savilian professors—namely, that of first holding an astronomical or geometrical professorship at Gresham College, London, and then moving to a Savilian Chair at Oxford; John Greaves and Christopher Wren were later to do the same.

By Savile's statutes, the astronomy professor was to carry out research as well as to teach. Bainbridge assembled an impressive set of instruments, several of which still survive in the Museum of the History of Science in Oxford, to be used both to teach astronomy and to conduct observations. He made the top room of the Schools tower on Catte Street into an observatory, and undertook a lengthy programme of observations over two decades. Indeed, he went much further in his conception of how to promote the research commitment charged by Savile, and drew up a grand design of coordinated simultaneous observations across the world for purposes of astronomy, cartography, and navigation, organizing astronomical expeditions in the 1630s to South America and the near East and setting in train a

The Oxford–Gresham connection

Gresham College was founded in London in 1597 under the will of the wealthy financier Sir Thomas Gresham, to provide public lectures in seven subjects, including geometry and astronomy. It is notable as the first such foundation and made a considerable impact, and indeed is still active today.

The first geometry professor was **Henry Briggs**, who went on to become the first Savilian Professor of Geometry at Oxford, and the first astronomy professor was the Oxford-educated **Edward Brerewood**. Until well into the eighteenth century the Gresham professors of astronomy and geometry were predominantly appointed from Oxford. This strong connection between Gresham College and Oxford mathematics was to play a significant part in the founding of the Royal Society.

The first five successors of Henry Briggs in the Gresham geometry Chair were all from Oxford—indeed, the first four were from Merton, Sir Henry Savile's college: **Peter Turner** from 1620 to 1631, when he returned to Oxford as the second Savilian Professor of Geometry; **John Greaves** from 1631 to 1643, when he became the second Savilian Professor of Astronomy, **Ralph Button** from 1643 to 1648—he returned to Oxford after the Civil War, but lost his position at the Restoration; **Daniel Whistler** from 1648 to 1657; and **Lawrence Rooke** of Wadham College who moved from the Gresham astronomy Chair to that of geometry in 1657 and died in 1662. After a short break when it was held by two Cambridge figures, Isaac Barrow and (briefly) Robert Dacres, it was another Oxford graduate, **Robert Hooke**, who held the Chair for a considerable period, from 1665 until his death in 1702.

Similarly, the first four Gresham astronomy professors were Oxford graduates: **Edward Brerewood** (Brasenose) from 1597 to 1613; **Thomas Williams** (Christ Church) from 1613 to 1619; **Edmund Gunter** (Christ Church) from 1619 to 1626; and **Henry Gellibrand** (Trinity) from 1626 until his death in 1637. The next two professors, Samuel Foster and Mungo Murray, were not especially connected with Oxford, but the next five astronomy professors were: **Lawrence Rooke** (Wadham) from 1652 to 1657; **Christopher Wren** (Wadham and All Souls) from 1657 to 1661; **Walter Pope** (Wadham) from 1661 to 1687; **Daniel Man** (Christ Church) from 1687 to 1691; and **Alexander Torriano** (St John's) from 1691 to 1713.

tradition to be pursued by Edmond Halley 40 years later. Although his grand design was not fulfilled, Bainbridge's energy and commitment were an inspiration to future generations of Oxford astronomers, and proved the soundness of Sir Henry Savile's choice for his first professor.

John Greaves, who became Savilian Professor of Astronomy in 1643, was a remarkable man, not only as an astronomer, but also as a traveller. Interested as he was in ancient astronomy and systems of measurement, he had tried to collate classical observations with modern ones, an enterprise which led him to undertake a voyage to the eastern Mediterranean and Egypt to determine independently the latitude of

Rhodes—the 'Greenwich meridian' of Ptolemaic astronomy. He also made a survey of the Great Pyramid. When one remembers that all of these places were in Muslim hands in the early seventeenth century, Greaves's work was no mean undertaking for a western Christian. But Greaves was evicted from his Chair and from Merton College in 1648, when the triumphant Parliamentary Visitors at the end of the Civil Wars deprived all Royalist dons of their posts, although he had the satisfaction of seeing his astronomical professorship pass to the Reverend Dr Seth Ward, a man whom he greatly respected.

Astronomy, experimental science, and the John Wilkins circle

Oxford's achievements in the seventeenth century were founded upon a rich culture of experimental science, and between 1648 and 1660 the city housed probably the most dynamic scientific community in Europe. Strong as Oxford's earlier astronomical reputation had been, it was after

Wadham College was built from 1610 to 1613 under the architectural influence of Savile's Fellows' quadrangle at Merton College. Under the wardenship of John Wilkins, it became the centre of Oxford experimental philosophy in the 1650s. In this engraving of 1738, Wilkins is shown at the front (displaying an image of Gresham College), with Thomas Sprat, Seth Ward, and Christopher Wren (displaying the Sheldonian Theatre).

In 1629 one of the Savilian professors (Henry Briggs or John Bainbridge) designed this early morning sundial on a buttress of Merton College Chapel.

Savilian professors of geometry, 1619–48

1619   Henry Briggs (1561–1631)

A Yorkshireman, born near Halifax, Henry Briggs was educated there and at St John's College, Cambridge. He was the first professor of geometry at Gresham College, founded in London in 1597. As someone with strong interests in astronomy and navigation, especially in the construction of tables, Briggs became fascinated by the potential of Napier's logarithms after their publication in 1614:

Naper, lord of Markinston, hath set my head and hands a work with his new and admirable logarithms. I never saw book, which pleased me better, and made me more wonder …

He travelled to Edinburgh in 1616 to stay with Napier for a month. In consultation with Napier he contributed greatly to the development of the tables, and his monumental *Arithmetica logarithmica* (1624) tabulates 30 000 logarithms to 14 decimal places. The quality of his mathematical insight may be judged from the fact that his calculation methods for logarithms involve a special case of the binomial theorem (for $n = \frac{1}{2}$) more than 40 years before Newton discovered the general theorem. An offshoot of his and Napier's work was to promote the general acceptance of decimal fractions.

Briggs was a long-standing friend of Henry Savile, who was also Halifax-born; they held similarly dismissive views about astrology and took a modern view in their appreciation of Copernicus. In 1616 Savile founded a lectureship in arithmetic at Merton College, which Briggs held, and he was chosen by Savile to become the first Savilian Professor of Geometry at Oxford. The respect in which he was held is clear from the memorial poem composed in Greek verse by another Fellow of Merton College:

Great Briggs who raced with stars, engirdled earth,
Euclid and Ptolemy in one matched not his worth …
E'en fate stayed not his art: in death enfurled
His soul communes with stars, his body spans the world.

1631   Peter Turner (1586–1652)

Peter Turner graduated BA (Christ Church, Oxford) in 1605, and in 1607 was elected a Fellow of Merton College. In 1620 he was appointed Gresham Professor of Geometry, in succession to the first professor, Henry Briggs, who had moved to the Savilian Chair of Geometry at Oxford. When Briggs died in 1631, Turner was again appointed his successor. Little is known about his work, although it is known that he lectured on magnetism and electricity. A strong royalist, he suffered imprisonment during the Civil War, and eventually was ejected from the Savilian Chair and his Merton fellowship, in 1648. He died in London, in poverty, three years later. According to Wood he was

a most exact latinist and Grecian, was well skilled in the Hebrew and Arabic, was a thorough pac'd mathematician, was excellently well read in the fathers and councils, a most curious critic, a politician, statesman, and what not … He wrote many admirable things, but being too curious and critical, he could never finish them according to his mind, and therefore cancell'd them.

John Bainbridge and John Greaves, the first two Savilian professors of astronomy.

Savilian professors of astronomy, 1619–49

1619 John Bainbridge (1582–1643)

John Bainbridge was born in Ashby-de-la-Zouch, and educated at the grammar school there, and at Emmanuel College, Cambridge (BA 1603, MA 1607, MD 1614). He was a physician in London when he came to Sir Henry Savile's attention through his *Astronomical description of the late comet* (of 1618), whereupon he was appointed the first Savilian Professor of Astronomy. His *Procli sphaera et Ptolomaei de hypothesibus planetarum* shows that he combined observational abilities with a concern for ancient astronomical texts. Like Briggs, he was a Copernican, and was probably the first person to lecture at Oxford on Kepler's astronomy, as well as the discoveries of Galileo. He was also in continual correspondence with continental scholars on developments in mathematics, and from at least 1624 was working on Arabic astronomical texts—on moving to Oxford he started to learn Arabic for this purpose. As Savilian professor he posted a notice on the gate giving the time and subject of his lecture, ending with the somewhat ill-formed Latin phrase *Lecturus de polis et axis* (meaning roughly 'I'm going to talk about poles and axes'): a wit is supposed to have added the graffito:

Dr. Bambridge
Came from Cambridge
   To read *de Polis et Axis*;
Let him go back again
Like as a dunce he came,
   And learn a new syntaxis.

Whatever his Latin abilities, Bainbridge played a crucial role in introducing the new astronomy to many generations of students, some of whom would later take part in the mid-century scientific developments in Oxford. Bainbridge was buried in Merton College Chapel, near Henry Briggs, and left his manuscripts to his friend Archbishop Usher; they are now in the library of Trinity College, Dublin.

1643 John Greaves (1602–52)

John Greaves was born in Hampshire, and studied at Balliol College, Oxford, from 1617 to 1621. He was elected to a fellowship at Merton College in 1624, where he became a friend of the Savilian professors Henry Briggs and John Bainbridge. There he learned oriental languages and studied ancient astronomical writings. In 1631 he was elected professor of geometry at Gresham College, in succession to Peter Turner who had resigned upon his appointment to the Savilian geometry Chair. He spent several years of the 1630s travelling in Europe and the Middle East, collecting Greek, Arabic, and Persian manuscripts and making observations. Elected Savilian Professor of Astronomy on Bainbridge's death in 1643, he was ejected in 1648 by the Parliamentary Visitors, and died in London four years later. Although circumstances were not conducive to his developing an extensive teaching role as Savilian professor, since his tenure coincided with the Civil War and its aftermath, Greaves contributed notably to the movement for preserving classical and oriental texts: he translated from an Arabic manuscript the *Book of lemmas* attributed to Archimedes, and left a number of astronomical and metrological texts recovered from ancient manuscripts.

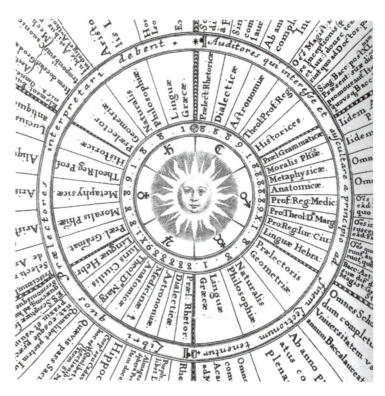

The University syllabus in 1635, designed by the Savilian Professor of Geometry, Peter Turner, presented the weekly lecture timetable in the form of what appears to be a Copernican universe.

John Wilkins' appointment in 1648 to the wardenship of Wadham College (by the same Commission that had evicted Greaves) that things developed as never before. It was Wilkins who invited Seth Ward, a Cambridge man, to Oxford, and gave him rooms in Wadham. Wilkins also attracted the chemist Robert Boyle to Oxford, while Christopher Wren's undergraduate career was spent under Wilkins' tutelage. Astronomers, anatomists, and chemists came to move in the Wadham circle. That great future pioneer of the epistemology of science, John Locke, who was a young don at Christ Church, came to join it, while his fellow collegian Robert Hooke became one of Wilkins' most illustrious protégés. Hooke was perhaps the greatest all-round experimentalist of this brilliant Wadham-based group; in addition to his other achievements he pioneered astronomical instrumentation and optical design, and in 1669 made one of the earliest attempts to measure a stellar parallax.

The members of this Oxford experimental group placed great stress upon instrumentation, for they came to realize that, if rational knowledge is gained through the senses, then a refinement of these senses would push back the limitations upon knowledge. The barometer, air pump, and pendulum clock had greatly improved mankind's ability to discern and quantify minute changes in nature, while Galileo's spectacular telescopic discoveries after 1610 were fundamental in challenging the cosmology of antiquity. Astronomy was, after all, the most highly developed of the sciences, with its clear division of the

Before coming to Oxford, Wilkins published a book popularizing the notion of a Sun-centred cosmology. On the frontispiece, Galileo and Copernicus discuss the new scientific discoveries.

sky into exact geometrical regions and its precise quantification of a complex series of long-term and short-term natural cycles. At a time, moreover, when chemistry, medicine, and botany depended upon largely speculative methodologies, astronomy already possessed a highly developed metrological basis, rooted in the division of the circle into 360 degrees by means of geometrical instruments. It was this close union between mathematics and practical measurement, therefore, which gave astronomy that investigative and predictive power which the other sciences strove so hard to emulate. With its already proven laws and established procedures, astronomy provided a lead for all those sciences seeking a rational, experimental foundation.

Copernicanism and Wilkins' journey to the Moon

John Wilkins himself had done a great deal to explore the implications of the new astronomy for English readers; as early as 1638, while still a 24-year-old scholar at Magdalen Hall, he had written his *Discovery of a new world in the moon*, which was a defence of Galileo. Wilkins had argued that, far from being fixed at the centre of the universe, the Earth was a planetary body which rotated around the Sun along with the other planets.

He later went on, in 1640, to suggest that a manned 'flying chariot' might travel from the Earth to the Moon, powered by machinery, and in his *Mathematical magick* of 1648 he explored the use of applied mathematics to revolutionize technology and improve all manner of machines. John Wilkins saw the planets as moving through an empty universe, now stripped of its legendary crystalline spheres and under the influence of a form of magnetic energy which radiated from the Sun in accordance with Keplerian proportions. When such a man came to hold high office in the University after 1648, as Warden of Wadham College, one can see how he himself acted as a magnet for those young scientific men who, after 1660, would found the Royal Society.

The distinction to which he had risen, and that to which he would soon rise further as Bishop of Chester, tells us a great deal about the intellectual liberalism of Oxford, for neither he, nor any of his friends, suffered in any way because of their scientific interests—quite the contrary, indeed. Seth Ward and Thomas Sprat both sat with Wilkins on the Episcopal Bench, while his lay protégés such as Christopher Wren, Thomas Willis, and Robert Hooke received official recognition for their scientific work. In spite of political turmoil, Civil War, and religious doctrinal battles going on in England during the 1640s and 1650s, astronomy and the natural sciences enjoyed a freedom which was virtually unique in Europe. This derived in part from the sincere religious

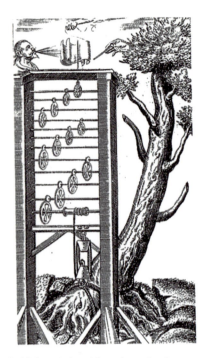

In *Mathematical magick; or, the wonders that may be performed by mechanical geometry* (1648), Wilkins showed how to uproot an oak tree by gearing up a breath of wind.

piety of most of the men involved in it, and in part from the way in which they used evidence from the natural and cosmic realms to demonstrate the grandeur and ingenuity of the divine creation, so that good science and good religion could complement rather than oppose each other.

Science and religion

In understanding the independent status of astronomy in the seventeenth century, we must remember that it made no statements that seriously threatened religious doctrine. The Earth's motion around the Sun, the physical composition of the planets, and the infinity of space, contradicted no Biblical texts in any serious way, and it took the Roman Catholic Church from 1543 to 1616 to get around to prohibiting Copernicus's book. Although the condemnation of Galileo in 1632 shocked Catholic intellectuals across Europe, it was largely political in character and a backlash against Galileo's partiality for controversy.

In England, however, we do not find any official ecclesiastical censure of science. Francis Godwin, whose speculations about a voyage to the Moon had first inspired Wilkins, became Bishop of Hereford, while many other clergy were willing to see the study of the natural world as a divine enterprise in its own right. While these men were not 'liberals', in the way in which the word is used today, they were heirs to a rich allegorical tradition of Christianity which could discern the difference between the poetic and the explicitly factual statements of Scripture.

Living as we do today in an age when crude 'fundamentalism' has acclimatized the western mind to a dramatic 'Bible versus science' juxtaposition, we sometimes fall into the trap of imposing these modern blinkers on to seventeenth-century people. By contrast, the Renaissance cosmological debate was conducted between men possessing a rich and sophisticated theological understanding.

No one more clearly epitomized the Christian 'virtuoso', or experimental scientist, than Robert Boyle. Although never an official member of the teaching establishment of the University, Boyle came to reside in Oxford as a private gentleman at the invitation of Dr Wilkins in the mid-1650s. From his laboratory in the High Street, he not only discovered the famous gas-pressure law which still bears his name and founded modern chemistry, but he also matured his theology of nature. Boyle saw all action in nature as occasioned by mechanical interrelations between fundamental particles, or atoms. But the relationships between atoms must, in themselves, be just as orderly as those between the planets, for they all bore the imprint of divine design. What nobler vocation could a gentleman pursue, therefore, than the exercise of God's gift of intelligence in the elucidation

Robert Boyle with his air pump. In 1663 the Danish ambassador witnessed the asphyxiation of a cat by the evacuation of the pump. This occasioned an anonymous verse describing the proceedings:

> Out of the glass the Ayre being screwed,
> Pusse died and ne'er so much as mewed.

A plaque in the High Street commemorates the site of Boyle's laboratory.

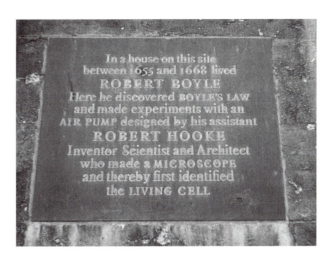

of those subtle truths which God had built into the fabric of the universe?

It was around 1657, when Boyle was searching for someone who could build a reliable air pump for him so that he could develop his vacuum and pressure experiments, that he came to employ Robert Hooke.

Robert Hooke and scientific instrumentation

Although Hooke had come up to Christ Church in 1653, at the age of 18, it was after his involvement as Boyle's assistant that his own experimental genius really flourished and helped to form two characteristics which were to lie at the heart of his mature attitude towards natural science.

The first of these characteristics was the application of mathematical laws to the physical properties of earthy substances, such as the volume and pressure of the air, for law-like operation was not unique to the heavens. Secondly, Hooke realized that new *quantitative* knowledge required specialized tools which refined human perception to gather exact data; in other words, without mathematical instruments, intelligible science was impossible. For the rest of his life, Hooke constantly sought to devise ways by which the forces and properties of nature could be made amenable to mathematical quantification. Inspired as he was by Galileo's dynamical researches, Hooke tried, whenever possible, to extend them into chemistry, physiology, and the invention of useful devices.

Astronomy underpinned Hooke's approach to natural enquiry. Being the most advanced of the sciences and already possessing the most elaborate instrumentation, it provided the natural model for mathematical investigation.

After becoming professor of geometry at Gresham College, London, in 1665, Hooke pioneered the development of instrumentation with particular stress upon the problems of astronomy. In 1669, he addressed himself to the pressing problem of the Earth's motion in space, and attempted to detect a stellar parallax with a new type of telescope. This was a *zenith sector*, a refracting telescope of 36-foot focal length, set up exactly in the vertical to point to the zenith. Hooke realized that the stellar parallax must be very small indeed, and further realized that previous attempts to measure it had been frustrated by the imprecise knowledge of the effect of atmospheric refraction upon starlight. By observing the star γ Draconis as it passed exactly overhead in London, however, Hooke argued that one could avoid entirely the atmospheric refraction problem, for zenith starlight passes through the air vertically, rather than obliquely as does the light from non-zenith stars. Yet, while Hooke's zenith sector was still too imprecise to measure the fraction of an arc second of even the largest parallaxes, his conception of the problem, along with its specialized instrumentation, was a *tour de force* of creative applied mathematics. It was also destined, over 60 years later, to provide the basic methodology by which Oxford's James Bradley discovered the *aberration of light*, and the Earth's axial nodding or *nutation*.

Hooke's other contributions to the solution of physical problems by the invention of specialized mathematical instruments were equally impressive. He developed and advocated the telescopic sight for astronomical measuring instruments, such as quadrants; this made it possible to measure star positions that were 40 times more accurate than could be obtained using the naked-eye sight. He also applied great attention to the perfection of the engraved scales of mathematical instruments, and was one of the first persons to recognize and publicize the mathematical properties of screw threads.

In 1667, he took up William Gascoigne's largely forgotten idea of the screw-micrometer as a device whereby very small fractions of a degree could be measured with precision. Although it was Richard Towneley who developed Gascoigne's idea of the screw of the eye-piece or 'filar' micrometer for telescopes, it was Hooke who developed the tangent screw as a general method of precision angle division.

While still at Christ Church, Hooke had become interested in the design of more accurate clocks and watches and, in this respect, his research into the physics of springs became fundamental. Like so many of Hooke's ideas, his work on springs functioned on a variety of practical and intellectual levels. Realizing that the energy given out by a spring is strictly proportional to the energy necessary to torque it up,

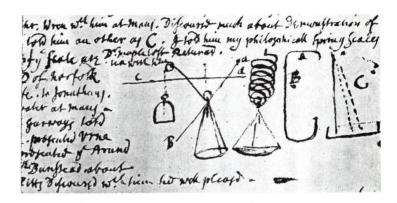

Robert Hooke's diary for 21 August 1678 records a visit to a coffee house with Sir Christopher Wren, where they exchanged information on their recent inventions, including Hooke's 'Philosophicall spring scales'.

he found a mathematical relationship which made possible an isochronous pulse that was just as accurate as that of a pendulum, yet, as the spring was independent of gravity, it did not have to be placed in a stable vertical position. The spring law which bears Hooke's name made possible the hairspring-regulated balance wheel of watches, a device which lay at the heart of all accurate portable timekeepers from the 1660s down to the development of quartz-crystal electronic watches in our own time.

Sir Christopher Wren

One of Hooke's lifelong friends was Sir Christopher Wren, whom he first met at Oxford and who was also to make fundamental contributions to astronomy, mathematics, and experimental physics before going on to an even more illustrious career as an architect.

It would be hard to review the life and career of Christopher Wren without feeling a twinge of envy. He was a youthful prodigy, whose gifts seemed only to increase with the passing of the years. Born into a distinguished family, his abundant talents, charm, and loyal friends (which included the Royal Family) sustained him through a healthy and active life terminated by a sudden death while dozing in an armchair at the age of 91.

Entering Wadham College in 1646 at the age of 14, two years before Dr Wilkins became Warden, he dazzled his contemporaries with his youthful talents and was elected a Fellow of All Souls College in 1653. Appointed Gresham Professor of Astronomy in 1657, he returned to Oxford to take up the Savilian Chair of Astronomy in 1661.

Like Hooke, Boyle, Wilkins, and the others, Wren developed a profound sense of orderliness of the natural world, and realized that this order was ultimately mathematical in character. One of the early Royal Society's projects was to produce a 'history' or accurate account of the weather, to establish its physical and hopefully predictive basis. In consequence, Christopher Wren and Robert Hooke produced designs for

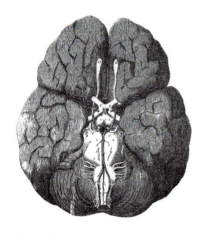

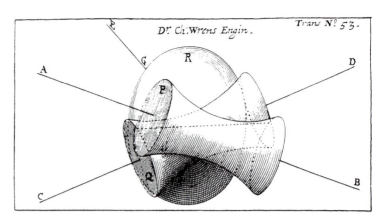

A fine draftsman, Christopher Wren drew the illustrations for the work on brain anatomy (*Cerebri anatome*, 1664) by his Oxford colleague Thomas Willis, the Sedleian Professor of Natural Philosophy.

Wren's blend of mathematical and practical insights is seen in his design of an engine for grinding hyperbolical lenses, shown to the Royal Society in July 1669.

several self-recording instruments to monitor temperature, pressure, and rainfall. His 'weather clock' of 1663 perhaps represents the first attempt ever made to record changing natural phenomena by a clock-driven pen-recorder device.

In addition to his work on instruments to measure and record nature, Wren possessed a first-class intellect for analytical mathematics. One of his greatest contributions in this area lay in the solution which he produced for the rectification of the cycloid.

A *cycloid* is the path taken by a point on the circumference of a rolling circle, such as a wheel. Its elegant shape, and some of its properties, had first been noted by Galileo, and a large number of European mathematicians, including Roberval, Descartes, Huygens, and Wallis, devoted their attention to the exploration of its properties. It was the fine proportions inherent in the cycloid which fascinated scholars; amongst other things, it made possible Christiaan Huygens' 'cycloidal cheeks', which were so important in controlling the swings of isochronal pendulums.

What was not yet known was how to rectify a cycloid—that is, to find a straight line of the same length as one arch of the curve. Wren wrote four tracts on the cycloid in which he demonstrated that the length of the cycloidal curve is exactly four times the diameter of the circle that generates the curve. He further developed a significant branch of curvilinear analysis for the study of cycloidal paths that formed an important link in the development of Newtonian fluxions and calculus.

Wren went on to pioneer researches into the laws of impact between colliding bodies, and in 1669 presented a paper to the Royal Society which resulted from these experiments. Using a frame of suspended balls (similar to the device popularly known as 'Newton's cradle'), he determined that the impact by which colliding balls hit and bounce off each other is proportional to their sizes and speeds. Eighteen years before the publication of Newton's *Principia*, Wren's

A bust of Sir Christopher Wren was carved by Edward Pearce in 1673, the year in which he resigned the Savilian Chair of Astronomy to concentrate on architectural work.

work provided a clear indication of how the as yet imprecisely perceived concepts of mass, distance, and velocity were coming to be understood by scientific men.

What was also coming to be realized at this period—at least by Hooke, Wren, Wallis, and the young Newton—was the way in which the laws that bind the motions of the planets are apparently the same as those which bind rolling, falling, and colliding terrestrial bodies. Although Wren, Hooke, Wallis, and (15 years later) Edmond Halley were moving towards partial solutions to this problem, it was Isaac Newton who eventually drew all the strands together and invented a coherent mathematical language whereby the laws of universal gravitation could be quantified and demonstrated. Yet Oxford's contribution to the development of gravitational theory should not be overlooked.

While we may no longer think of astronomy and physiology as mathematically related disciplines, they did possess many potential connections in the seventeenth century. William Harvey's discovery of blood circulation in 1628 opened up a possibility for the mathematical investigation of living things, for if the heart was a pump, then it must exert mechanical pressure on the structure of the body. In the way that Robert Boyle was attempting to explain chemical reaction by means of atoms, so Thomas Willis of Christ Church was trying to understand the 'physico-mechanical' relations of the human body. Wren's universal interests drew him to anatomists as well as mathematicians in the scientific circles of the 1650s and 1660s, while his artistic and observational gifts made him an ideal anatomical illustrator. It was Wren who made most of the drawings for Willis's *Cerebri anatomi* (1664), which was the first 'modern' treatise on the anatomy of the human brain.

Once it was realized that the blood circulates under the mechanical pressure of the heart, and also passes up the carotid arteries to the brain, drugs could be administered intravenously, rather than mixed with food. In addition to his own work on the cardiac and circulatory systems of dogs and other animals, Wren was one of the first persons to administer an intravenous drug, when in 1656 he and two medical friends injected his dog with a mixture of wine and opium and noticed how quickly it fell asleep, in contrast with the drug's much slower action when taken orally.

Wren's scientific intellect, like that of Hooke, travelled across many fields of investigation, and a common approach imbued all their investigations. That approach consisted of realizing that the whole of nature, from planets via colliding bodies to circulatory systems, obeys laws that are capable of being understood mathematically. It was this omnipresence of mathematics, combined with that good taste which was itself an expression of divine reason, which made it possible for Wren to devote the remaining three-fifths of his life to architecture.

Architecture and mathematics

It would be inconceivable in the twentieth century that two academic astronomers should be appointed to re-design the principal buildings of a capital city in the event of its being suddenly destroyed. Yet this is precisely what happened in London after the Great Fire of 1666. Christopher Wren, Oxford's Savilian Professor of Astronomy, was appointed Surveyor to the King, to be responsible for new Royal buildings, such as churches and St Paul's Cathedral. Robert Hooke, Gresham College Professor of Geometry and Curator of Experiments of the Royal Society, was similarly appointed Surveyor to the City of London, to re-design the streets and most of the civic and commercial buildings. While Wren already possessed a reputation as an architect, with the Royal assistant-surveyorship to his credit, not to mention the design of the as yet incomplete Sheldonian Theatre, Hooke seems to have been an architectural novice. No matter how one looks at it, however, neither man had received a professional training in architecture.

The remarkable and innovative Sheldonian Theatre (1664–7) was designed by Christopher Wren while he was Savilian Professor of Astronomy at Oxford.

Yet both men succeeded brilliantly in their tasks. While Wren came to give up science for architecture, resigning his Savilian Chair in 1673, Hooke continued to conduct fundamental scientific research at the same time as redesigning the city's buildings. What made it so easy for these astronomers suddenly to produce exquisite architecture was the same geometrical sense, combined with original creative gifts, which made it possible for them to study pneumatics, impact, physiology, and instrumentation. In these men, a sense of the mathematical harmonies inherent in all things was combined with a natural eye for beauty, enhanced and refined by an Oxford classical education.

Every seventeenth-century astronomer knew his Euclid, and from geometrical theorems it was a natural step to the concrete geometry of structures. Men who had encountered the classical Roman architectural writer Vitruvius as part of their cultural training, and who had easy access to the printed works of Palladio and Serlio and the buildings of Inigo Jones (another gifted amateur), would have had little difficulty in devising a presentable building.

Classical and Renaissance architecture possessed strongly formulaic elements, and once one had mastered the naturally tasteful relations of the cube and the half-cube, one possessed a formula with which to fill a given space. With knowledge of the client's requirements, the nature of the site, and the amount of money available, the facade, fenestration, internal arrangement of rooms, and decoration could be computed with relative ease. What made Hooke's and Wren's buildings so exceptional was not the mere classical pastiche that one would expect from simply following the rules, but their daring originality.

Wren's innovative Sheldonian Theatre and St Paul's dome, and Hooke's sadly destroyed Physicians' College and Bethlehem (Bedlam) Hospital with its 520-foot facade, were masterpieces by any standard. Wren was probably the greater architect, but he devoted the greater part of his life to architecture while Hooke continued as a 'full-time' scientist. Even so, Hooke's buildings were often confused with Wren's by later generations, and it was not until the chance discovery of Hooke's *Diary* from a private sale in 1891 that the true provenance of some 'Wren' buildings became known.

The most appropriate building upon which both men expended their energies was the Royal Observatory in Greenwich. Although it was designed by Wren (as Royal architect) in 1675, Hooke was intimately involved and probably laid out the ground plan. We do not know of the part played by Hooke in the planning of the observatory's fabric, but he certainly designed the original set of instruments which went into it. These were devised in accordance with Hooke's own theories of optics, mechanics, horology, and scale graduation, and while their experimental status sometimes exasperated John Flamsteed (the first Astronomer Royal) when he used them, they came to form the prototypes for other instruments used in observatories across Europe.

The Royal Observatory not only provides us with a splendid example of an astronomical building designed by two distinguished Oxford astronomers, but is also the supreme instance of great architecture done on a shoe-string. Built from second-hand bricks and timber from the demolished Tilbury Fort, and erected in record time, the Royal Observatory cost £520 9s—which was £20 9s above the estimate. For comparison, Christopher Wren's 'Great Model' of the proposed St Paul's Cathedral, an exquisite 18-foot-long miniature domed edifice still preserved in St Paul's crypt, cost £505 6s.

Oxford science in the middle and late seventeenth century seemed to abound with gifted men who could turn their hands to anything and make original contributions. Indeed, just as Christopher Wren and Robert Hooke could switch from astronomy to architecture, so John Wallis switched from divinity to mathematics, simply because he was given the Savilian Chair of Geometry. Once established, he became one of Oxford's greatest pure mathematicians.

Further reading

For modern studies of John Wilkins, see Barbara Shapiro's *John Wilkins 1614–1672: an intellectual biography*, University of California, 1969, and Allan Chapman, 'A world in the Moon: John Wilkins and his lunar voyage of 1640', *Quarterly Journal of the Royal Astronomical Society* **32** (1991), 121–32. For Wilkins' collected works, see the two volumes of J. Wilkins, *Mathematical and philosophical works*, London, 1802, in the Frank Cass and Co. reprint edition.

Margaret Espinasse's *Robert Hooke*, Heinemann, London, 1956, is still the only biography of Hooke, although *Robert Hooke, new studies* (edited by Michael Hunter and Simon Schaffer), Boydall Press, Woodbridge, 1989, provides new scholarly insights; see also the *Diary of Robert Hooke 1672–1680* (edited by H. W. Robinson and W. Adams), London, 1935.

Sir John Summerson's *Sir Christopher Wren*, Collins, London, 1953, provides a good account of his life and architecture, while much of his scientific work is published in *Parentalia: or memoirs of the family of the Wrens ... but chiefly Sir Christopher Wren* (edited by Sir Christopher Wren, junior), London, 1750. Wren's mathematical and astronomical work is discussed in J. A. Bennett's *The mathematical science of Christopher Wren*, Cambridge, 1982. Robert Plot's *Natural history of Oxfordshire*, London, 1705, in the Minet reprint edition, contains some of the earliest published studies of Wren's designs for the Sheldonian Theatre, Oxford.

R. E. W. Maddison's *The life of the honourable Robert Boyle F.R.S.*, Taylor and Francis, 1969, provides a valuable account of the Oxford scientific community of the mid-seventeenth century. Anthony V. Simcock likewise examines the early scientific community in *The Ashmolean Museum and Oxford science 1683–1983*, Museum of the History of Science, Oxford, 1984. Robert T. Gunther provides one of the most complete accounts in Volumes II and XI of *Early science in Oxford*, Oxford, 1923, 1937. Lively and readable gossip about the lives of many early Oxford scientists is contained in John Aubrey's *Brief lives* (edited by O. L. Dick), Secker and Warburg, London, 1975.

Other articles of interest are Allan Chapman's 'Gresham College: scientific instruments and the advancement of useful knowledge in seventeenth-century England' [the 1997 annual invitation medal lecture], *Bulletin of the Scientific Instrument Society* **56** (1998), 6–13, and his essay articles on 'Great Britain, astronomy (1500–1980)', 'Oxford University, astronomy (1500–1980)', and 'Cambridge University, astronomy (1500–1980)' in *History of astronomy, an encyclopedia* (edited by J. Lankford), Garland, 1997.

Johannis Wallis S. T. D.

Geometriæ Professoris Saviliani, in Celeberrima
Academia Oxoniensi,

OPERUM MATHEMATICORUM

Volumen Tertium.

QUO CONTINENTUR

CLAUDII PTOLEMÆI ⎫
PORPHYRII ⎬ Harmonica:
MANUELIS BRYENNII ⎭

ARCHIMEDIS ⎰ Arenarius, &
⎱ Dimensio Circuli;

Cum EUTOCII Commentario:

ARISTARCHI SAMII, de Magnitudinibus & Distantiis
Solis & Lunæ, Liber:

PAPPI ALEXANDRINI, Libri Secundi Collectaneorum,
hactenus desiderati, Fragmentum:

Græce & Latine Edita, cum Notis.

ACCEDUNT

EPISTOLÆ nonnuilæ, rem Mathematicam spectantes;

ET

OPUSCULA quædam MISCELLANEA.

OXONIÆ,
E THEATRO SHELDONIANO, An. Dom. MDCXCIX.

CHAPTER 6

John Wallis

Raymond Flood and John Fauvel

The introduction of John Wallis into the University of Oxford was occasioned by politics. During the Civil War, the University had been the royalist headquarters, and in the reckoning afterwards most college heads and Fellows were deposed. Both Savilian professors, Peter Turner, the professor of geometry, and John Greaves, the professor of astronomy, were expelled in 1648 for royalist sympathies, and the Parliamentary Commissioners replaced them by Wallis and Seth Ward, respectively. Like Ward, the 33-year-old Wallis was a Cambridge man; unlike him, he had been open in his support for the revolutionary cause during the Civil War.

It might strike us as remarkable that prior to taking up the Savilian Chair Wallis had little mathematical experience and enjoyed no public reputation as a mathematician. As he himself wrote,

In the year 1649 I removed to *Oxford*, being then *Publick Professor of Geometry*, of the Foundation of S^r. *Henry Savile*. And *Mathematicks* which had before been a pleasing diversion, was now to be my serious Study.

His previous involvements had been in theology and administration, which had included the secretaryship to the Assembly of Divines—a body set up during the Civil War to devise a new form of church government. However, his mathematical skills had become evident when he deciphered secret codes for Cromwell's intelligence service in 1642; and on reading William Oughtred's *Clavis mathematicae* ('The key to mathematics') in 1647 he was so enthralled that he began to explore the subject for himself and rediscovered Cardano's method of solving cubic equations. These interests, together with the impression he had made on influential people through his work for the Assembly of Divines, were sufficient to bring him to the Savilian Chair of Geometry. A more far-sighted mathematical appointment on flimsier evidence is difficult to imagine. Wallis's appointment to the Savilian Chair, which he held until his death 54 years later, marked the beginning of an intense period of activity which established Oxford as the mathematical powerhouse of the nation, for a time at least, and Wallis himself as the most influential of English mathematicians before the rise of Newton.

John Wallis's *Collected works*, published at the end of his life from the Sheldonian Theatre designed by his pupil Christopher Wren, attest to a remarkable career promoting both mathematical studies and publishing at Oxford. The third volume contained his contributions to the Oxford ancient texts project: his editions of works by Ptolemy, Archimedes, Aristarchus, Pappus, and others.

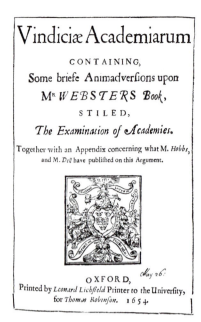

Vindiciæ Academiarum

CONTAINING,

Some briefe Animadverſions upon
Mr WEBSTERS Book,

STILED,

The Examination of Academies.

Together with an Appendix concerning what M. *Hobbs*,
and M. *Dell* have publiſhed on this Argument.

OXFORD, *May 26.*
Printed by *Leonard Lichfield* Printer to the Univerſity,
for *Thomas Robinſon.* 1654.

When the puritan reformer John Webster
attacked the universities for neglect of
practical subjects and ignorance of recent
mathematics, Seth Ward vigorously rebutted
Webster's 'buzzing Discourse', the 'puerility
of his Arguments', and his 'crude and jejeune
Animadversions'.

Oxford in the 1650s

The Savilian statutes obliged Wallis to give, besides lectures on Euclid's *Elements*, the *Conics* of Apollonius and the works of Archimedes, introductory courses in practical and theoretical arithmetic. These elementary lectures gave rise to his *Mathesis universalis* (1657) which is valuable for his promotion of the power of appropriate symbolism and careful examination of the development of algebraic notation and techniques, as well as for the insight it gives us into the level of preparation of students attending mathematics classes at Oxford.

We can also learn something of mathematics teaching in 1650s Oxford from the response made by Seth Ward to an attack on the universities from the puritan reformer John Webster. In a lively pamphlet, *Vindiciae academiarum*, Webster had claimed, among many other things, that recent mathematics was neglected in the universities, and any advances in the subject had been made by private individuals elsewhere:

but that some private spirits have made some progress therein, as Napier, Briggs, Mr. Oughtredge, and some others, it had lain as a fair garden unweeded or cultivated, so little have the Schools done to advance learning, or promote Sciences.

Ward had no difficulty in pointing out the absurdity of such a charge when Briggs, for one, had been a professor at Oxford. He went on to describe the mathematics and astronomy teaching provided, noting the benefits of appointments made under the new political ascendancy before alluding to the recent work of John Wallis which had not yet been published.

Arithmetick and *Geometry* are sincerely & profoundly taught, Analyticall Algebra, the Solution and Application of Æquations, containing the whole mystery of both those sciences, being faithfully expounded in the Schooles by the professor of Geometry, and in several Colledges by particular Tutors ... These Arts he mentions, are not only understood, and taught here, but have lately received reall and considerable advances (I meane since the Universities came into those hands wherein now it is) particularly Arithmetick, and Geometry, in the promotion of the Doctrine of *Indivisibilia*, and the discovery of the naturall rise and mannagement of Conic Sections and other solid places.

Indeed, it is clear from *Mathesis universalis* that, as well as exploring his theoretical interests and creating a research school to advance mathematical learning, Wallis was teaching Oxford mathematics at a practical, down-to-earth level which met the urgings of more extreme reformers such as Webster better than they realized. This was supported by a network of college tutors, as indicated by Ward's remark about 'particular Tutors'. Some wealthy students, furthermore, employed in addition a private instructor specially for mathematical coaching, although the Savilian professors themselves generally gave private tuition free as it was part of their duties under the Savilian

Several mathematicians of 1650s Oxford learned their trade from the Surrey rector William Oughtred, whose posthumous papers were published in Oxford as well as his influential algebra text *Clavis mathematicae*.

GUILELMI OUGHTRED
ÆTONENSIS,

Quondam collegii Regalis
In Oughtred (W.)

CANTABRIGIA Socii.

Opuscula MATHEMATICA hactenus inedita.

OXONII,

E THEATRO SHELDONIANO,
Anno ꓕꓷ 6 7 7; excudit

statutes to make themselves available and 'of easy access to the studious who would consult them on mathematical subjects'. Ward's radical approach to teaching Euclidean geometry was reported by John Aubrey, that Ward 'did draw his Geometricall Schemes with black, red, yellow, green and blew Inke to avoid the perplexity of *A*, *B*, *C*, etc.' Other tutors too gave freely of their time to those interested to learn. In short, there was a range of teaching provision through which students could learn various aspects of mathematics, from basic arithmetic and geometry to astronomy and the use of instruments.

The pedagogical concerns of the Savilian professors are exemplified in their enthusiastic promotion of the text from which both had learned their algebra, Oughtred's *Clavis*. This work, originally appearing in Latin in 1631, had been translated into English in 1647 by Robert Wood, an Oxford mathematician who later became for a short time mathematical master at Christ's Hospital. A new Latin edition was published in Oxford in 1652, with the support of Ward and Wallis, and containing as an appendix a work on sundials translated by the 20-year-old Christopher Wren, a student at Wadham College 'from whom we may expect great things' in the prescient words of Oughtred's preface. Further Oxford editions, in Latin and in English, followed in later decades. Although by the end of the century many came to feel that the *Clavis* had long passed its usefulness, Wallis's enthusiasm for promoting Oughtred's work and ensuring its continued availability was supported by Isaac Newton, who wrote in the 1690s:

Mr Oughtred Clavis being one of ye best as well as one of ye first Essays for reviving ye Art of Geometricall Resolution & Composition I agree wth ye Oxford Professors that a correct edition thereof to make it more usefull & bring it into more hands will be both for ye honour of or nation & advantage of Mathematicks. Is N.

Arithmetica infinitorum

After his appointment to the Savilian Chair, Wallis studied all the major mathematical literature in the Bodleian Library, at that time an achievable task, and the mathematical library for the Savilian professors (now absorbed into the Bodleian). He encountered the works of Torricelli in which Cavalieri's method of indivisibles (which Ward referred to as the 'Doctrine of *Indivisibilia*') was used. This was of considerable interest to him because he thought the method could be used to find the area of the circle. It occupied him for over three years and led to his work *Arithmetica infinitorum*, which placed him in the forefront of mathematicians of his time and remains one of the most influential works in the history of mathematics.

Dedicated to William Oughtred, whose teaching and approach had made a great impression on Wallis, *Arithmetica infinitorum* appeared at a critical time in the development of mathematics, and was full of fresh discoveries and insights. In it, Wallis arithmetized the *Geometria indivisibilibus* of Cavalieri who, using a laborious pairing of geometric indivisibles, had found the area under any curve of the form $y = x^n$ in the case that n is a positive whole number. Wallis, however, associated numerical values to the indivisibles which allowed him to extend the result successfully to the case when n is fractional. This was a great advance on anything previously obtained and Wallis—always as interested in the process as in the result—gave a full account of his methods in order, he said, to throw open the very fount to his readers, rather than imitate the methods of the ancients, who strove to be admired rather than understood and whose methods were 'harsh and difficult, so that few would venture to approach them'. The very word 'interpolation' in its mathematical sense was introduced by Wallis in this work.

It was here, too, that Wallis produced his celebrated formula for what we would call $4/\pi$ and the reciprocal of what we would write as $\int_0^1 (1 - x^2)^{1/2}$; namely,

$$\frac{4}{\pi} = \frac{3}{2} \cdot \frac{3}{4} \cdot \frac{5}{4} \cdot \frac{5}{6} \cdot \frac{7}{6} \cdot \frac{7}{8} \cdot \frac{9}{8} \cdot \frac{9}{10} \cdots$$

Wallis passed this result to the Oxford graduate and amateur mathematician William Brouncker, later to be first president of the Royal Society, who produced another remarkable expression in the form of a 'continued fraction' (another term first introduced in *Arithmetica infinitonum*):

$$\frac{4}{\pi} = 1 + \frac{1^2}{2+} \cdot \frac{3^2}{2+} \cdot \frac{5^2}{2+} \cdot \frac{7^2}{2+} \cdots$$

Among the mathematicians influenced by *Arithmetica infinitorum* was, crucially, Isaac Newton. It was through his study of this work in 1664–5 that Newton came to discover the binomial theorem, as he recalled in a letter to Henry Oldenburg, secretary of the Royal Society:

At the beginning of my mathematical studies, when I had met with the works of our celebrated Wallis, on considering the series by the intercalation of which he himself exhibits the area of the circle and the hyperbola …

What appealed to Newton was the fundamental engine of discovery in Wallis's work, the exploration and recognition of pattern. D. T. Whiteside has drawn attention to the connection between this aspect of *Arithmetica infinitorum* and Wallis's decoding activities during the Civil War:

Essentially Wallis in his interpolation approach sets up the pattern of tabulated instances in a two-dimensional array, and then compares individual instances with surrounding ones in a search for general aspects of the pattern—much as the decoder uses context checks in trying to abstract a meaning from the pattern of symbols before him.

Jobannis VVallis,
GEOMETRIÆ PROFESSORIS
SAVILIANI,

ORATIO INAVGVRALIS:

IN

Auditorio Geometrico, *Oxonii*, habita;
ultimo die Menfis Octobris, Anno
Æræ Chriftianæ 1649. quum publicam
Geometriæ Profeffionem aufpicatus eft.

OXONII,
Typis *Leonardi Lichfield* Academiæ Typographi.
Impenfis *Tho. Robinfon.* 1657.

Wallis's inaugural lecture as Savilian Professor of Geometry was given on the last day of October 1649 in the Geometry Lecture Room at the east end of the 'Schools quadrangle' which now forms part of the Bodleian Library.

Not only was Wallis's career set in motion by his cryptological skill, which seems to have characterized his mathematical style as well, but he continued to serve the State in a code-breaking capacity to the end of his life. This was not widely bruited, for obvious security reasons, and remains one of the lesser-known aspects of Wallis's work. Indeed, he wrote from Oxford in 1690 to protest about a reported leak of information, by pointing out to his Whitehall contact the basic principles of secret decoding work:

why I doubt whether it be advisable to make it so publike (that I decipher y^e French Kings letters) is, because it can then hardly be avoided but that

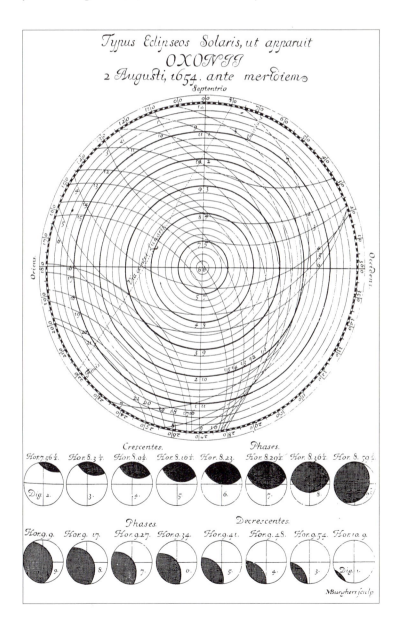

A solar eclipse was observed at Oxford on 2 August 1654 by John Wallis, Christopher Wren, and Richard Rawlinson.

Hobbiani Puncti

DISPVNCTIO

OR

The Vndoing of Mr *Hobs's*

Points:

In Anſwer to M. Hobs's

ΣΤΙΓΜΑΙ,

Id eſt,

STIGMATA HOBBII.

By JOHN WALLIS, Profeſſor in Geometry.

OXFORD.

Printed by *Leonard Lichfield*, Printer to the Univerſity for *Tho. Robinſon,* Anno 1657.

THOMÆ HOBBES

QVADRATVRA CIRCVLI,

CVBATIO SPHÆRÆ,

DVPLICATIO CVBI;

Confutata.

Authore *JOHANNE WALLIS* S. T. D

Geometria Profeſſore SAVILIANO.

OXONIÆ·

Typis LICHFIELDIANIS Acad. Typograph. *Impenſis* THO. GILBERT, A. D. 1669.

Title pages of Wallis tracts against Hobbes.

---

this will some way or other come to .y^e. French Kings knowledge from some of his correspondents here; which will be attended at lest with one or both of these inconveniences, viz: A greater care to prevent the intercepting of such letters; And, a change of the ciphers they now use (which they have allready changed more than once, for I have allready nine or ten of their ciphers by mee) for others more difficult, & (perhaps) in superable.

Newton was not the only mathematician to benefit from studying *Arithmetica infinitorum*; others had already absorbed the techniques and were making use of them. A good example is William Neile, a student at Wadham College who at the age of 19, in 1657, used the techniques to rectify (find the length of) the semi-cubical parabola $ay^2 = x^3$. Wallis gave an account of Neile's work in his *Tractatus duo* (1659), together with an improvement of it by William Brouncker, and was later embarked on a priority dispute when the French mathematician Pierre Fermat published essentially the same rectification in 1660 and appeared to be claiming originality in not only the rectification of the curve but also the very possibility of rectifying any curve. In fact, shortly after Neile's work, another promising young Oxford mathematician, Christopher Wren, had succeeded in rectifying another curve, the cycloid—work which was also published by Wallis in *Tractatus duo*. A notable feature of this episode was Wallis's pride in the achievements of the research school, in effect, which he was building up at Oxford.

This was not the first time that Wallis had had occasion to quarrel with Fermat, who had responded to *Arithmetica infinitorum* rather ungraciously, complaining that the new symbolism was unnecessary and lengthier than classical techniques and claiming to have already established the quadratures that Wallis had found. Lengthy correspondence followed which Wallis eventually published as *Commercium epistolicum* (1658). Further quarrels with another Frenchman, Blaise Pascal, left Wallis with a somewhat jaundiced view of French mathematicians which found its way into the great history of algebra he wrote a quarter of a century later.

Wallis the controversialist

Foreign scholars were not the only ones with whom Wallis carried on lengthy controversies. For nearly a quarter of a century, from the mid-1650s onwards, Wallis was engaged in a virulent dispute with an Oxford scholar of an earlier generation, Thomas Hobbes, who had been an undergraduate at Magdalen Hall (later incorporated into Hertford College) in the first decade of the century. A friend of Francis Bacon, as well as of Galileo, Descartes, and Mersenne, Hobbes was one of the outstanding intellectual figures of the age. Aubrey's report of

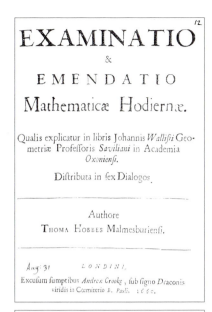

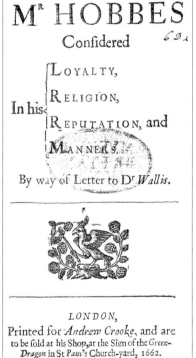

Title pages of Hobbes tracts against Wallis.

how Hobbes came to mathematics in his middle years reveals how lacking his early Oxford days must have been in this respect:

He was 40 years old before he looked on Geometry; which happened accidentally. Being in a Gentleman's Library, Euclid's *Elements* lay open, and 'twas the *47 El. libri I* [the 'theorem of Pythagoras']. He read the Proposition. *By G—*, sayd he (he would now and then sweare an emphaticall Oath by way of emphasis) *this is impossible!* So he reads the Demonstration of it, which referred him back to such a Proposition; which proposition he read. That referred him back to another, which he also read. *Et sic deinceps* that at last he was demonstratively convinced of that trueth. This made him in love with Geometry.

Hobbes' new-found enthusiasm for mathematics subsequently pervaded his philosophical approach and style of writing, and he also gained a mathematical and scientific reputation respected by his European friends. His materialist and anti-clerical *Leviathan* (1651) created widespread controversy even before he aroused the ire of both Ward and Wallis through attacks on the post-Civil War state and performance of the universities, which he saw as riddled, somewhat as Webster did, with priestcraft and outmoded learning. The publication of his *De corpore* in 1655 was the occasion for both Wallis and Ward to prepare fierce attacks on many aspects of Hobbes' work. The titles of Wallis's pamphlets, and Hobbes' rejoinders, tell their own story.

This notorious dispute, which lasted until after Hobbes' death in 1679, is complicated to evaluate. It is true that Hobbes' view of mathematics was quite different from that of Wallis, and also true that Hobbes did not approach Wallis's sophisticated understanding of modern, or indeed ancient, mathematics. But the contest went on too long to be only about Hobbes' mathematical competence. Even though Hobbes rather unwisely claimed to have squared the circle, he also made some telling and acute observations on Wallis's mathematical style and arguments. Overall, the issues went deeper. One thing which plainly worried both Wallis and Ward was Hobbes' claim to have grounded his materialistic philosophy upon mathematics; if this were widely believed or accepted, the association could only damage public sympaty for the subject which the Savilian professors, both men of the established church, were trying to promote. More generally, as Simon Schaffer has argued, this was a dispute over 'the propriety of the separate authority in civil society of teachers, mathematicians, or divines'. But the tone in which the dispute was conducted leaves a strong impression of the forthright nature of Wallis's vigorous personality.

Wallis was blessed with a formidable intellect, a prodigious memory, a robust constitution, and was a kind and affectionate father. He also possessed a highly contentious nature, and managed to create many enemies, remaining all his life a man of short temper and robust dialogue. The antiquary Anthony à Wood (himself scarcely less irascible) recorded a trivial but apparently typical example, when in the early 1680s Wallis borrowed a set of keys from Wood. On Wood asking for their return, two years later, 'he told me he loved not to be expostulated with, that I was in drink that I talked so with him, so that if I had cringed and licked up his spittle, he would let me have had the key. He pointed to the door, and bid me be gone with his 3 corner cap.'

Mathematical language

Besides his exploration of patterns in indivisibles, another important contribution that Wallis made in the 1650s was what Ward described as 'the discovery of the naturall rise and mannagement of Conic Sections'. Wallis published his investigations of this, called *De sectionibus conicis, nova methodo expositis*, in 1656, bound in the same volume as the *Arithmetica infinitorum*. The conics—parabola, ellipse, and hyperbola—and their properties had been known from antiquity, but up to now the curves had been viewed as sections of a cone, something reached through three-dimensional geometry. Wallis's 'new method' was to regard them as plane curves with no reference to the cone, after the initial derivation, and to produce their properties through the use of the techniques of algebraic analysis introduced by Descartes.

A feature of Wallis's work which merits special note is his enthusiastic and thoughtful adoption and extension of the new symbolism of Oughtred and Descartes, evident throughout his works of this period

PARS PRIMA.

PROP. I.

De Figuris planis juxta Indivisibilium methodum considerandis.

SUppono in limine (juxta Bonaventuræ Cavallerii *Geometriam Indivisibilium*) Planum quodlibet quasi ex infinitis lineis parallelis conflari : Vel potius (quod ego mallem) ex infinitis Parallelogrammis æque altis; quorum quidem singulorum altitudo sit

totius altitudinis $\frac{1}{\infty}$, sive aliquota pars infinite parva; (esto enim ∞

nota numeri infiniti;) adeoque omnium simul altitudo æqualis altitudini figuræ.

Est item (propter tangentem) $DT \overline{>} DO$ (hoc est, DT æqualis vel major quam DO; illud quidem si D, P, coincidant; hoc, si secus) & $DTq \overline{>} DOq$, hoc est $\frac{f^2 \pm 2fa + a^2}{f^2} p^2 > \frac{-d \pm a}{d} p^2$; & (utrumque multiplicando in $df2$ &

In his 1655 book on conic sections, Wallis introduced two lasting mathematical symbols: for infinity and for 'greater than or equal to'.

and another reflection of his interest in cryptography. He improved on their notation and avoided, too, some of the more particular and less helpful Oughtredian excesses. In *De sectionibus conicis* Wallis introduced symbols that are still in current use, for infinity and for 'greater than or equal to'. This work was not widely appreciated, perhaps overshadowed by the *Arithmetica*, and was too much for some readers—Hobbes, perhaps not a disinterested commentator, described it as 'covered over with the scab of symbols'. Even where useful notations were yet to be introduced, such as those for fractional and negative indices (first displayed by Newton in 1676), Wallis went far towards laying the groundwork. While not actually using the notation explicitly, he wrote, in *Arithmetica infinitorum*:

$$\text{`}\frac{1}{x}\text{ whose index is } -1\text{'} \quad \text{and} \quad \text{`}\sqrt{x}\text{ whose index is }\frac{1}{2}\text{'}.$$

Wallis's concern for mathematical symbolism is but one facet of a lifelong exploration of issues of language and communication. His first and indeed most successful book was not a mathematical treatise but an English grammar (*Grammatica linguae anglicanae*, 1653). Part of his energies in the 1660s were devoted to teaching two deaf and dumb persons to speak—although this led to yet another controversy when Dr William Holder, who had also worked with one of the young men, felt that Wallis had overlooked his contribution in reporting on the treatment to the Royal Society.

The Royal Society

The Royal Society was founded on Wednesday 28 November 1660, at Gresham College in London. A group met after a lecture given by the Gresham astronomy professor, Christopher Wren, and 'Something was offered about a design of founding a College for the Promoting of Physico-Mathematicall Experimentall Learning'. Of the 12 people at this meeting who are thought of as the founding members of the Royal Society (the Royal Charter was granted later, in 1662), seven—Robert Boyle, William Brouncker, Jonathan Goddard, William Petty, Lawrence Rooke, John Wilkins, and Christopher Wren—had Oxford associations, and another, Sir Paul Neile, was the father of John Wallis's research associate William Neile. The Oxford influence on the founding of the Royal Society is clear, although John Wallis later drew attention to London meetings at Gresham College from 1645 onwards which also contributed both to the vigour of 1650s Oxford science and to the tradition of meeting and discussing mathematical and scientific matters which culminated in the founding of the Royal Society.

Although not at that first meeting, Wallis was an early and active Fellow of the Royal Society, often by correspondence since his Oxford

William Brouncker, the first president of the Royal Society, is the left-hand figure of the frontispiece of Thomas Sprat's 1667 history of the Society. Wallis encouraged Brouncker's mathematical researches and published some of his results in books of the 1650s.

Oxford's Sheldonian Theatre exemplifies the creative tension between antiquity and innovation that characterized the Wallis era. Designed on an ancient Roman model by the Savilian Professor of Astronomy, Christopher Wren, the expanse of the Sheldonian's flat ceiling caused a sensation.

duties precluded very frequent visits to London. His research associate William Brouncker, discoverer of the continued fractions expansion of π, was the first president of the Society, and what has been described as 'a semi-official mathematical sub-committee' was formed of Brouncker, Wren, and Wallis, to take an interest in and resolve various issues. A good example of the work of the young Society, and Wallis's contribution within it, is the publication of the papers of Jeremiah Horrocks, the great Liverpool-born astronomer who had died, tragically young, in 1641. Horrocks was a student near-contemporary of Wallis at Emmanuel College, Cambridge, but his mathematical genius matured more rapidly. He carried out some remarkable theoretical and observational work, including the recognition of the elliptical orbit of the Moon, and the first accurate prediction and observation of the transit of Venus on 24 November 1639, before his death a little over a year later. Although many of his papers were dispersed and lost, some survived and in the 1660s the Royal Society procured them; Wallis edited them and saw them through the press.

Along with many of his contemporaries in the Royal Society, Wallis had remarkably broad interests. The pages of the Society's *Philosophical*

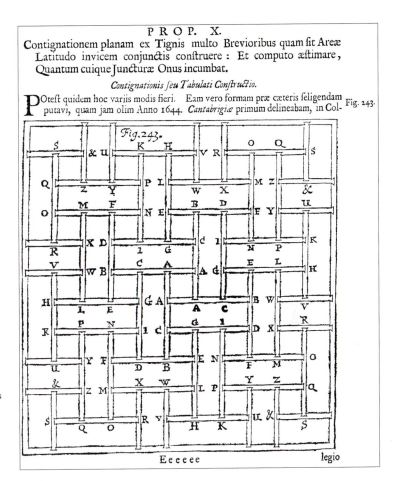

PROP. X.

Contignationem planam ex Tignis multo Brevioribus quam fit Areæ
Latitudo invicem conjunctis conftruere : Et computo æftimare,
Quantum cuique Juncturæ Onus incumbat.

Contignationis feu Tabulati Conftructio.

POteft quidem hoc variis modis fieri. Eam vero formam præ cæteris feligendam
putavi, quam jam olim Anno 1644. *Cantabrigiæ* primum delineabam, in Col- Fig. 243.

Fig. 243.

Eeeeee   legio

A model for the flat roof structure of Christopher Wren's Sheldonian Theatre, designed by John Wallis, was one of the sights of 1660s Oxford. The mathematics of the interlocking beams had been worked out by Wallis some years earlier in an innovative calculation involving the solution of an unprecedented number of simultaneous equations.

Transactions reflect the range of things he had views on and wanted to communicate: unusual arch-work he found in a disused kitchen in Northamptonshire; the relation between flesh-eating and conformation of the intestines; the isthmus that used to link Dover and Calais; the effect of music on its hearers in ancient times—as well as topics reflecting his interest in the history of mathematics, such as Nicolaus of Cusa's awareness of the cycloid and who invented the mariner's compass. In 1685 Wallis supported his argument that memory is better at night by reporting that he calculated the square root of 3 in his head to 20 decimal places (arriving at the answer 1.73205,08075,68877,29353) and retained it in his mind before writing it down the next day.

Besides contributing considerably to the activities and influence of the Royal Society, Wallis played a full role in Oxford life too during the 1660s, continuing to promote mathematics at every opportunity. On the evening of 11 July 1663, he lectured at Oxford on Euclid's parallel postulate, and presented a seductive argument purporting to derive it from Euclid's other axioms. The argument assumed, as Wallis himself observed, that similar figures can take different sizes. Wallis found this assumption very plausible, and if it were so, then the parallel postulate would be a consequence of the other axioms of Euclid. It does, however, imply a remarkable result: in any geometry in which the parallel postulate were false, similar figures have to be identical in size as well as in shape, and so scale copies can never be made. It also has the consequence, as Lambert observed a century later, that there would have to be an absolute measure of length. This argument by Wallis is the first mature western attempt to derive the parallel postulate as a theorem.

An interesting example of the lasting influence of Wallis's mathematics on the development of Oxford is the small but crucial part his work played in the construction of the Sheldonian Theatre. This remarkable building for University ceremonial functions, donated by the Archbishop of Canterbury and former Warden of All Souls College, Gilbert Sheldon, was designed in the 1660s by Christopher Wren, the Savilian Professor of Astronomy. It follows the design of the classical theatre of Marcellus in Rome, an open D-shaped arena with tiered seats. Oxford not being by the Mediterranean, however, there was need for a roof, but without load-bearing columns which would destroy the simplicity and spaciousness of the classical concept. To construct a roof spanning 70 feet, Wren turned to some mathematics already worked out by John Wallis in the 1650s, enabling an architect to calculate the loading, and thus the feasibility, of an interlocking beam structure that needed support only where its edges rested on the walls. The strength of the result is attested by the fact that the roof spaces were used for many years as a book store by the University Press, which operated out of the Sheldonian Theatre until well into the eighteenth century.

Oxford ancient texts project

By mid-century the Bodleian Library, founded in 1598, had become one of the world's leading libraries. It now housed major collections of manuscripts, thanks to Sir Thomas Bodley's policy, carried on by his successors, of encouraging generous benefactions, as well as acquisitions throughout Europe and the Middle East. John Greaves, for example, Seth Ward's predecessor in the Savilian astronomy Chair, was one of those who travelled and collected manuscripts—in his case, in Alexandria and Constantinople—for Archbishop Laud's donations to the Bodleian. When the political climate changed in Oxford, with the new generation of parliamentary appointees such as Ward and Wallis, steps were taken to make the contents of the library more open and widely useful by preparing a subject catalogue. This was urged both by members of the Wadham group and by Gerard Langbaine, the Keeper of the University Archives. Langbaine co-opted leading scholars to work on the cataloguing, including Ward, Wilkins, Willis, and Wallis, but the task was more onerous and time-consuming than expected and the momentum trickled away.

Langbaine died in 1658 (through catching his death of cold in the Bodleian Library, it is said) and John Wallis succeeded him as University Archivist. This appointment provoked yet more controversy: the Savilian statutes forbade a Savilian professor from taking on an added role such as this, some argued, and there was much suspicion of unfair practice by Wallis to secure the post. An outbreak of the usual pamphleteering and recriminations ensued, but Wallis retained the position. The purpose of the archives was to provide evidence of the University's property, jurisdiction, rights, and privileges, and, as in everything else he touched, Wallis set to with energy and in 1664 drew up a 'Repertorie of Charters, Muniments, and other Writings, belonging to the University of Oxford, under the care of the Custos Archivorum'. Wallis's catalogue of the University archives remained the standard catalogue for over two and a half centuries, until a new one was drawn up in the 1930s.

Wallis's zest for ancient documents sprang from the same desire for fundamental understanding, with a tension between tradition and innovation, which characterized his mathematics. In his autobiography he described his state of mind upon entering the Savilian Chair:

And (herein as in other Studies) I made it my business to examine things to the bottom; and reduce effects to their first principles and original causes. Thereby the better to understand the true ground of what hath been delivered to us from the Antients, and to make further improvements of it.

By the 1670s, the enthusiasm which he and others had for rooting the progress of the subject in knowledge of the past began to form itself into a definite project for collecting, editing, and publishing ancient manuscripts. Many of these were by now to be found in the Bodleian and other libraries. The leading light in this project was Edward Bernard, Christopher Wren's deputy and successor in the Savilian astronomy Chair. John Fell, too, the great Dean of Christ Church who was keen to promote the activities of the newly activated University Press, published a programme in 1672 of works he proposed should be printed, including the 'ancient Mathematicians Greek and latin in one and twenty volumes', to be based on fresh collations of the manuscripts. Down the years, various such plans were produced, and several Savilian professors became involved in bringing out editions of the great Greek mathematicians. The most splendid volumes to emerge from this enterprise were those of Euclid (by David Gregory in 1703), and Apollonius (by Edmond Halley in 1710), both of monumental magnificence. John Wallis worked on the texts of Archimedes,

The memorial to Edward Bernard in St John's College.

Aristarchus, Pappus, and Ptolemy. Various editions of things continued to come out in dribs and drabs during the eighteenth century, culminating, in a sense, in a lavish Oxford edition of Archimedes in 1792, more than a century after the project had been set in hand.

Several motivations lay behind this activity, ranging from individual career reasons to considerations about the role of Oxford in the world of learning, but beliefs about the right way to progress mathematical and scientific research played an important role. Bernard, in 1671, explicitly balanced the promotion of new research with awareness of the achievements of the past:

Books and experiments do well together, but separately they betray an imperfection, for the illiterate is anticipated unwittingly by the labours of the ancients, and the man of authors deceived by story instead of science.

Wallis took to heart the lessons of past manuscript losses, which had been a preoccupation of Oxford scholars since at least the sixteenth century, and sought to ensure that the work of his contemporaries, too, was not lost to the world. His generous concern in promoting the work of colleagues by publishing the results of his research collaborators, in the pages of *Arithmetica infinitorum* onwards, indicates the strong commitment that continued to the end of his life. He wrote to Isaac Newton in 1695, in bolder language than anyone else dared use, reproaching him for his continued reluctance to share his discoveries explicitly with the world:

I can by no means admit your excuse for not publishing your Treatise of Light & Colours. You say, you dare not *yet* publish it. And why *not yet?* Or, if not now, when then? You adde, least it create you *some trouble.* What trouble *now,* more then at another time? ... Mean while, you loose the Reputation of it, and we the Benefit. So that you are neither just to yourself, nor kind to the publike.

And as late as 1701 Wallis was urging John Flamsteed, the Astronomer Royal, to the same effect: 'I am not likely to live so long as to see your Observations published; but, however, I would not have the publike loose them.'

Wallis had a highly charged sense, too, of the importance of Oxford and its University Press. According to him,

The Art of printing was first brought into England by the University, and at their Charges; and here practiced many years before there was any printing in London, and we have been in the continual possession of it ever since ...

on which Peter Sutcliffe has commented that Wallis 'was wrong in every historical particular, but sound in his convictions'. Perhaps an inflated view of Oxford's historical role was necessary for Wallis to do the good he did in promoting the interests of the University and its

Press. He lost no opportunities in attracting printing tasks to Oxford, and had a professional understanding of the demands of mathematical typography. As part of his strategy to persuade Isaac Newton to release some of his work for publication, he wrote to him, on 30 May 1695:

I think Oxford the most convenient place for it; Because here we have most of the Cuts allready, & furniture fit for it; & our Compositors are acquainted with this kind of troublesome work; which to others unacquainted with it will seem strange. And Mr Caswell or I will see to the correcting of the Press.

A treatise of algebra

Wallis's last great mathematical work, *A treatise of algebra, both historical and practical*, was written in English and published in 1685, in his seventieth year. Of all his vast output, it was this work that over the next hundred years was most widely read, and it remains an extraordinary work in which both Wallis's strengths and weaknesses come to the fore. The idea of presenting a field of modern mathematics so intimately bound up with an account of its development was itself a novel one—this is the first substantial history of mathematics in the English language—although in tune with the spirit in which the Royal Society and others sought to classify and index knowledge, and describe past discoveries, in order to provide a firm foundation for moving forward and making fresh discoveries. In many respects the history is well done, with a full and judicious survey of the achievements of past centuries on the basis of wide reading of the sources available to him, notably in the first 13 of its 100 chapters which trace the history of the subject up to the time of Viète.

When he came to the mathematics of his own century, Wallis's past difficulties with French mathematicians seem to have influenced his judgement in a rather startling way. The reader was left in no doubt that the achievements normally attributed to French mathematics after Viète had a solidly English basis. Not only was it Thomas Harriot who made the discoveries thought to be due to Descartes, Wallis asserted, but Descartes had plagiarized the results from Harriot, who

hath taught (in a manner) all that which hath since passed for the *Cartesian* method of *Algebra*; there being scarce any thing of (pure) *Algebra* in *Des Cartes*, which was not before in *Harriot*; from whom *Des Cartes* seems to have taken what he hath (that is purely *Algebra*) but without naming him.

This claim, with the further substantiation that Wallis offered over several chapters, did not impress all readers (particularly the French), although it was a little toned down by the time of the Latin edition in 1693. The controversy somewhat distracted attention from the final 25 chapters, concerned with the new methods of analysis by indivisibles

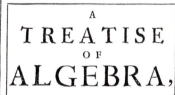

Savilian professors of astronomy during the Wallis era

1649   Seth Ward (1617–89)

Born in Buntingford, Seth Ward graduated BA (Sidney Sussex, Cambridge) in 1637, and was elected a Fellow of Sidney Sussex in 1640. He went to study privately with William Oughtred, and on returning to Cambridge introduced Oughtred's *Clavis mathematicae* to Cambridge mathematics teaching. In 1643 he was chosen mathematical lecturer in the University of Cambridge. In 1644 he was removed from his fellowship on political grounds, but within a few years had taken an oath of allegiance to the Commonwealth. When in 1648 John Greaves realized that he was about to lose his Savilian Chair, he arranged for Seth Ward to be elected his successor. At Oxford Ward worked on the theory of planetary motion, and set up an observatory at Wadham College, as well as defending the new learning in the universities and participating vigorously in the activities of the circle around John Wilkins. In 1661 he resigned to further his career within the Church of England, and by 1667 was Bishop of Salisbury.

1661   Christopher Wren (1632–1723)

Born in East Knoyle, Wiltshire, Christopher Wren was educated at Westminster School and Wadham College, Oxford (BA 1651). Elected a Fellow of All Souls College in 1653, and Gresham Professor of Astronomy in 1657, he was appointed Savilian Professor of Astronomy upon Seth Ward's resignation. Upon his appointment as Surveyor of the Royal Works in 1669, Wren nominated Edward Bernard as his deputy in the Savilian Chair, and gave it up altogether in 1673. Throughout his distinguished architectural career Wren maintained his scientific interests, serving as president of the Royal Society in 1680–2.

1673   Edward Bernard (1638–96)

Born near Towcester, Edward Bernard was educated at Merchant Taylors' School and St John's College, Oxford (BA 1659). He studied a variety of oriental languages and also, enthused by Arabic treatises in the Bodleian Library, mathematics under John Wallis. He visited Leyden in 1668–9 to study oriental manuscripts and returned full of enthusiasm for editing ancient mathematical writers. In 1669 he was appointed to deputize for Wren in the Savilian Chair of Astronomy, and became his successor in 1673. In 1676 Charles II appointed Bernard as tutor to his sons, the dukes of Grafton and Northumberland, who were living in Paris. In the words of the *Dictionary of National Biography*,

His retiring disposition and erudite pursuits rendered him an object of ridicule in gay society, and he resumed his antiquarian studies at Oxford, after about a year's absence, saddened by his novel experiences, though consoled by the acquisition of many rare books.

Although he did some work on astronomical tables of the fixed stars, and observed the solar eclipse of July 1684 in Oxford, the bulk of his studies were concerned with ancient knowledge and the availability of manuscripts. For some years he tried to divest himself of the Savilian Chair, and in 1691 resigned to take up a rich living at Brightwell, nine miles from Oxford. After his death his books were bought by the Bodleian Library for £140, and his manuscripts for a further £250.

1691   David Gregory (1659–1708)

Born in Aberdeen, David Gregory was educated at Marischal College, Aberdeen, and the University of Edinburgh, where in October 1683, a month before he graduated MA, he succeeded his uncle James Gregorie (who had died in 1675) as professor of mathematics. Becoming increasingly unhappy at Edinburgh, where he refused to swear an oath of allegiance to the new British monarchs (William and Mary) as required by the Scottish parliamentary commissioners, he sought to further his career in England. In the contest to fill the Savilian astronomy Chair in 1691, following the resignation of Edward Bernard, Gregory secured strong and successful support from both Isaac Newton and the Astronomer Royal, John Flamsteed. Gregory proved a loyal disciple of Newton, besides entering with enthusiasm into the ancient texts project.

and infinite series, summarizing his *Arithmetica infinitorum*, together with an account of more recent work in this area such as Newton's binomial theorem and further discussion of the work of Newton and others.

Much fresh and important mathematics was introduced by Wallis in the pages of *A treatise of algebra*, both in the English version of 1685 and even more so when it appeared in Latin in the second volume of his collected works in 1693; this new edition contained an account, essentially written by Newton himself, of the method of fluxions, including the first appearance in print of his 'dot' notation. These volumes showed no slackening in Wallis's ability to provoke controversy even as he promoted mathematics; the third volume of his collected works, which appeared in 1699, put in print for the first time the two long letters of 1676 in which Newton told Leibniz about his work on infinite series, with some guarded remarks about the calculus. It was the publication of these letters by Wallis that fanned (not undeliberately) the embers of a growing controversy about priority in discovering the calculus which soured the later years of both Newton and Leibniz.

Final years

During Wallis's lengthy tenure of the Savilian Chair of Geometry there were four occupants of the astronomy Chair, each of whom made important and memorable contributions to Oxford life and studies. His last Savilian colleague, David Gregory, carried on several of the traditions that Wallis had fostered, notably the ancient texts project, as well as bringing fresh ideas from outside Oxford and invigorating the study of the physical sciences.

Gregory's appointment to the vacant Chair of astronomy, on the resignation of Edward Bernard, was the culmination of a remarkably skilful campaign. On the face of it, two home-grown applicants might have been thought stronger candidates. John Caswell (eventually to be Gregory's successor) was a familiar Oxford figure, and John Wallis's protégé and collaborator; Wallis published Caswell's *Brief (but full) account of the doctrine of trigonometry* at the end of his *A treatise of algebra* in 1685. But although interested in the post, he was discouraged, according to John Flamsteed, by the machinations of the other Oxford candidate, Edmond Halley. Halley, however, had made such an enemy of Flamsteed, the Astronomer Royal, that David Gregory, the candidate favoured by Isaac Newton, gained the appointment. Newton's recommendation of Gregory, a Scotsman whom he had hardly met, was astonishingly warm:

I do account him prudent sober industrious modest & judicious & in Mathematiques a great Artist. He is very well skilled in Analysis &

John Wallis, by Sir Godfrey Kneller (1702). Samuel Pepys paid for this picture and presented it to Oxford. It now hangs in the Examination Schools.

Geometry both new & Old. He has been conversant in the best writers about Astronomy & understands that science very well. He is not only acquainted wth books but his invention in mathematical things is also good. He has performed his duty at Edinburgh wth credit as I hear & is respected the greatest Mathematician in Scotland & that deservedly so far as my knowledge reaches. For I take him to be an ornament to his Country & upon these accounts recommendable to the Electors of the Astronomy Professor in Oxford now vacant.

Although Gregory was certainly able and knowledgeable, this encomium is also a tribute to a shrewd courtship in which he flattered and charmed the older man. He had recognized early on that Newton's patronage would be invaluable in developing an academic career in late seventeenth-century England, and worked hard to ensure this. In return, as it were, he was a vehicle for introducing and promoting Newtonian ideas at Oxford.

Wallis retained his vigour to the end. In 1699, one of his Oxford colleagues had written to Samuel Pepys, the former president of the Royal Society and a friend of Wallis for over 30 years, 'He says 83 is an incurable distemper. I beleive Death will no more surprise him than a Proposition in Mathematicks.' Some measure of the impression Wallis left on contemporaries may be seen in the remarkable full-length portrait of him at the age of 86, by the court painter Sir Godfrey Kneller, which now hangs in the Examination Schools. This painting—in which the aged Wallis, swathed in scarlet like some Renaissance prince–prelate, stares out at the viewer in cold disdain—was commissioned by Pepys for presentation to the University of Oxford. In response, the University presented a diploma of thanks to Pepys, which drew attention to

the perfection wherewith the Very Mathematicks, as well as Your unexampled Generosity, seem represented in the Picture of our most learned Professor. 'Tis an Argument of the highest Genius to bee capable of putting a just Value on the great Dr Wallis, who was never known to say or do a vulgar thing. He soars into lofty Regions of Learning, and traceth out most Sublime Paths to the Mathematicks, being next a-kin to those Heavens he measures, and those Starrs whose Numbers are obvious to his Arithmetick, and out of the reach of all but a truly penetrating Eye.

Further reading

The most comprehensive discussion of Wallis's work remains J. F. Scott's *The mathematical work of John Wallis*, Taylor and Francis, 1938, but a more up-to-date and reliable essay is that of Christoph Scriba in the *Dictionary of scientific biography*. The three volumes of John Wallis's *Opera mathematica* have been published in facsimile by Georg Ohms Verlag, Hiderheim, 1972.

A full and broad account of the context of mathematical and scientific studies over this period at Oxford, by Mordechai Feingold, forms Chapter 6 of *The history of the University of Oxford*, Vol. IV, *Seventeenth-century Oxford* (edited by Nicholas Tyacke), Clarendon Press, Oxford, 1997, pp. 359–448.

A Description of the Passage of the Shadow of the Moon, over England, In the Total Eclipse of the SUN, on the 22d Day of April 1715 in the Morning.

CHAPTER 7

Edmond Halley

Allan Chapman

While John Wallis drew acclaim for the promotion of pure mathematical research in seventeenth-century Oxford, the University also fostered several men who established a great tradition in both theoretical and observational astronomy. Edmond Halley was perhaps the most famous, and certainly the most widely published, English astronomer of his day. His impact was felt not only in Oxford, but nationally and internationally, and he was the first of three consecutive Oxford men to hold the office of Astronomer Royal.

In 1656, the year of Halley's birth, astronomy was developing at a speed unprecedented in its history. In the 46 years since Galileo's first telescopic discoveries, observational planetary astronomy had come into being and had effectively challenged traditional evidence for the geocentric cosmology. New instruments and techniques of observation, such as the micrometer and pendulum clock, had been developed while Halley was still a schoolboy at St Paul's School, London, and the first, albeit unsuccessful, attempts were made to measure the distances of the stars by Christiaan Huygens in Holland. At Oxford, too, things were moving fast. Christopher Wren developed an international reputation in the Savilian Chair of Astronomy during the 1660s, with those studies in dynamics that provided important links between the work of Kepler and Newton, before receiving a knighthood in 1673 and moving to an even more illustrious career as an architect.

Coming up to The Queen's College in 1673, Halley astonished the University even at the age of 17, where he began to make observations using a fine private collection of astronomical instruments provided by his London merchant father. His instruments incorporated such recent technical innovations as telescopic sight and micrometers, as well as telescopes for observing surface detail on the Moon and planets. Astronomy was a current preoccupation in Oxford: the year he went up, John Wallis published his edition of the papers of Jeremiah Horrocks, the Lancashire astronomer who had died young in 1641 and can be seen as Halley's predecessor as an innovative observational astronomer. Within two years, Halley had made contact with John Flamsteed, the recently appointed Astronomer Royal, and began to

Halley's concern with the public dissemination and understanding of science can be seen in the mass eclipse observation that he organized while Savilian Professor of Geometry at Oxford. Calculating that on 22 April 1715 London would experience its first total eclipse since 20 March 1140, he distributed maps all over England to solicit observations of local eclipse conditions, as well as to reassure citizens that it was a natural event, portending no harm to the new monarch, the Hanoverian King George I.

Halley went up to The Queen's College, Oxford, in 1673. He left in 1676, without taking his degree, in order to make his expedition to St Helena.

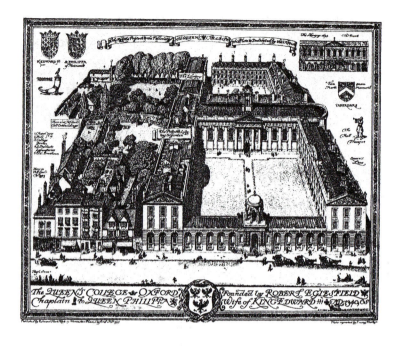

collaborate with him, both through making observations at Oxford and by travelling to assist at Greenwich. In the Royal Society's *Philosophical Transactions* for 1675, Flamsteed remarked that 'Edmond Halley, a talented young man of Oxford, was present at these observations and assisted carefully with many of them.'

Observing the southern stars—an undergraduate achievement

Edmond Halley's undergraduate career was cut short and he left Oxford without taking his degree. However, this disruption was occasioned by an adventure that was to secure him an international reputation by the age of 22. In 1676, the 20-year-old Halley sailed to St Helena (an island 16° south of the Equator), with a set of new instruments, to map and explore the skies of the southern hemisphere.

It was Halley's work with Flamsteed, we may presume, which had awakened his mind to an opportunity to contribute to astronomical work in a spectacular way. The foundation of the Royal Observatory at Greenwich in 1675 had recognized the official need to re-map the heavens at what everyone expected to be public expense. But the new observatories at Greenwich and Paris were concerned with making *better* catalogues of the long-familiar northern heavens. Halley's proposal to do the same for the southern skies was supported by several influential patrons, using both his Oxford and his family connections. Behind the letter from King Charles II which asked the East India

Sir Joseph Williamson, Charles II's secretary of state, was a Fellow of Queen's College and supported Halley's expedition to St Helena. He was president of the Royal Society from 1677 to 1680 (the first three presidents of the Royal Society, Brouncker, Williamson, and Wren, were all educated at Oxford).

While still an undergraduate, Halley published in the Royal Society's *Philosophical Transactions* his observations of the occultation of Mars by the Moon, as seen from Oxford on 21 August 1676.

Mr. Edmund Hally's *Observations,concerning the same Occultation* of Mars *by the Moon, made at* Oxford, *Anno* 1676. Aug.21.*P.M.*

Temp.Corr.

h. ' "		
11.43.30	THe center of Mars from the Nearest limb of the Moon,	719½ =12.40
11.49. 2	*Again,*	571 =10. 3
11.54.58	*Again,*	409 == 7.12
12. 3.25	*The center of Mars from the North Cusp of* ☽,	1118 =19.41
12.10.28	*The gibbous part of Mars touched the Moons limb.*	
12.10.42	*Mars was wholly covered, being distant from the Cusp,*	963 =17.14
12.40.00	At this time a *Halo* encompassed the Moon, in whose Circumference was *Saturn,* the *Pleiades, Capella,*and the following of the foot of *Perseus.*	
13.10.41	*Mars did emerge, I suppose, his Center.*	
13.12.45	*Mars was distant from the Northern horn of* ☽,	1018=17.55
13.31.10	*Mars passed over a point noted in the Telescope.*	
13.33.15	*The Southern limb of Ætna passed by the same point.*	
13.34.00	*The lucid limb passed over the same point.*	
13.52.35	*The Moons diam.observed,*1698==30'.1". *alt.*☽ 31°. *circ.*	
13.57.52	*Mars from the Northern horn of the Moon,*	2042=36. 5
14. 2.53	*Mars from the Southern horn of the Moon.*	2266=40. 3

Having carefully considered the Moons Parallaxes in the observations of this Occultation at *Dantzick* and *Greenwich,* I find from the *Immersion* the difference of Meridians between *Greenwich* and *Oxford* 4'.57"; between *Greenwich* and *Dantzick* 1ʰ. 14'. 50": By the *Emersion* the first of those differences is found 4'. 59", the latter 1ʰ.14'.41":which near agreement shews the Exactness of all the Observations.

Company to transport Halley and a friend to St Helena there lay a web of influence including such luminaries as Sir Joseph Williamson, the secretary of state who had close Queen's College connections, Lord Brouncker, the president of the Royal Society and another old Oxford figure, and the eminent practical mathematician Sir Jonas Moore.

While Halley was not the first person to map the southern skies, for navigational charts showing them had been commercially available for 150 years, he was the first to do so to the new level of accuracy made possible by telescopic measuring instruments. Using a large sextant equipped with telescopic sights, a quadrant, pendulum clock, and micrometers, he obtained right ascension and declination angles to within about 20 arc seconds, as opposed to the several minutes' error parameters of existing charts. He also marshalled together several miscellaneous groups of stars to form new constellations, one of which he diplomatically christened Robur Caroli, *Charles's Oak*, a patriotic gesture which did not go unnoticed in the Palace of Whitehall.

In addition to mapping the southern skies, Halley was fascinated by the mysterious objects which they contained and which were not

In 1676 Halley left his undergraduate studies at Oxford to make an astronomical expedition to St Helena, a trip which is still remembered there.

visible in European latitudes. These objects included nebulae and star clusters, which were classes of objects that would interest Halley for the rest of his long life. He discovered the star clusters of the constellation of Centaurus—perhaps the first of a long list of deep-space objects to be subsequently discovered in the southern hemisphere and of significance to the modern astrophysicist.

When his *Catalogus stellarum Australium* was published, following his return home in 1678, Halley found that his reputation as an astronomer was made. The Royal Society elected him one of its youngest ever Fellows at the age of 22, he was acclaimed in Paris, and Charles II ordered Oxford University to grant Halley his MA degree by royal *mandamus*, or command, rather than troubling him to be examined for it. This may well be the first university degree conferred in explicit recognition of research achievement.

It also says a great deal about the young Edmond Halley's standing as a practical astronomer, not to mention his potential as a diplomat, that the Royal Society chose to send him to Danzig in 1679 to adjudicate in an acrimonious dispute between Robert Hooke and the 68-year-old Johannes Hevelius. Hooke, who had done more than any single astronomer to develop and advocate the use of telescopic sights for astronomical measurement, severely censured Hevelius for failing to use them, and an unsavoury volume of abuse had arisen. That a young man of 23 should have been sent to Danzig to comment upon the observing procedures of the elder statesman of European astronomy may seem remarkable in itself, but that he accomplished his task with so much success and amiability speaks volumes for Halley's diplomatic as well as scientific skills.

The establishment of Halley's career

While Oxford had launched Halley's astronomical career, capping it with an honorary MA, he was to hold no formal status or appointment in the University for a further 26 years, although he remained on familiar terms with the place and was always regarded as an Oxford man. Moreover, many areas of research which he initiated over these years were brought to maturity after his return to Oxford as Savilian professor in 1704. However, it was in the Royal Society, at the heart of metropolitan and international English science, that he largely moved in the interim. Halley was not a man to be attracted to routine teaching, and at this stage of his career he would have found the University too narrow and provincial for his intellectual energies. It was through his involvement with the Royal Society that he was able to exercise those talents with which he was best endowed—research, administration, and diplomacy.

Halley's temperament and social skills also played a strong role in his success. He was fortunate in combining one of the most original scientific minds of his time with an astute social sense, an affable disposition, a strong constitution, a keen sense of humour, and financial shrewdness. Although he was the son of a prosperous London merchant, his private means do not appear to have been sufficient, and he seems to have needed to earn his living. Indeed, Halley's financial position presents the historian with something of a puzzle, for while he was capable of undertaking extraordinary acts of generous patronage, he frequently sought paid work and had the reputation for possessing a sharp eye for a profitable venture. He was an entrepreneur by instinct.

Edmond's marriage in 1682 to Mary Tooke, the daughter of an Exchequer auditor, was singularly successful. It probably brought a dowry, and lasted for 54 years—an astonishingly long time, and against all the statistical odds in an age when a person was lucky to live beyond the age of 40.

Flamsteed, Hooke, and Newton

When one looks at the leading figures on the English astronomical scene in the 1680s, it is clear that Edmond Halley was an exception, at least in terms of temperament. John Flamsteed, the Astronomer Royal, was irritable and often resentful. Robert Hooke, Curator of Experiments at the Royal Society and Gresham Professor of Geometry, tended to be defensive and quarrelsome, while Isaac Newton, Lucasian Professor of Mathematics at Cambridge, shared most of these traits with the extra one of reclusiveness. They also tended to hypochondria, and filled their private memoranda with gruesome details of their real and imaginary ailments, although all three lived to 'modern' age spans of between 68 and 85. What was more, each man actively disliked the others, and sometimes upset Royal Society meetings with acrimonious back-biting.

Halley had a very different temperament. Unlike Flamsteed, Hooke, and Newton, who spent their entire adult lives within the narrow circle of south-east England, Halley was already an acclaimed traveller at 23, and would travel relentlessly around England, Europe, and the Atlantic until his mid-60s. He enjoyed rudely sound health of body and mind, which perhaps irritated his more sickly and obsessive colleagues. Added to this, he possessed a love of good company and merry jests which Flamsteed and Newton especially thought unfitting in a grave philosopher. His long and happy marriage can only have further distanced him from Newton and Hooke, who came increasingly to manifest the eccentricities of elderly bachelors.

Halley's relationships with these men were diverse and complicated. After an initial period of admiration during Halley's youth, Flamsteed became the persistent *bête noire* of his life. The Astronomer Royal detested Halley's jocundity and easy success. As a man who felt ignored

Royal Society portrait of 'young Halley' by Thomas Murray: Edmond Halley in about 1687, at the time he paid for the publication of Newton's *Principia*, holding a diagram relating to his work on the number of roots of cubic and quartic equations.

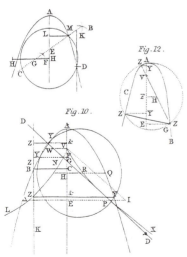

Diagrams from Halley's work on the numbers and limits of roots of cubic and quartic equations, published in the *Philosophical Transactions* of the Royal Society in 1687.

by the establishment (as, to some extent, he was), Flamsteed clearly resented Halley's knack of playing the successful scientific courtier, his picking up occasional sinecures, his friendship with Tsar Peter of Russia, his senior naval commission, and his diplomatic missions abroad for the Royal Society and Queen Anne. He successfully ruined Halley's otherwise well-favoured application for the Savilian Chair of Astronomy in 1691, to which David Gregory was appointed, when he informed Oxford that Halley would 'corrupt the youth of the University' with his scandalous ideas and conversation. But Flamsteed did have some genuine grounds against Halley, who in turn was far from generous to his old encourager.

Halley and *Principia*

Poles apart in terms of temperament, Halley and Newton enjoyed a successful relationship that spanned four decades. Both men had diverse and roving intellectual interests—albeit for different subjects on most occasions—and were first-rate mathematicians with interests in

FRANCISCI WILLUGHBEII Armig.
DE

HISTORIA PISCIUM

LIBRI QUATUOR
LONDINENSIS editi.

Juſſu & Sumptibus SOCIETATIS REGIÆ

In quibus non tantum De Piſcibus in genere agitur, Sed & ſpecies omnes, tum ab a-
lis traditæ, tum novæ & nondum editæ bene multæ, naturæ ductu ſervante
Methodo diſpoſitæ, accurate deſcribuntur.

Earumque effigies, quotquot haberi potuere, vel ad vivum delineatæ, vel ad
optima exemplaria impreſſa, Artifici manu elegantiſſime in æs inciſæ, ad de-
ſcriptiones illuſtrandas exhibentur. Cum Appendice Hiſtorias & Obſervationes
in ſupplementum Operis collatas complectente.

TOTUM OPUS

Recognovit, Coaptavit, Supplevit,
Librum etiam primum & ſecundum integros adjecit
JOHANNES RAIUS e Societate REGIA.

OXONII,
E THEATRO SHELDONIANO, Anno Dom. 1686.

Francis Willughby's *History of fishes*, financed
by the Royal Society and printed at the
Sheldonian Theatre, was not only the
occasion for Halley's subsidizing of Newton's
Principia but became currency for the
impoverished Society: in 1687 Halley was paid
his wages as clerk to the Royal Society with
50 copies of the *History of fishes* (and 20 more
for arrears of salary).

astronomical instrumentation. Both men shared lifelong interests in the process by which physical motion in the heavens could be expressed in precise geometrical terms, thereby linking external observed phenomena in nature with the predictive and analytical capacities of systematic thought. But Halley also had the genius to recognize the even greater mathematical genius of Newton, to urge him to write *Principia mathematica*, and then pay for the costs of publication out of his own pocket because the Royal Society was currently broke, having extended its finances overmuch in publishing Francis Willughby's *History of fishes*.

Indeed, without Halley the *Principia* might never have seen the light of day. Without Halley's discussions and constant encouragement to make his work public, Newton might well have kept the inverse-square law and all its consequences in an undeveloped and unpublished state for the rest of his life. No doubt Robert Hooke would have been happier if this had happened, for Newton's failure to acknowledge ideas first given to him by Hooke created one of the bitterest controversies to wrack the early Royal Society. It also showed clearly where Halley's personal allegiances lay, and inevitably put him beyond the pale for Hooke. Newton's behaviour drove another wedge, likewise, between Halley and Flamsteed. Flamsteed felt anger about his apparent lack of gratitude to the Royal Observatory, after supplying Newton with numerous observations important to the development of his lunar theory in the 1690s. In his private papers, Flamsteed frequently spoke of Halley as 'Newton's Creature'.

On the other hand, the friendship between Halley and Newton could not always have been easy to maintain, for the earnest Newton also admonished Halley for his levity and his reservations about the scientific accuracy of the book of Genesis. The continuity of their relationship tells us much about Halley's qualities of friendship, tact, and diplomacy.

The dimensions of the solar system

Like all seventeenth-century astronomers, Halley's mental processes were essentially geometrical, in so far as he thought in terms of complementary angles, precise proportions, and the exact division of natural and mathematical wholes. One reason why astronomy was considered to be a perfectible science in the seventeenth century stemmed from the belief that all of its convolutions were ultimately reducible to conic sections. Kepler's laws at the beginning of the seventeenth century, Jeremiah Horrocks's studies in the middle of it, and Newton's work at the end, all emphasized the unity of 'matter and intellect'. All that was needed to complete the picture was to give exact weightings to the proportions.

The crucial element in the quest for exact dimensions of the solar system was to determine with great accuracy the astronomical unit, or distance between the Earth and the Sun. Halley realized in 1677 that a precise and technically simple method was available. Observing a transit of the planet Mercury across the Sun's disc from St Helena, it occurred to him that the exact position of the planet upon the Sun and its exact time of transit would be different for observers in different places on the Earth's surface. The method was perfect in theory: to dispatch astronomers to various parts of the globe to make simultaneous observations of the easily predicted transits. On their return the respective values would be collated with the latitudes and longitudes of their observing stations, a parallax angle could then be obtained, and from it the exact distance of the planet.

Halley came to realize that, while Mercury transits were quite frequent, the planet subtended too small a diameter upon the Sun for the contacts to be observed reliably. Venus promised to be a more likely candidate: its apparent diameter in transit is relatively large, and its motion is slower and easier to observe. The problem lay in the rarity of the transits of Venus. Only two astronomers had seen a transit of Venus up to Halley's time—the Englishmen Jeremiah Horrocks and William Crabtree, who saw the transit of 1639. No astronomer alive in 1680 could hope to see the next predicted transits of 1761 and 1769 (they occur in pairs eight years apart, with approximately 120 years between the pairs).

It says much about Halley's extraordinary far-sightedness that he began to work out the criteria necessary to observe the mid-eighteenth-century transits, in a series of papers to *Philosophical Transactions*. When the transits did occur, after Halley's death, they provided data for calculating a radial distance for Venus, and hence an astronomical unit, which is very close to the one accepted today.

Comets

In addition to his attempts to establish these constants of the solar system, another interest was to secure Halley's fame and make his name a household word—the study of cometary orbits. His interest in comets combined many of the great astronomical issues of the day: the interpretation of observations to form cosmological models, the laws of planetary motion, and the status of historical evidence conjoined with modern observation in the framing of predictive laws.

Halley's interest in comets seems to have been triggered by the brilliant appearance in 1680 of one of the first major comets to have its path through the sky measured with the newly invented telescopic sight and micrometric instruments. At that time there was a great deal of dispute as to the shape of the orbits: a wide variety of curves, and even straight lines, were suggested as possible paths for comets. Flamsteed at Greenwich was the first to suggest that another bright comet, which appeared a few weeks later in late December 1680 and into

Halley published his cometary researches, as one of the first acts of his Savilian Chair, with a historical introduction that attests to his implicit support for Savile's tradition of grounding current scholarship in its historical roots.

1681, with its tail pointing in the opposite direction, was the same as the first 1680 comet which had passed behind the Sun and was now receding.

After 1695, when Halley had Newton's techniques of orbital and moving-body analysis at his disposal, he began to apply himself seriously to comets as astronomical bodies. He started off by hypothesizing that comets move in parabolic orbits—a shape favoured by Newton. But if this were the case, then the same comet would never come back, for the parabola is an 'open', non-returning, orbit—unlike the 'closed' ellipse. Fitting observations of recent comets to elliptical orbits, however, he found that the comet of 1682 might be the same as the bright one which had been observed in 1607, 1533, 1456, 1380, and 1305. Taking account of the perturbations of Jupiter, he predicted that it would return in December 1758.

As with the long-term proposals for measuring the transits of Venus, Halley was alerting future astronomers to an event which he knew he would not live to see. *Halley's comet* became his memorial when the German amateur Georg Palitsch sighted the returning comet on Christmas Day 1758, in the region of sky that had been predicted. Halley's comet was more than merely an interesting discovery. It was one of the first vindications of Newtonian theory, demonstrating the power of the new science to explain ancient classes of hitherto mysterious phenomena, and showing them to be subject to the same predictable laws as the Sun, Moon, and terrestrial bodies.

Geomagnetism and geophysics

Edmond Halley was one of the first scientists to pay serious attention to the Earth as a physical member of the solar system in its own right, approaching the problem in an attempt to find a law-like basis for geomagnetism.

Throughout the seventeenth century, in the wake of the publication of William Gilbert's *De magnete* (1600), magnetism had been an object of great scientific interest as a natural phenomenon which bore directly upon the art of navigation. Henry Gellibrand of Trinity College, Oxford, who first found a fascination with geometry when attending Sir Henry Savile's lectures, became interested in the gradual changes through which the Earth's magnetic field moved, and began one of the first systematic analyses of the problem after becoming professor of astronomy at Gresham College, London, in 1627. By the end of the seventeenth century, it was known that the Earth's magnetic field moves with time, that there is a vertical 'dip' in addition to the lateral field, and that both the vertical and horizontal fields vary over time and veer dramatically away from north and south in different parts of the world.

By the 1690s, Halley had developed a theory to explain the magnetic field and its variations. He proposed that beneath the Earth's spinning

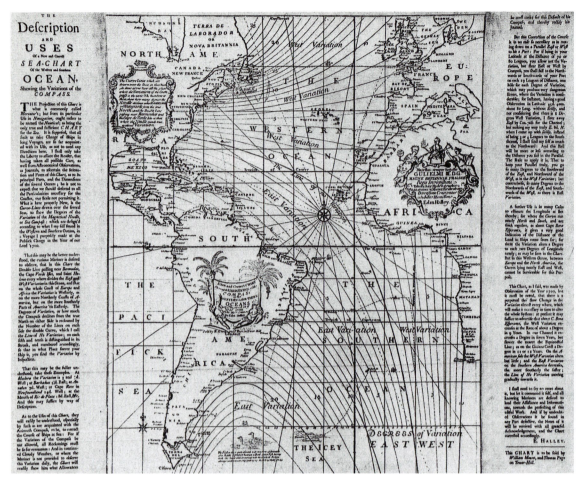

Halley's chart of magnetic variation over the Atlantic introduced the important technique of using lines to join points with the same value.

crust a core rotates at a slightly different velocity. Both inner core and outer shell possess magnetic poles, so that observed variations in the terrestrial magnetic field can be explained in terms of the changing synchronicity of the four poles. He also suggested that the subterranean gap between the core and outer shell is filled with an 'effluvium', and in 1716 he proposed the first association between the aurora borealis and the geomagnetic field. Halley suggested that when this charged luminous effluvium escapes into the air, the motion and display of the resulting aurora is governed by the rotating Earth and its magnetic field.

In pursuit of his geomagnetic interests and their possible application to the discovery of a technique for finding longitude at sea, Halley succeeded in getting himself commissioned captain in the Royal Navy in 1698, to command *HMS Paramore* on a survey expedition of the Atlantic to plot the changing magnetic variations with relation to astronomical coordinates. Following his return to England in 1701, he set about publishing his findings, and produced a map which came to be regarded as something of a wonder in its own right. Although his magnetic navigation scheme proved unworkable, it was of great importance geomagnetically. When the German scholar Conrad von Uffenbach visited England in 1710 and spent several weeks in Oxford, he was delighted to purchase a set of Halley's magnetic charts for nine shillings, even though he was neither an astronomer nor a geographer.

A presentational technique of enduring utility that emerged out of the *Paramore* voyage was Halley's conception of joining locations possessing the same magnetic deviation with a connecting 'isogonic' line, so that they could be conveniently traced across the chart. This method of showing equal quantities on a map subsequently entered into international usage, to show contours, depths, and all manner of cartographic, geographical, and other details, in addition to magnetic ones.

Halley as a historian of astronomy

Many of the great astronomers of the seventeenth century prefixed the publication of their own achievements with the history of the science leading up to their own time. (In Chapter 6 we saw a parallel pattern in the mathematical work of John Wallis, notably in his great *A treatise of algebra*.) Riccioli, Hevelius, and others had given credit to their predecessors, and especially their near contemporaries such as Tycho Brahe, by way of setting the scene for their own work, while John Flamsteed's monumental *Historia coelestis Britannica* (1725) begins with a major history of astronomy.

All of these histories were intended to show a forward progress of achievement. Halley, on the other hand, displayed a different approach towards historical evidence, seeing it as a body of information that could be used to amplify modern understanding. In this respect he was following in a tradition of Oxford astronomy stemming from Sir Henry Savile and the first Savilian professors. Halley was an assiduous collector of astronomical data from previous centuries. Besides the famous historical studies upon which his cometary work depended, he collected evidence of ancient eclipses as a way of checking the new lunar theory based on Newtonian gravitation, as well as ancient inscriptions recording astronomical events such as that from Palmyra.

But he was also aware of the great divide which separated scientists of his century from previous astronomers: a series of inventions in his own day had transformed the quality and quantity of data almost beyond recognition. The telescope, pendulum clock, and micrometer allowed astronomers to see and measure things in the sky that had been quite beyond the bounds of perception only a few decades before. In addition to his awareness of the significant observations made in the past, Halley also saw his own work and that of his contemporaries as forming terms of reference for future researchers.

Halley and the public understanding of science

Concerned as we are with the public understanding of science, it is of interest that Halley recognized a duty of scientists to educate the public. He relentlessly informed the readers of *Philosophical Transactions* on a wide range of natural events. His preparatory work on the forthcoming transits of Venus was matched by his detailed publications on eclipses, meteors, meteorology, and natural phenomena. In 1716, for instance, he published a paper on the strange brightness of Venus seen on 10 July in that year, in a direct attempt to explain a long-term characteristic of the planet's orbit currently being exploited by astrologers and scaremongers. For, as Halley emphasized, 'It may justly be reckoned one of the principal uses of mathematical sciences, that they are in many cases able to prevent the Superstition of the unskillful vulgar.' An awareness of the great cyclical patterns of nature, which he could substantiate from both his historical and mathematical interests, gave him a capacity to educate the public in the new science.

The Royal Society's *Philosophical Transactions* was read by a wide range of intellectually inclined people, and not just active scientists. By writing for it, the only scientific magazine of its time, one obtained access to a broad range of intelligent readers. A wide penumbra of journalists, playwrights, and hacks further combed its pages for unusual material that could be recycled in newspapers and broadsheets. All the mad 'projectors' or scientists whom Jonathan Swift parodied in the Lagado Academy in *Gulliver's travels* (1726), for instance, came from barely distorted reports of what appeared in *Philosophical Transactions* or Royal Society projects. Although Halley might have occasionally opened himself to satirical usage from his articles, he was also able to get his ideas, and those of contemporary scientific thinking, straight into the public domain. Two things which clearly qualified Halley for this role were his well-attested gift as a speaker and his clarity as a prose writer.

Savilian Professor of Geometry

Edmond Halley was 48 years old when he obtained what might be called his first proper academic post, being elected in 1704 to succeed John Wallis (who died in October 1703) in the Oxford Chair of geometry. Few front-rank scientists established their reputations in circumstances more removed from the security of the ivory tower than did Edmond Halley, or came eventually to hold a high academic appointment from a more eccentric background. But by now Halley's astronomical and scientific reputation stood too high to be dismissed, even by the acidic epistles of the Astronomer Royal. A letter of Flamsteed's, of December 1703, reveals his irritation at the turn of events: 'Dr Wallis is dead—Mr Halley expects his place—who now talks, swears and drinks brandy like a sea captain.'

On moving into (one half of) his predecessor's house, Edmond Halley had an observatory built on to the roof in 1705, which his colleague David Gregory wrote 'will be very convenient and indeed useful to the university, and what Sir Henry Savile did expect from them'.

APOLLONII PERGÆI

DE

SECTIONE RATIONIS

LIBRI DUO

Ex Arabico MSᵗᵒ. Latine Verfi·

ACCEDUNT

Ejufdem de SECTIONE SPATII

Libri Duo Reftituti.

Opus Analyfeos Geometricæ ftudiofis apprimè Utile.

PRÆMITTITUR

Pappi Alexandrini Præfatio

ad VIIᵐᵘᵐ Collectionis Mathematicæ,

nunc primùm Græce edita :

Cum Lemmatibus ejufdem Pappi ad hos
Apollonii Libros.

Opera & ftudio Edmundi Halley
Apud Oxonienses
Geometriæ Profefforis Saviliani.

OXONII,

E Theatro Sheldoniano
Anno MDCCVI.

*Ariftippus Philofophus Socraticus, naufragio cum ejectus ad Rhodienfium
litus animadvertiffet Geometrica fchemata defcripta, exclamaviffe ad
comites ita dicitur,* Bene fperemus, Hominum enim veftigia video.
Vitruv. Architect. lib.6.Præf.

(left) The first part of Halley's edition of the works of the classical Greek mathematician Apollonius was published in Oxford in 1706.

(right) The frontispiece of Halley's edition of the main works of Apollonius, published in 1710, continued (with an appropriate choice of geometrical figures in the sand) the tradition of reminding readers of the story of Aristippus, introduced to an Oxford audience in Henry Savile's lectures 140 years before (see page 53).

Back in Oxford after a quarter of a century, Halley set about securing his reputation within the University with a widely admired inaugural lecture:

Mr Hally made his Inaugural Speech on Wednesday May 24, which very much pleased the Generality of the University. After some Complements to the University, he proceeded to the Original and Progress of Geometry, and gave an account of the most celebrated of the Ancient and Modern Geometricians. Of those of our English Nation he spoke in particular of Sir Henry Savil; but his greatest encomiums were upon Dr Wallis and Mr Newton, especially the latter, whom he styled his Numen etc. Nor could he pass by Dr Gregory, whom he proposed as an example in his Lectures; but not a word all the while of Dr Bernard …

Over the next few years Halley justified the geometrical nature of his Chair by wisely taking up a suggestion from Henry Aldrich, Dean of Christ Church, that he take up a task left unfinished by his predecessor, John Wallis. He prepared what was to become the definitive edition of Apollonius's *Conics*, including in the project a minor work of Apollonius, 'The cutting-off of a ratio' (which had survived in an Arabic manuscript) and a reconstruction of the fragmentary *Sectio spatii* (the cutting-off of an area). Always the astute judge of men and situations, Halley must have realized that a scholarly edition of a major classical geometer would silence those critics who accused him of

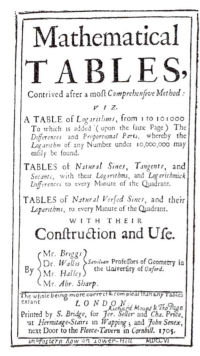

Mathematical TABLES,

Contrived after a moſt *Comprehenſive Method :*

v i z.

A TABLE of *Logarithms,* from 1 to 101000 To which is added (upon the ſame Page) The *Differences* and *Proportional Parts,* whereby the *Logarithm* of any Number under 10,000,000 may eaſily be found.

TABLES of *Natural Sines, Tangents,* and *Secants,* with their *Logarithms,* and *Logarithmick Differences* to every Minute of the Quadrant.

TABLES of *Natural Verſed Sines,* and their *Logarithms,* to every Minute of the Quadrant.

W I T H   T H E I R

Conſtruction and Uſe.

By { Mr. *Briggs,* Dr. *Wallis,* Mr. *Halley,* Mr. *Abr. Sharp.* } Savilian Profeſſors of Geometry in the Univerſity of *Oxford.*

The whole being more correct & compleat than any Tables extant

L O N D O N :

Printed by S. Bridge, for *Jer. Seller* and *Cha. Price,* at Hermitage-Stairs in *Wapping* ; and *John Senex,* next Door to the *Fleece-Tavern* in *Cornhill.* 1705.

in Poſtern Row on Tower-Hill   MDCCVI

A 1705 book of tables indicates the cumulative contribution of Oxford mathematicians to a major seventeenth-century development. In the dedicatory letter to Edmond Halley the compiler, Henry Sherwin, remarked on 'It being universally acknowledged that by the Professors of Geometry in the Savilian Chair (viz. Mr. Briggs, Dr. Wallis, and Yourself) the logarithmical art hath received its greatest improvements.'

being a gadfly and an opportunist, and secure him the respect of the academic establishment.

Halley worked on part of the *Conics* in conjunction with David Gregory, his Savilian co-Professor of Astronomy, who died in 1708. The work was eventually published in 1706 and 1710 and clearly won the University's mark of approval, for Halley was rewarded with an honorary DCL (Doctor of Civil Laws) degree and a warm citation commending his public services from the University's Chancellor, the Duke of Ormonde. After this success, he prepared an edition of the *Spherics* (the geometry of the surface of a sphere) of the first-century Alexandrian astronomer Menelaus, translating it into Latin from Arabic and Hebrew manuscripts. This work was printed at the time but not published, for reasons now obscure, until 1758, long after Halley's death.

During his Oxford professorial years up to 1720, Halley made many contributions to Oxford life. His reputation attracted distinguished foreign visitors to Oxford, and he lectured on mathematics as befitted his Chair—some of his lectures, on solutions of polynomial equations, were published. It was during these years too that Halley came to make his most profound and intellectually far-reaching contributions to astronomy. He was to develop further the Newtonian impact upon the University, already commenced by David Gregory, while his induction to the Savilian Chair coincided with the publication of his fundamental cometary studies of 1705. But his real contributions to astronomy during 1704–20 were to be made in his papers on the physical composition of deep space, stellar distribution, and the proper motions of stars. During his professorial years many of Halley's long-standing research interests, such as geomagnetism and planetary dynamics, came to maturity. It is truly remarkable how Halley continued to maintain a veritable outpouring of original ideas and discoveries in the physical sciences in spite of the diverse demands placed upon him.

The proper motions of the stars

In 1710 Halley commenced a detailed study of the star catalogue of Ptolemy (compiled in the second century AD, but containing data that were even then up to 600 years old), and was struck by several discrepancies of star positions when compared with observations of the same stars nearer to his own time. Even when one discounted such factors as refraction and latitude, the errors were too great to ignore. In particular, Sirius, Aldebaran, and Arcturus displayed systematic position shifts of up to 30 arc minutes over 1800 years of observations.

From these studies, Halley drew the correct conclusion that the so-called 'fixed stars' are not fixed at all, but move independently in space. Although he could not prove it at the time, he correctly surmised that many dimmer stars also display 'proper motions' which were too small to be detected with available techniques. Since 1718, when Halley announced his results in *Philosophical Transactions*, many stars have been found to display these proper motions.

Halley's discovery of the existence of independent proper motions was one of his most portentous discoveries, lying as it does at the heart of all modern theories of stellar distribution. It was also characteristic of his originality of approach that he made it from a painstaking comparison between ancient and modern observations, rather than in the observatory.

Nebulae and stellar distribution

Between 1715 and 1720, at the height of his creativity as Oxford's Savilian professor, Edmond Halley published a series of short papers in *Philosophical Transactions* which might be considered to lay the foundation of modern astrophysics. In these papers, he inquired into the nature of the matter which occupies deep space, how it is distributed, the distance to which it extends, and why some things emit light and others do not. They provide wonderful demonstrations, not only of the fertility of the 60-year-old Halley's scientific imagination, but also the degree to which he could use contemporary physical evidence to construct plausible models, and state clearly for the first time some questions for which modern astrophysicists are still seeking solutions.

In 1715–16, Halley published a study of six variable stars, inquiring why some undergo regular and dramatic changes in luminosity. He then went on to discuss nebulae, addressing the question of why they glow at all. He was especially interested in the paradox of why nebulae without central stars succeed in emitting light anyway, and went on to posit the existence of a self-luminous 'medium' which was somehow capable of glowing without the presence of an illuminating star. This was a concept to which William Herschel would return in 1791, and which is an acknowledged ingredient of much cosmological theory. He then went on, from a narrow yet cogent body of observed data, to posit the immense distances of these nebulae. If they are apparently so large, starless, dim, and lacking any detectable parallax, they must not only be vastly remote, but much bigger than the individual stars: 'perhaps no less than our whole solar system'.

In 1720, he published a pair of what might justifiably be called the briefest 'profound' papers in the history of astronomy. These papers, 'Of the infinity of the sphere of the fix'd stars' and 'Of the number, order and light of the fix'd stars', occupied no more than four printed pages between them in Volume 31 of *Philosophical Transactions* (1720), although they struck to the heart of the problem of the physical construction of deep space. In the first, Halley used Newtonian gravita-

A portrait of Halley presented to the Bodleian Library by the artist, Thomas Murray, in 1713. The painting may mark the award of his Oxford doctorate of civil laws.

tional criteria to argue in favour of an infinite universe, and in the second he reached the same conclusion on optical and symmetrical considerations. If stars are equal in size and equidistantly 'planted' throughout space (as eighteenth-century concepts of order and symmetry pre-supposed), then that would explain why first-magnitude stars are greatly outnumbered by those of the second magnitude, and they in turn by the third, and so on. Telescopes revealed stars which were even dimmer than those visible to the naked eye, and which increased in number in direct relationship to their dimness. In short, the dimmer the magnitude of the star, the more of its class there were. The universe must therefore extend infinitely far in all directions, as each improvement in optical power succeeded in yielding progressively richer harvests of progressively dim stars.

This infinite distribution of glowing stars, augmented with the luminous 'medium' of the nebulae, posed an even greater problem: with so much ambient light in the universe, why is the sky dark at night? Halley was not the first astronomer to recognize this paradox of the infinite universe, but he was the first to give it a succinct articulation, although his explanation was incorrect. It was not until a century later, in 1823, that Heinrich Olbers made the classic formulation of this 'paradox', by whose name it is known today.

Astronomer Royal

The first Astronomer Royal, John Flamsteed, died on the evening of 31 December 1719, having occupied the office for over 44 years. Neither he nor Halley had cared much for each other since Halley's days of youthful promise, and Flamsteed had done undisputed mischief to Halley's career, most notably in 1691 when his remarks swayed the electors at Oxford to award the Savilian professorship of astronomy to David Gregory. Yet Halley, in turn, had seized some opportunities to get his own back, most especially in 1712 when he was prevailed upon by Newton to edit Flamsteed's observations as *Historia coelestis in libri duo*. Not only did Halley's editorial rigour leave a good deal to be desired— the printed volumes contained many numerical errors—but he added a preface, written by himself without consulting the author, in which he attacked Flamsteed for sluggishness, secretiveness, and lack of public spirit. This 'pirated edition' of Flamsteed's life's work drove home the final division between Halley and Flamsteed. It was therefore an enormous affront to the memory and achievement of Flamsteed when, early in 1720, Halley was appointed Astronomer Royal to succeed him. Flamsteed's widow, Margaret, was enraged, and sold all of her husband's instruments (which were the Astronomer Royal's personal property) to ensure that Halley would never lay hands upon them.

A collection of eighteenth-century astronomical instruments, photographed in the 1890s. In front of Halley's quadrant is James Bradley's zenith sector.

Halley probably sought the vacant Astronomer Royal's job, not only for the modest £100 per annum salary that it carried, but because he could hold it in tandem with his Oxford professorship. By the pluralist standards of the Hanoverian age, it was considered natural for a talented and energetic man to accumulate offices, and it was not necessary for Halley to resign his Oxford Chair on account of his new appointment. Yet Halley did not treat Greenwich as a sinecure, but threw himself into a determined programme of research there. It says a great deal for Halley's apparent health and vigour that he was offered this arduous job in the first place, bearing in mind that he was 64 years old and would be 69 before his new set of instruments was ready. More extraordinary was his declared intention of devoting his office to the observation of the motion of the lunar coordinates through a full 18-year cycle. Most extraordinary of all, he succeeded in accomplishing this task, and was to be Astronomer Royal for nearly 23 years.

For decades, Halley and other astronomers had been recommending a method of finding the longitude at sea by the Moon's position against the fixed stars. To make this method work, it was necessary for astronomers to map the highly complex 18-year lunar cycle to a much higher degree of precision. One suspects that Halley's enthusiasm for the challenge was not only the undoubted kudos that a solution would bring, but also the incentive of the £20 000 reward offered under the act of 1714. His critics in the 1720s always accused him of chasing after the longitude prize. But to win both the kudos and the money—both dear to Halley's heart—he needed an astronomical establishment better equipped than Oxford's.

When Halley took up residence at Greenwich in 1720, he moved into an observatory without instruments. Being shrewder and more courtly than Flamsteed, he got the government to pay for them—a fortunate precedent for future holders of the office. Even so, he had to start from scratch, and with his generous £500 official grant he set about designing, ordering, and supervising the construction of what would become by 1725 a prototype set of meridian astronomy instruments. He was also fortunate in securing the services of George Graham FRS, the great clockmaker and foremost mechanician of the age, to construct them.

Halley's plan was to observe the Moon's place in right ascension (or east–west position) by timing its meridian passage through a transit telescope against a good regulator clock, which seems to have been in operation by 1721. By 1725, Graham had completed a much more ambitious instrument for Halley—a great quadrant of eight-feet radius with scales of novel design and telescopic sights. With this instrument, fixed on a stone meridional wall, Halley could read declinations, or vertical angles between celestial bodies. This basic set of instruments was to be of fundamental importance, not only to Halley's work, but to the design of all subsequent large observatory instruments. The transit, clock, and mural quadrant (or later, circle) became essential components of any serious observatory from the 1720s to the early twentieth century.

Halley never obtained lunar coordinates that were sufficiently accurate to find the longitude reliably, or to win the £20 000 prize. This did not derive from any slackness or lack of zeal on his part; it was said that Halley never missed a single lunar transit that was visible at Greenwich in 18 years—a formidable achievement for any man, let alone one entering his 80s. His failure derived from the fact that it was not then possible to measure the Moon's position against the fixed stars with sufficient accuracy, even on Graham's superlative instruments. Not until the 1760s, with further improvements in mechanics and precision instrument making, would it be possible to measure down to the one or two arc seconds that are necessary.

Halley was the outstanding Oxford astronomer of his day, but he was not alone. Newtonianism, astronomy, and the new science were widely discussed in the University, most particularly by David Gregory, John Keill, John Desaguliers, and James Bradley. These developments are discussed in the next chapter.

Further reading

The articles on Halley in the *Dictionary of national biography* and the *Dictionary of scientific biography* provide useful information. For general histories of Oxford and science, see the three volumes of C. E. Mallet, *The history of Oxford University*, Oxford, 1924, and Volumes II and XI of Robert T. Gunther's *Early science in Oxford*, Oxford, 1923, 1937.

An important recent biography of Halley is Alan Cook's *Edmond Halley: Charting the heavens and the seas*, Clarendon Press, Oxford, 1997. Earlier biographies include those by Colin Ronan, *Edmund Halley—genius in eclipse*, New York, 1969, and London, 1970, and Angus Armitage's *Edmond Halley*, London, 1966. Halley's work on geomagnetism is published in *The three voyages of Edmond Halley in the Paramore, 1698–1701*, edited by N. J. W. Thrower, Hakluyt Society, London, 1981, while E. F. MacPike's *Correspondence and papers of Edmond Halley*, Oxford, 1932, provides useful primary material.

The great bulk of Halley's own researches were published in the *Philosophical Transactions of the Royal Society* between about 1680 and 1740. John Flamsteed's accounts of Halley are to be found in *The 'Historia coelestis Britannica' of John Flamsteed, 1725*, edited and introduced by Allan Chapman, National Maritime Museum Monograph **52**, London, 1982. Halley's work on the transit of Venus preparations is published in Harry Woolf's *The transits of Venus*, Princeton, 1959; see also C. A. Murray, 'The distance of the stars', *The Observatory* **108**, no. 1087 (December 1988), 199–217.

On 8 May 1992, the Royal Astronomical Society held a special meeting to mark the 250th anniversary year of Halley's death; see the *Quarterly Journal of the Royal Astronomical Society* **34**, no. 2 (June 1993), 135–55.

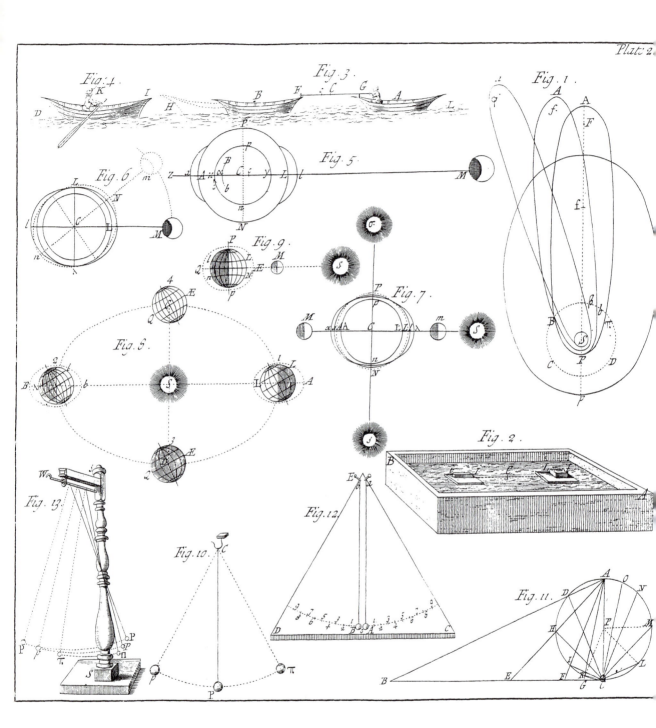

Plate 2.

Fig. 4.

Fig. 3.

Fig. 1.

Fig. 6.

Fig. 5.

Fig. 9.

Fig. 7.

Fig. 8.

Fig. 2.

Fig. 13.

Fig. 10.

Fig. 12.

Fig. 11.

CHAPTER 8

Oxford's Newtonian school

Allan Chapman

At the beginning of the eighteenth century, Oxford University became the centre for a group of mathematicians and scientists who were active in the development and promotion of the ideas contained in Newton's *Principia*. At a time when Newton had ceased to reside in Cambridge and was working on his lunar theory from London, it was in Oxford that his ideas were most actively fostered and disseminated. The principal members of this group of Oxford Newtonians were David Gregory, John Keill, John Theophilus Desaguliers, John Whiteside, and James Bradley. Although none of them possessed outstanding scientific originality as researchers, they probably did more than anyone to propagate Newtonianism in the period.

David Gregory was a nephew of the great Scottish mathematician James Gregorie and the custodian and developer of many of his illustrious uncle's ideas. He was also a vigorous teacher in his own right, whom Newton strongly and successfully recommended for the Savilian Chair of Astronomy when it fell vacant in 1691. On his arrival in Oxford, Gregory became (until his early death in 1708) the University's first active resident Newtonian. His inaugural lecture, on 21 April 1692, sketched the history and methods of celestial physics with particular emphasis on Newton's work. Indeed, the second half of the lecture, a catalogue of results established in *Principia*, must have been the fullest public account of Newton's work ever heard in Oxford.

John Keill, David Gregory's former Edinburgh pupil who followed him to Oxford, was a man of similar capacity. A high churchman (unusual for a seventeenth-century Scot), Keill argued against the 'atheistic' philosophies which were coming out of the continent, and emphasized the compatibility of Newtonianism and Christianity. He was especially interested in Newtonian explanations for the coherence of matter and its relation to experimental philosophy, or physics. After working in various lecturing capacities in Oxford, Keill eventually succeeded to the Savilian Chair of Astronomy in 1712, a post which he retained until his death in 1721. It was Keill who introduced the teaching of Newtonian ideas through courses on experimental philosophy in the 1690s. In the testimony of Desaguliers, Dr John Keill

Newtonian experimental philosophy—'to explain and prove experimentally what Sir Isaac Newton has demonstrated mathematically'—was developed and promoted in eighteenth-century Oxford. Desaguliers' *Course of experimental philosophy* gave several plausible demonstrations of Newton's third law of motion, involving boats, magnets, tides, and pendulums.

David Gregory's inaugural lecture as Savilian Professor of Astronomy was a strong statement of the value of geometry in understanding the world, and of the peculiarly English contribution to improving natural philosophy along these lines, fostered especially at the University of Oxford.

Oratio Inauguralis

a Davide, Gregorio M: D: Astronomiæ professore Saviliano, in Auditorio Astronomico Oxoniæ, Habita vigesimo primo die Mensis Aprilis Anni 1692, quum publicam Astronomiæ professionem auspicatus est.

Insignissimo Domino Vice Cancellario. Reliquiaq Academiæ Proceres honorandi. Academici Doctissimi.

Professionis Astronomicæ munus in Academiâ Oxoniensi suscepturum decet in ipso principio, gratâ Oratione Excellentissimi viri D: Henrici Sabilij Memoriâ litare, qui Astronomiæ insomnibus illis portarum mundi vigilibus hæc otia fecit, et il Lustrissimis professorum Savilianorum electoribus devinctissimum, me profiteri, qui ea nobis quoque propria esse voluerunt.

Cum primis autem animo meo gratissimum est, post magnam ætatis partem in alijs academijs exactam, potuisse me in Academiam Oxoniensem principi Geometrarum D: Johanni Wallisio Collegam et Literatissimo D: Edwardo Bernardo successorem cooptari.

was the first who publickly taught Natural Philosophy by Experiments in a mathematical Manner: for he laid down very simple Propositions, which he prov'd by Experiments, and from those he deduc'd others more compound, which he still confirm'd by Experiments; till he had instructed his Auditors in the Laws of Motion, the Principles of Hydrostaticks and Opticks, and some of the chief Propositions of Sir Isaac Newton concerning Light and Colours. He began these Courses in Oxford … and that Way introduc'd the Love of the Newtonian Philosophy.

John Theophilus Desaguliers, of Christ Church, was of Huguenot descent. By the end of the first decade of the eighteenth century, he was actively lecturing in experimental philosophy in Oxford, and was appointed a lecturer at Hart Hall (now Hertford College). Desaguliers' great talent lay in the high-level popularization of Newtonian physics within the University, which he did with great effect by means of a collection of elaborate models and machines that were especially useful for communicating abstract principles such as waves or the dynamics of

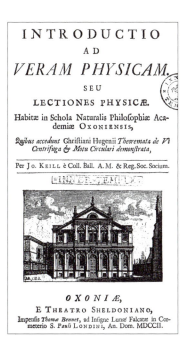

ASTRONOMIÆ
PHYSICÆ & GEOMETRICÆ
ELEMENTA.

Auctore DAVIDE GREGORIO M. D.
Astronomiæ Professore Saviliano OXONIÆ,
& Regalis Societatis Sodali.

OXONIÆ,
E THEATRO SHELDONIANO, An. Dom. MDCCII.

INTRODUCTIO
AD
VERAM PHYSICAM.
SEU
LECTIONES PHYSICÆ.
Habitæ in Schola Naturalis Philofophiæ Aca-
demiæ OXONIENSIS,
Quibus accedunt Chriftiani Hugenii *Theoremata de Vi
Centrifuga & Motu Circulari demonftrata,*
Per JO. KEILL è Coll. Ball. A. M. & Reg. Soc. Socium.

OXONIÆ,
E THEATRO SHELDONIANO,
Impensis *Thomæ Bennet,* ad Infigne Lunæ Falcatæ in Cœ-
meterio S. *Pauli* LONDINI, An. Dom. MDCCII.

Books by David Gregory and John Keill,
published from the Sheldonian Theatre,
helped to position and promote the teaching
of physical science in the post-Newtonian era.

comets to non-mathematical academics. In 1712, Desaguliers moved to London, where his lecture demonstrations and writings made him famous. He even entered courtly circles, after impressing King George I with his lectures, and went on to demonstrate a relationship between the rational laws of Newtonian mechanics and constitutional monarchy.

John Whiteside was Keeper of the Ashmolean Museum in Oxford, housed in the building in Broad Street which now contains the Museum of the History of Science. After Desaguliers had left for London, Whiteside used its lecture room to deliver a very successful Oxford course in experimental physics, with a privately owned collection of apparatus that was valued at £400 in 1723. He was also a skilled astronomer and communicated some of his astronomical results to the Royal Society via Edmond Halley. Halley had been discussing the possible substance out of which the glowing 'effluvium' of the aurora borealis might be formed, and in 1716 he cited an experiment by John Whiteside as evidence. In an article in the *Philosophical Transactions*, Halley recalled watching Whiteside ignite gunpowder *in vacuo* in the laboratory of the Ashmolean Museum. The room had been deliberately darkened, and Halley argued that the faint luminescence that remained in the evacuated glass vessel could be the same as the glowing effluvium of the Northern Lights.

Gregory, Keill, Desaguliers, and Whiteside brought important pedagogical gifts to Oxford. All four were effective teachers, and their promotion of the Newtonian philosophy, in a form which could be

Balls take the same time to roll down a
cycloidal slope no matter where they start,
just as a pendulum moving in a cycloidal arc
takes the same time to swing whatever the
amplitude of the vibration: an illustration
from Desaguliers' *Course of experimental
philosophy*.

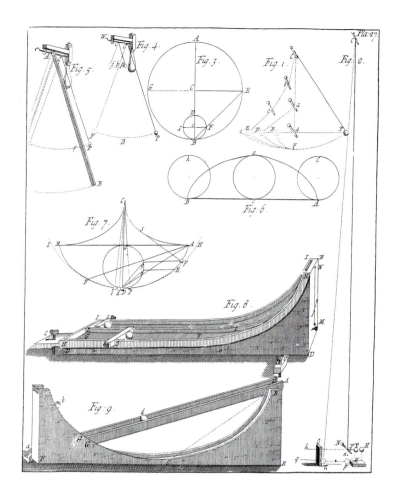

comprehended by largely non-scientific audiences, must have done a
great deal to assure its acceptance among the educated laity, which constituted the academic and undergraduate body. When one remembers
the overall Christian context of their teaching, they must have played
an important part in showing that the new science was compatible
with Anglicanism, by demonstrating the logic and order of the creation. The incorporation of the Newtonian philosophy into the cultural fabric of the Georgian intelligentsia owed a great deal to Gregory,
Keill, and men like them. In London, especially, it owed a very great
deal to Desaguliers. But in terms of intellectual influence, Oxford's
most creative astronomer after Halley was James Bradley.

James Bradley

If Halley was the greatest theoretical astronomer to come out of
Oxford, then Bradley might well be considered as its greatest practical
observer. His influence upon the subsequent history of the science was

The Ashmolean Museum was used for lectures in Newtonian science throughout the eighteenth century.

profound: as late as the 1830s, the great German astronomer Friedrich Bessel regarded Bradley's observations as laying the foundations upon which all subsequent work was built. Bradley developed the hitherto largely practical *art* of observation and raised it to an exact science in its own right, so that theoretical problems in astronomy could obtain solutions and predictable bases from deliberately planned courses of observations. While he was not the first astronomer to do this—Tycho Brahe, Hevelius, and Flamsteed had attempted to work along similar lines—he was the first to do it successfully and systematically over nearly 40 years, creating parameters of accuracy that are still accepted today.

His uncle was one of the leading amateur astronomers of the day, Dr James Pound, an Oxford graduate and rector of Wanstead in Essex. Dr Pound knew Halley well, and introduced his nephew to him when James was establishing himself as a Balliol College luminary and rising clergyman. Bradley became very much Halley's protégé and collaborator. In 1716, when he was 23, Bradley seized upon an invitation from Halley to make positional observations of Mars and some nebulae, which earned him the Savilian Professor of Geometry's approval.

The conspicuously talented Bradley now had two careers opening before him, both of which he could have developed in tandem (like many contemporary men of science) had he been less scrupulous. In addition to astronomy and his election to the Royal Society in 1718,

James Bradley, the embodiment of Oxford's Newtonian tradition, was Keill's successor as Savilian Professor of Astronomy and Halley's successor as Astronomer Royal.

Bradley's clerical career was flourishing. The following year he became vicar of Bridstow, Monmouthshire, along with receiving other marks of promise and patronage from the Bishop of Hereford. As these appointments were largely sinecures, he had plenty of opportunity to travel regularly to Essex to work in his uncle's observatory, as well as to cultivate scientific company in Oxford and London.

In 1721, he was elected Savilian Professor of Astronomy, following the death of John Keill and in the wake of Halley's appointment as Astronomer Royal the year before. Bradley was required to resign all of his clerical appointments and he refused to accept any future ones. It is not easy for us today to appreciate the magnitude of sacrifice which this decision entailed. A clerical career, to a man of Bradley's obvious talent and character, could easily have whisked him up the ladder to a bishopric by early middle age, with a seat in the House of Lords and all the power and influence enjoyed by senior clergy in the Georgian age. Halley, it is true, had never taken holy orders, although he did retain his Savilian professorship, as well as enjoying other sinecures when Astronomer Royal, such as a half-pay Royal Navy captaincy. But to forgo the opportunity of a clerical career for the modest annual salary of a Savilian professor (£138 5s 9d throughout the 41 years of Bradley's holding the Chair) must have seemed surprising to his contemporaries.

The aberration and nutation

In 1725, Bradley collaborated in an attempt to measure the stellar parallax of the zenith star γ Draconis with a zenith sector—a long telescope that hung vertically in the meridian. If a position shift could be found, the distance of the star could then be calculated trigonometrically. Within a few days he indeed detected a motion in the position of γ Draconis: but it was too large, and in the wrong direction to that expected. As a result, Bradley decided to commission a more versatile zenith sector from George Graham. With this, he was able to observe a wider range of zenith stars and found that not only γ Draconis, but all of the adjacent stars, partook in the same motion. Bradley came to realize that the position shift of all stars in a given direction must be related to the Earth's moving in that particular direction of space, during a particular part of its orbit. This suspicion was confirmed when, several months later, the stars began to move back, so that within a whole year they exhibited a regular extreme motion amounting to 40 arc seconds. This 'aberration of light' was a masterpiece of painstaking observation and evidence-interpretation, and won Bradley international acclaim when he published it in 1728, providing as it did the first clear demonstration that the Earth must move around the Sun.

Using the same zenith star, γ Draconis, in 1727, he also discovered another motion of the stars which could not be explained by parallax,

The lecture course on experimental philosophy changed little in content over many years. This syllabus is either for Bradley's course in the 1740s or for the similar one given by his precursor John Whiteside in the 1720s.

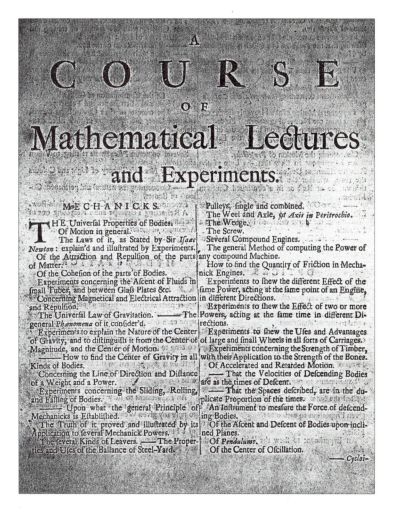

A

COURSE

OF

Mathematical Lectures

and Experiments.

MECHANICKS.

THE Universal Properties of Bodies.
Of Motion in general.
The Laws of it, as Stated by Sir *Isaac Newton*: explain'd and illustrated by Experiments.
Of the Attraction and Repulsion of the parts of Matter.
Of the Cohesion of the parts of Bodies.
Experiments concerning the Ascent of Fluids in small Tubes, and between Glass Plates &c.
Concerning Magnetical and Electrical Attraction and Repulsion.
The Universal Law of Gravitation. ———The general *Phænomena* of it consider'd.
Experiments to explain the Nature of the Center of Gravity, and to distinguish it from the Center of Magnitude, and the Center of Motion.
——— How to find the Center of Gravity in all Kinds of Bodies.
Concerning the Line of Direction and Distance of a Weight and a Power.
Experiments concerning the Sliding, Rolling, and Falling of Bodies.
——— Upon what the general Principle of Mechanicks is Established.
The Truth of it proved and illustrated by its Application to several Mechanick Powers.
The several Kinds of Leavers. ———The Properties and Uses of the Ballance of Steel-Yard.

Pulleys, single and combined.
The Weel and Axle, or *Axis in Peritrochio*.
The Wedge.
The Screw.
Several Compound Engines.
The general Method of computing the Power of any compound Machine.
How to find the Quantity of Friction in Mechanick Engines.
Experiments to shew the different Effect of the same Power, acting at the same point of an Engine, in different Directions.
Experiments to shew the Effect of two or more Powers, acting at the same time in different Directions.
Experiments to shew the Uses and Advantages of large and small Wheels in all sorts of Carriages.
Experiments concerning the Strength of Timber, with their Application to the Strength of the Bones.
Of Accelerated and Retarded Motion.
——— That the Velocities of Descending Bodies are as the times of Descent.
——— That the Spaces described, are in the duplicate Proportion of the times.
An Instrument to measure the Force of descending Bodies.
Of the Ascent and Descent of Bodies upon inclined Planes.
Of *Pendulums*.
Of the Center of Oscillation.

——— Cycloi-

nor by precession of the equinoxes, nor aberration. After systematically observing the motion over many years, he realized that the Earth exhibited a slight axial 'nodding', caused by the Moon's gravitational pull. Suspecting that this cycle would close once the Moon had completed its own 18-year cycle, he withheld publication until 1748. From the complete set of observations Bradley discovered that his predictions were precisely confirmed. He named the new motion 'nutation', from the Latin *nutare*, to nod.

The discovery of aberration and nutation were milestones in the development of precision astronomy in the eighteenth century, and brought home to astronomers the physical truth of Newton's laws, for both phenomena accorded precisely with criteria implicit within Newton's *Principia* of 1687. Furthermore, Bradley's published figures for the original quantification of aberration and nutation come to within a decimal fraction of a single arc second of those accepted today—a remarkable tribute to the meticulous quality of his observations.

James Bradley as university teacher

While Bradley's resignation of his clerical prospects upon election to the Savilian professorship might appear single-minded in some respects, the Chair carried with it possibilities that went well beyond its £138 annual salary. The professor could also charge the students who attended his courses, and when Bradley took on a set of lectures in 'experimental philosophy' (that is, physics) in the 1730s, following the death of Whiteside, his financial rewards were handsome indeed.

Bradley was not only a scientist of genius; he was also an inspired and gifted teacher of the mathematical sciences who developed a significant following. He came to average 57 students per course, who paid him three guineas (£3.15) per head, and among his surviving papers in the Bodleian Library, the names of over one thousand individuals can be traced.

The Newtonian context of Bradley's Oxford teaching is clearly emphasized in a detailed set of notes taken by an anonymous student who attended his 20-lecture experimental philosophy course in 1747. This student notebook gives a remarkable insight into the teaching of physics in eighteenth-century Oxford, and shows how a scientist of international standing presented the subject to a non-specialist audience. On several occasions, Bradley gave acknowledgement to Dr Keill, and one realizes that the structure of the course, with its divisions dealing with optics, mechanics, hydrostatics, and so on, derived from earlier courses by Keill and the London lecturer–demonstrator Francis Hauksbee. In a predominantly arts university, where no undergraduate would have received a specialist training in mathematics or science, Bradley and his predecessors aimed to present the intellectual core of 'classical' Newtonian physics, not through a mathematical approach, but by demonstration using apparatus and philosophical argument.

In his first lecture, James Bradley 'mentioned the three rules laid down by the great Sir Isaac Newton as proper to be observ'd by all who are desirous of making any proficiency in natural knowledge', which related to the laws of cause and effect. Newton's laws of motion were discussed, as were his optical and prism experiments, and, of course, gravitation. A Newtonian context was provided for the lectures on mechanics, hydrostatics, and pneumatics, so that a student would conclude the course with a non-mathematician's appreciation as to how great mathematical laws bind nature into a unified whole.

By 1747, Bradley was the living embodiment of the Newtonian tradition, being Halley's acknowledged successor in both Oxford and

The Old Ashmolean building (now the Museum of the History of Science) still bears the marks of the students waiting to attend lectures in the mid-eighteenth century.

Greenwich and, through Halley, a man who had been part of Newton's circle in the Royal Society over the last decade of Newton's life. This intimate situation comes out in many of the anecdotes recorded in the 1747 lecture notes: the young men who attended Bradley's lectures must have realized that, when himself young, their professor had known the then old men who had created so much of modern science. Bradley told his students, for example, how Halley's improvements to the diving bell provided such an abundance of sweet air at the sea bed, that Halley 'drank a Bowl of Punch with as much ease & pleasure as upon the surface of the Earth'. He related how it had been Robert Boyle in Oxford who first realized that the 'Torricellian tube' or barometer could be used to predict the weather. He also provided a distinctly more sophisticated variant of the legendary 'falling apple' story of the discovery of gravity, gleaned perhaps from Newton himself:

The Manner of its being discovered was thus, Sr. Isaac Newton sitting in his Garden saw something, probably a Leaf, fall from a Tree, which described a curve Line, this put him upon trying how the same would fall in vacuo, & by a Series of Propositions found Gravity in Bodies.

Most of the men who attended Bradley's courses appear to have been BAs preparing for their Master of Arts degrees and holy orders—postgraduate students, in effect. After he became Astronomer Royal, Bradley divided his time between Greenwich and Oxford. When one adds together his various salaries and fees by 1750, one finds that he was making around £650 per annum, with free residences provided in Greenwich and Oxford; this was over twice the personal revenues received by the Bishop of Oxford. Bradley had succeeded in turning himself into someone who was in effect Oxford's first professional scientist, inasmuch as he made a very handsome living out of physics and astronomy without the need for a clerical benefice. Perhaps his decision to renounce clerical advancement, developing a lucrative profession out of what had hitherto been reckoned as a part-time job, was not so foolish after all.

Bradley as Astronomer Royal

In 1742, James Bradley succeeded Halley as Astronomer Royal, in accordance with Halley's wishes. He was also permitted to retain his Savilian professorship and continued to discharge his Oxford duties with diligence. It was a wise appointment on the government's part: not only was Bradley the former Astronomer Royal's Oxford protégé, but he had also made critical examinations of the Greenwich instruments. Equally wise was the permission which enabled Bradley to remain as Savilian Professor of Astronomy at Oxford, thereby making it possible to use insights gained from his Greenwich researches to enhance his Oxford teaching. As a conscientious and energetic man, Bradley came to divide his time between research at Greenwich and the teaching of his now famed courses at Oxford. It would be hard to find a better relationship between teaching and research or greater collaboration between a major scientific establishment and the University than that developed by Bradley between 1742 and 1762.

Moreover, upon moving into the Royal Observatory as Astronomer Royal, Bradley's first priority was to test and correct all of Halley's instruments. He succeeded in extracting a £1000 grant from the Admiralty to re-equip the observatory in 1748; this was double the size of Halley's grant, and that in a period of virtually zero inflation—perhaps the largest single government grant to British science up to that point. In this way, Bradley might be considered a pioneer in the business of University and government liaison for the financing of scientific work. Although the new Royal Observatory instruments were not directly used in Oxford teaching, the design principles upon which they were based and used formed an integral part of his subsequent Oxford courses, along with the researches into solar system gravitation and longitude finding in which they were employed.

By the time of his death in 1762, James Bradley had not only revolutionized the art of astronomical observation, but had turned it into an exact intellectual activity in its own right. Bradley's published Greenwich observations set a wholly new standard, both of accuracy and procedure, which all subsequent astronomers were obliged to follow. Considering Bradley's work on the aberration, nutation, instrument design, and error correction, one becomes aware of how he succeeded most dramatically in bringing the practical and intellectual dimensions of astronomy together, as no-one had done before. His procedures inspired both instrument makers and users to new heights of aspiration, which ultimately made possible the precise tabulation of the lunar orbit, the finding of the longitude, and the measurement of the first stellar parallax by Bessel in 1838.

Nathaniel Bliss

Nathaniel Bliss was Halley's successor as Savilian Professor of Geometry.

A young clergyman who was very much influenced by Bradley's lectures was Nathaniel Bliss. Rather curiously, his name does not appear on the list of Pembroke College men who attended Bradley's courses, and one wonders how many other people attended without their names appearing in the surviving lists. Dr Bliss was elected to succeed to the Savilian professorship of geometry when Halley died in 1742. We know that Bliss made observations of Jupiter's satellites in 1742 and of the bright comet of 1745. We also know that he had helped Bradley in some of his work at the Royal Observatory. But Bliss is a tantalizing figure in the sense that he left few enduring achievements or records behind him, although he was clearly a figure of considerable significance and energy in his day. Not only was he a vigorous Oxford teacher, but he was sufficiently weighty in national scientific terms to be elected Astronomer Royal in succession to Bradley in 1762.

One presumes that Bliss must have been hale and hearty to be appointed to such a demanding scientific job at the age of 62, while still retaining his Savilian Chair. The Admiralty, no doubt, was expecting another 20-odd years of work out of him, as they had received from Halley after his appointment to Greenwich at the age of 64 in 1720. But if this was the hope, their Lordships were disappointed, for poor Dr Bliss died only two years later. As he had completed no major researches in that time, his Greenwich observations were not published until 1805, when Thomas Hornsby included them as a supplement to his edition of the observations of James Bradley.

We must, however, assume that Bliss continued to teach in Oxford after his appointment as Astronomer Royal, for on 21 May 1765, his enterprising widow launched what one assumes to have been a continuation of his popular lectures, delivered by Bradley's Savilian astronomical successor, Thomas Hornsby, 'for the Entertainment of ladies and others'. Mrs Bliss was clearly the driving force behind the special lecture given in the Bodleian Library tower above the Schools quadrangle, for her name heads the published prospectus as organizer, taking precedence over Professor Hornsby who merely delivered it, and stipulating that the admission fee would be by ticket for half a crown. One assumes that the ladies who attended were well off, for in 1765 a half-crown would have maintained a labourer's entire family for a couple of days.

Nathaniel Bliss was the third Savilian professor in succession to become Astronomer Royal. It is unfortunate that his untimely death after only two years in that office has left him so overshadowed by Halley and Bradley.

Conclusion

Astronomy was actively cultivated in Oxford, as both a theoretical and practical subject, from the early seventeenth century. The founding of the Savilian Chairs in 1619 had provided a framework within which it could be officially promoted on a didactic level, although not as a specific degree subject. In addition to whatever original contributions to the subject the professors made in practical terms, they interpreted changing astronomical ideas, be they Copernican, Newtonian, or cosmological, for the comprehension and cultural polish of a largely non-scientific body of gentlemen and a handful of ladies.

In terms of vigour, staying-power, and intellectual brilliance, Edmond Halley was the most illustrious of these early professors, building upon a reputation established in youth and continued through Oxford to the Royal Observatory. But if Halley was Oxford's bright comet, his protégé, James Bradley, brought about a union between the theoretical and practical branches of astronomy which altered and enhanced the character of astronomy as an exact science. After Bliss's death in 1764, the directorship of the Greenwich Royal Observatory passed into the hands of a succession of Cambridge men, and thus out of our story, although Oxford responded by establishing an astronomical observatory of its own which, within a decade, was regarded by visiting foreign scientists as the best equipped and most visually beautiful in Europe, as we see in Chapter 10.

Further reading

All of James Bradley's published papers, and some previously unpublished ones, along with the only complete *Life of Bradley*, are included in Stephen Peter Rigaud's *Miscellaneous works and correspondence of the Rev. James Bradley ...* (edited by P. Rigaud), Oxford, 1832; see also Robert Grant's *A history of physical astronomy*, London, 1852, and Allan Chapman, 'Pure research and practical teaching: the astronomical career of James Bradley, 1693–1762', *Notes and Records of the Royal Society* **47** (2) (1993), 205–12. A detailed analytic study of Bradley and the aberration appears in John Fisher's M.Sc. thesis, 'James Bradley and the new discovered motion,' University of London, 1993.

For further background material, see Lesley Murdin's *Under Newton's shadow: astronomical practices in the seventeenth century*, Adam Hilger,

Bristol, 1985, and the chapters by A. G. MacGregor and A. J. Turner, 'The Ashmolean Museum', and G. L'E. Turner, 'The physical sciences', in *The history of the University of Oxford*, Vol. V, *The eighteenth century* (edited by L. S. Sutherland and L. G. Mitchell), Clarendon Press, Oxford, 1986, pp. 639–58 and 659–81. For the technical details of Bradley's observing procedures, see Allan Chapman's *Dividing the circle*, Chichester, 1995.

The articles on James Bradley and Nathaniel Bliss in the *Dictionary of national biography* also provide useful information. For general histories of Oxford and science, see the three volumes of C. E. Mallet, *The history of Oxford University*, Oxford, 1924, and Volumes II and XI of Robert T. Gunther's *Early science in Oxford*, Oxford, 1923, 1937.

LINEÆ
TERTII ORDINIS
NEUTONIANÆ,
SIVE
Illustratio Tractatus D. *Neutoni*
De Enumeratione *Linearum*
Tertii Ordinis.

Cui Subjungitur,

Solutio Trium Problematum.

R Bates A M d coll yisu

Authore Jacobo Stirling, è Coll. Ball. *Oxon.*

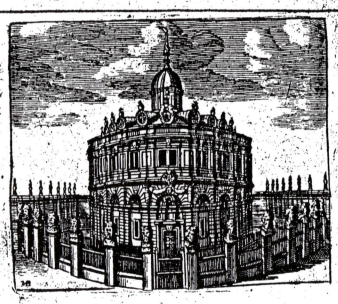

OXONIÆ,

E Theatro Sheldoniano, Impensis *Edvardi Whistler*
Bibliopolæ *Oxoniensis,* MDCCXVII.

Soc Reg Lond

CHAPTER 9

Georgian Oxford

John Fauvel

Oxford mathematics was in good shape at the beginning of the eighteenth century. The Savilian Professor of Geometry, John Wallis, was coming to the end of a long and distinguished career. His young and energetic colleague, the Savilian astronomy professor David Gregory, had been in post since 1691, and was soon to be joined by the equally energetic Edmond Halley, elected to the Savilian geometry Chair upon Wallis's death in 1703. There seemed no reason why Oxford mathematics should not build upon the Wallis legacy and flourish as much in the new century as it had done in the previous one. Gregory and Halley made a strong team, working together to strengthen Oxford's mathematical and astronomical development. They lived next door to each other, in the two halves of John Wallis's house in New College Lane, and shared interests both in astronomy and in the editing of ancient texts.

David Gregory was an enthusiastic teacher, and distributed his plan for making himself available to students who wished to learn mathematics:

> If any number of schollars desire him to explain to them the Elements, or any other of the Mathematicall Sciences if they are allready acquainted with the Elements, he will allow that company such a time as they among themselves shall agree upon, not less than an hour a day for three days in the week; in which time he will go through the said Science, explaining every proposition and illustrating it with such examples, operations, experiments, and observations as the matter shall require, until all the company fully apprehend and understand it.

The syllabus of seven suggested courses, distributed both privately and in a 1707 almanac, *Mercurius Oxoniensis*, gives an idea of the ambition of Gregory's intentions, although it is not clear how many students took up his offer.

But this bright start to mathematics teaching in eighteenth-century Oxford clouded over somewhat with the early death of Gregory, from consumption, in 1708. In this chapter we follow the story of mathematical activity in Oxford over the next century or so, in so far as that can be distinguished from astronomy and Newtonian science.

James Stirling's brilliant first book, *Newtonian lines of the third order*, was also the last book on contemporary research mathematics to be published in Oxford during that century.

Memorial to David Gregory in the University Church.

David Gregory's proposed mathematics courses

1. The first six with the Eleventh and Twelfth Books of Euclid's Elements.

2. The Plain Trigonometry, where is to be shewed the construction of naturall Sines, Tangents, and Secants, and of the tables of Logorithmes, as well of naturall numbers as of Sines, etc. The Practical Geometry, comprehending the descriptions and use of instruments and the manner of measuring heights, distances, surfaces, and solids.

3. Algebra, wherin is taught the method of resolving and constructing plain and solid problems, as well arithmeticall as geometricall; to which will be subjoyned the resolution of the indetermined arithmetical (or diophantaean) problemes.

4. Mechanicks, wherein are laid down the principles of all the sciences concerning motion; the five powers commonly so called explained, and the engines in common use reducible to those powers described.

5. Catoptricks and Dioptricks, where the effects of mirrours and glasses are shewed; the manner of Vision explained; and the machines for helping and enlarging the sight, as telescopes, microscopes, etc., described.

6. The Principles of Astronomy, containing the explication of all the most obvious Phaenomena of the Heavens from the true System of the World, and the Generation of the Circles of the Sphear thence arising. Here also is to be taught the doctrine of the Globes and their use, with the problems of the first motion by them resolved. After this is to be demonstrated the Sphaericall Triginometry and the application thereof to Astronomy shewed in resolving the problemes of the Spere by calculation, and the construction of the tables of the first motion depending on this.

7. The Theory of the Planets, where the more recondite Astronomy is handled: that is, the Orbits of the Planets determined by observation; the tables for their motions described, and the method of constructing them taught, and the use of these tables shewed in finding the Planets' places, the Eclipses of the Luminares, etc.

Gregory's successor was John Caswell, a veteran Oxford teacher, geodesist, and instrument designer, to the disappointment of John Keill, Gregory's friend and fellow Scot. Keill had already failed to secure the Sedleian Chair of Natural Philosophy in 1704. His friends sensed a plot, and read Keill's failure as the result of Whig low church scheming to prevent a Tory high church appointment. Jonathan Swift, for one, saw the political dimension clearly, writing to a friend in 1709:

> You know, I believe, that poor Dr Gregory is dead, and Keil sollicites to be
> his successor. But Party reaches even to Lines and Circles, and he will
> hardly carry it being reputed a Tory, wch yet he wholly denyes.

Deny it as he might, Keill was a tough fighter for the causes in which he believed—he is mainly remembered today for his assiduity in stirring the

Eighteenth-century Oxford has often been thought of as sunk in decadence and academic slumber. An advertisement for the mathematics lectures of the Savilian Professor of Geometry in mid-century disproves the more extreme of such charges, while the similarity to David Gregory's syllabus of half a century earlier suggests that the structures of undergraduate mathematics provision remained unchanged.

N. BLISS M.A. *Savilian Professor of Geometry*, proposes to explain the Elements of the most useful Mathematical Sciences, at his House in *New-College* Lane, in the following *Classes* or *Courses*.

I. ARITHMETICK *Vulgar* and *Decimal* with its Application to common Affairs, as well as to the other Parts of the Mathematicks.

II. The first *Six Books* with the *Eleventh* and *Twelfth* of EUCLID's *Elements*.

III. ALGEBRA, wherein will be taught the Method of resolving the several kinds of *Equations*, illustrated by a great Variety of useful and curious Problems, as well *Arithmetical* as *Geometrical*.

IV. PLAIN TRIGONOMETRY, wherein will be shewn, the Construction of the *Natural Sines, Tangents* and *Secants*, and the Table of *Logarithms*, as well of the *Natural Numbers*, as of the *Sines, Tangents* &c. with the Use of the *Logarithmic Tables* in the Solution of the several Cases of *Plain Trigonometry*. To which will be added the *Practical* GEOMETRY, comprehending the Description and Use of *Instruments*, and the Manner of measuring *Heights, Distances, Surfaces*, and *Solids*.

V. SPHERICAL TRIGONOMETRY, with its Use in the Resolution of the most common Problems of the *Sphere*; together with the Method of Projecting the several Cases *Stereographically*. To which will be added the full Description and Use of both the *Celestial* and *Terrestrial Globes*, and the Method of Solving the same Problems by them, which were solved by *Spheric Trigonometry*.

VI. The Elements of the CONIC SECTIONS, with the Demonstration of such of their Properties as are of most frequent Use, together with the Mensuration of the Superficial and Solid Content of the *Cone* and its *Frustums* and *Sections*. To which may occasionally be added the Method of Projecting the Sphere *Orthographically*, exemplified in the Construction of *Solar* and *Stellar Eclipses*.

It is proposed that the Number of *Scholars* in each of these *Classes* or *Courses*, be not less than *Six*, or more than *Ten*; to whom he will read three Days in a Week, and not less than an Hour each Day, explaining the Propositions, and illustrating them with Examples, and such Observations as the Matter shall require until the Company apprehend and understand it: And each *Person* shall have full Liberty to propose such *Doubts* or *Scruples* as he pleases.

For the *Text* to be explain'd, and to give Occasion for necessary *Digressions*, a printed Book will be used, if there be any that is proper; in other Cases every *Scholar* shall have Liberty to transcribe his *MS. Notes*, if he pleases.

It is computed that any one of these *Classes* or *Courses* will require about *three Months*; and any *Gentleman* may go through any *one* or *all* of them as he pleases, paying *two Guineas* at the Beginning of each Course, and *half a Guinea* more for every *Month* the Course shall continue longer than *three*.

priority dispute between Newton and Leibniz whenever it showed signs of abating. Caswell did not hold the post very long, but died in 1712, whereupon Keill was at last appointed to the Savilian Chair of Astronomy.

A straw in the wind of changing political fortunes, and their repercussions on mathematical appointments, was the suicide in 1712 of

John Wallis's grandson William Blencowe, who had succeeded Wallis as decipherer to the government. Recording the event in his diary, the robustly Jacobite Oxford antiquary Thomas Hearne described Blencowe as a 'proud, Fanaticall Whigg' who was 'discontented because the Whiggs were turn'd out, as they deserv'd'. John Keill was appointed government decipherer in place of Blencowe, but, according to Hearne, 'was turn'd out upon the D. of Brunswick's coming to the Crown' (that is, when George I came to the throne).

The late Stuart years saw the best mathematician that Oxford managed to produce in the eighteenth century. James Stirling was also a Scot, from a prominent Jacobite family, and studied at Balliol College from 1711 to about 1715. He lost his scholarship in 1715 on refusing to take the oath of allegiance to the new Hanoverian King George I, and never graduated. In July 1716, Hearne recorded in his diary that:

One Mr Sterling, a Non-juror of Bal. Coll. (and a Scotch Man), having been prosecuted for cursing K. George (as they call the Duke of Brunswick), he was tryed this Assizes at Oxford, & the Jury brought him in not guilty.

Nonetheless, he was recognized as the brightest young mathematician around. John Keill wrote to Newton of Stirling's success in solving a problem of orthogonal trajectories set by Leibniz for the English mathematicians, and in 1717 Stirling's first book *Lineae tertii ordinis Neutonianae* was published at Oxford. This was a commentary on Isaac Newton's remarkable classification and enumeration of cubic curves, a subtle work which most contemporary mathematicians had failed to appreciate. In what D. T. Whiteside has described as 'the turning-point in contemporary appreciation of the essential accuracy and technical accomplishment' of Newton's work, Stirling

produced the first proficient exposition and percipient critique of Newton's tract, elaborating its presuppositions, proving most of its undemonstrated theorems and arguments, and adding four (of the six) minor species of cubic which Newton himself had there overlooked ... This 'illustration' of Newton's tract set the style for later eighteenth-century accounts of the analytical geometry of cubic and higher plane curves—notably those by Leonhard Euler, Gabriel Cramer and Edward Waring—which were in turn to lead directly into the yet more complex ramifications of the topic developed by Plücker, Cayley and a host of others in the nineteenth century.

Newton himself subscribed for two copies of this book, and tried to help Stirling later in his career. With financial help from Newton, Stirling spent the next few years in Venice, from which he returned in 1724, teaching in London before returning to Scotland. The major part of his later life was spent as manager of Lanarkshire lead mines, and he did little mathematics after the mid-1730s. After Stirling's departure, we hear little about Oxford mathematics for 40 years or more.

AN
ESSAY
ON
The *Usefulness* of
MATHEMATICAL LEARNING,
IN
A Letter from a *Gentleman* in the CITY to his *Friend* in
O X F O R D.

Printed at the THEATER in *Oxford* for *Anth. Peisley* Bookseller, 1701.

Authorship of this anonymous essay, reprinted several times over the century, has been variously attributed. Written when Wallis and Gregory were still flourishing, it reflects the confidence in mathematical studies of early eighteenth-century Oxford.

Writing to Archbishop Wake about the Savilian Chair left vacant by the death of John Keill in 1721, the Provost of Oriel College George Carter commented, in terms that suggested his message was familiar, 'Mathematics are studied, Your Grace knows very well, but little amongst us, especially by such as I should be glad to recommend to this Place.'

The reasons for any diminution in Oxford mathematical activity are complex. Some responsibility must be placed upon the senior academics. The suitability of Oxford dons was, for example, fiercely attacked in the same year by Nicholas Amhurst (who was perhaps not wholly disinterested, having been expelled from his college fellowship at St John's two years earlier): 'I have known a profligate *Debauchee* chosen Professor of *Moral* Philosophy; and a Fellow, who never looked upon the Stars, sober in his life, Professor of *Astronomy.*' The latter reference would seem to be to John Keill—it is not known with what justice. Furthermore, the Savilian Chair of Geometry over this period was held by Edmond Halley, who had moved to Greenwich in 1720 to take up the post of Astronomer Royal and by this stage of his life took little interest in promoting non-astronomical mathematics, or indeed in teaching at Oxford.

Throughout these years, the two Savilian Chairs seem to have been treated as essentially interchangeable in terms of qualification or inclination. The main interests of Halley's successor Nathaniel Bliss (Savilian professor from 1742 to 1765) lay in astronomy also—and later two geometry professors in succession, Abraham Robertson and Stephen Rigaud, moved over to the astronomy Chair on the death of its incumbent. John Smith, another Scot (from Maybole in Ayrshire) was Savilian Professor of Geometry for over 30 years, from 1766 to

7–8 New College Lane was the residence of the Savilian professors from 1672 to 1854. Those who occupied one or other of the two houses on the site include John Wallis, David Gregory, Edmond Halley (for whom the University built the observatory on the roof of the left-hand house), John Keill, James Bradley, Nathaniel Bliss, Thomas Hornsby, Joseph Betts, John Smith, Abraham Robertson, Stephen Rigaud and Baden Powell.

OBSERVATIONS

ON THE

USE AND ABUSE

OF THE

Cheltenham Waters,

IN WHICH ARE INCLUDED

Occasional Remarks on different Saline
Compositions.

By J. SMITH, M. D.

Savilian Professor of Geometry in the University of
OXFORD.

CHELTENHAM:

Printed and Sold by S. HARWARD; sold also at his
Shops in GLOCESTER and TEWKESBURY; by MURRAY,
ELMSLY, and CADELL, in LONDON; and by the
Booksellers in OXFORD and BATH. 1786.

Price One Shilling and Six-pence.

The seventh Savilian Professor of Geometry's
contribution to learning.

1797, without being a mathematician of any sort—he taught anatomy and chemistry—and without being in residence for the time he spent as a physician in Cheltenham. It was not until the election of Robertson in 1797 that an active mathematician once again filled the Savilian Chair.

On a more institutional or ideological level, it is worth reflecting that the status of mathematics, in particular the Newtonian mathematics that so preoccupied the Oxford of Gregory and Keill, was but a cog in the wider political issues of the age. Oxford was identified throughout this period with high church and Tory opposition to the Hanoverian dynasty. While not as far to the romantic right as was a Jacobite such as James Stirling, both Gregory and Keill were noted Tories. Newtonianism, as a general approach to understanding the world, became increasingly identified, however, with the Whig oligarchy, and it could be that political and theological tensions contributed to the way that Oxford mathematical activity faded without the strong promotional personality of a Wallis, a Gregory, or a Keill.

We have but few hints about mathematical teaching or research in Oxford between 1720 and 1760, although the story for astronomy and natural philosophy is rather different, as reported in Chapters 8 and 10. In any case, it is clear from the reissue by the University Press of Keill's textbooks for students, covering Euclid and trigonometry in 1731 and again in 1747, and from other scattered evidence, that some undergraduate mathematical activity persisted. But as the century wore on, the reputation of Oxford as sunk in decadent slumber became so well established that many commentators in later years—well into the next century—could hold it up as an accurate portrayal, regardless of whether their remarks were currently true. Adam Smith is a case in point. In a celebrated charge he claimed in 1776 that 'In the university of Oxford, the greater part of the publick professors have, for these many years, given up altogether even the pretence of teaching.' The relaxed academic atmosphere of Oxford was an old theme of Smith's—in a letter of 1740 he wrote 'it will be his own fault if anyone should endanger his health at Oxford by excessive study'.

The student experience 1760–1800

It cannot be claimed that the eighteenth century was the University's most creative research period, nor that the teaching there—or anywhere in England, for that matter—produced mathematicians the equal of the best elsewhere in Europe. However, mathematical tuition was available for those who wanted it, and for some of the time achieved what was intended, to form a natural if unobtrusive part of a liberal education. Precisely because the mathematics teaching was not very ostentatiously promoted, this is a good opportunity to watch how

In 1737 William Hogarth drew Oxford scholars attending a lecture on *Datur vacuum*.

mathematics fitted in to the overall gentleman's education that Oxford provided at this period—an Oxford which Dame Lucy Sutherland has memorably described as a 'small but complex society, set in an equally small and beautiful, but very dirty and unsalubrious, town'. Charles James Fox, later a prominent Whig politician, was an undergraduate at Hertford College in the 1760s and seems to have worked quite hard, considering that for young men of his class, the so-called *gentlemen commoners*, Oxford was more like a finishing school, with no especial obligations of study or attendance. He wrote to a friend in 1765: 'I read here much, and like vastly (what I know you think useless) mathematics. I believe they are useful, and I am sure they are entertaining, which is alone enough to recommend them to me.' Fox's nephew, Lord Holland, commented years later that his uncle 'had a wonderful capacity for calculation, and a great aptitude, no doubt, for all branches of mathematics; but I have often heard him regret that he had applied so little to them; and ascribe his neglect of them to the superficial manner in which they were taught at Oxford.'

Superficial or not, we get some perception of the pace of Oxford mathematical studies at the time from a letter written to Fox by his tutor William Newcome. Newcome was either entirely unruffled by the fact that his charge spent months at a time living it up in Paris or London, or felt that a gently sardonic tone was the best way of gaining his occasional attention. Discussing the mathematics tutorials at Hertford College, he wrote to Fox, during one of his periodical absences:

I expect you will return with much keenness for Greek and for lines and angles. As to trigonometry, it is a matter of entire indifference to the other geometricians of the college (who will probably continue some time here) whether they proceed to the other branches of mathematics immediately, or wait a term or two longer. You need not, therefore, interrupt your amusements by severe studies, for it is wholly unnecessary to take a step onwards without you, and therefore we shall stop until we have the pleasure of your company.

We gain a similarly laid-back impression from Fox's friend James Harris, afterwards first Earl of Malmesbury. Harris was at Merton College from 1763 to 1765. and recalled in his later years that:

The two years of my life I look back to as most unprofitably spent were those I passed at Merton … My tutor, an excellent and worthy man, according to the practice of all tutors at that moment, gave himself no concern about his pupils. I never saw him but during a fortnight, when I took it into my head to be taught trigonometry.

From the tutor's perspective, it must have been hard to encourage learning on any subject, even had he wished to, in a system where there were marked gradations between students, some of whom had no intention of taking a degree, and where degrees were awarded for oral exercises of various kinds sometimes described as 'farcical' by contemporaries, with no written component until the following century.

A satirical poem of the period described the mathematics which the gentlemen commoners were careful not to get too involved with.

> I rise about nine, get to breakfast by ten,
> Blow a tune on my flute, or perhaps make a pen …
> While Law, Locke and Newton and all the rum Race
> That talk of their modes, their Ellipses and Space,
> The seat of the Soul and new Systems on high
> In holes as abstruse as their mysteries lie.

Students who were not in this social class had cause to take their learning opportunities more seriously. The 15-year-old Jeremy Bentham was a commoner at The Queen's College, and reported to his father on the experimental philosophy lectures of the Savilian Professor of Geometry, Nathaniel Bliss, in March 1763:

We have gone through the Science of Mechanics with Mr. Bliss, having finish'd on Saturday; and yesterday we begun upon Optics; there are two more remaining, viz: Hydrostatics, and Pneumatics. Mr. Bliss seems to be a very good sort of a Man, but I doubt is not very well qualified for his Office, in the practical Way I mean, for he is oblig'd to make excuses for almost every Experiment they not succeeding according to expectation; in the Speculative part, I believe he is by no means deficient.

Young Bentham also attended lectures of the Savilian Professor of Astronomy, Thomas Hornsby (see Chapter 10). In 1767 he wrote to his father seeking permission to attend, because this would involve Bentham *père* in the further expense of a guinea for the lecture fee: 'the Course will last no longer than a Week', Bentham observed, with Hornsby lecturing every day. It seems clear that Adam Smith's charges against the Oxford professoriate have to be taken with a pinch of salt.

All the same, donnish perceptions of mathematics were not unmixed. Dr Markham, Dean of Christ Church, told an undergraduate in 1768 that 'only classical and historical knowledge could make able statesmen, though mathematics and other things were very necessary for a gentleman'. But Dean Markham's Christ Church also used mathematics as part of the disciplinary system: an undergraduate rusticated for some now-forgotten offence was ordered to master, among other things, Books V, VI, XI, and XII of Euclid's *Elements*, besides working all the examples in the first part of Maclaurin's *Algebra*. We happen to know a lot about the undergraduate experience over the eighteenth century at Christ Church because the collections books from 1699 onwards have been preserved there. These are lists of authors read by individual students, with some schemes of required readings for classes. They show little evidence of mathematical reading up to 1760, but thereafter a sharp rise in mathematical books—mostly Euclid's *Elements* in Keill's 1701 version and later editions, and Maclaurin's *Algebra* (from 1748 and later editions).

THE

ELEMENTS

OF

EUCLID,

WITH

DISSERTATIONS,

INTENDED

To affift and encourage a critical examination of thefe Elements, as the moft effectual means of eftablifhing a jufter tafte upon mathematical fubjects, than that which at prefent prevails.

VOL. I.

By JAMES WILLIAMSON, M.A.

FELLOW OF HERTFORD COLLEGE.

Sed nil dulcius eft, bene quam munita tenere
Edita doctrina Sapientum templa ferena. LUCR.

OXFORD.

PRINTED AT THE CLARENDON PRESS.

M DCC LXXXI.

A conscientious example of Oxford historical scholarship in the late eighteenth century.

In 1772, an aristocratic undergraduate named James Cochrane circulated a leaflet complaining about the mathematics lectures: in particular, he noted that after a whole term the lecturer had reached only Proposition 7 of Book I of Euclid's *Elements*. The response of the lecturer, Samuel Love of Balliol, gives us some insight into the state of Oxford mathematics teaching in the 1770s. After pointing out that he was standing in for someone else, so had started the lecture course only half-way through the term, he claimed that, since the mathematical lecture was for all undergraduates, the mixture of abilities meant he could not go faster than one Euclidean proposition per lecture. Indeed, once he had to cancel the lecture when not enough of the class of 20 or 30 turned up, twice he didn't appear because he was detained elsewhere, and 'the Execution of the Criminal interfered with another Lecture'. Considering all these obstacles to holding the lecture at all, he did well, he claimed, to lead the students as far as Proposition I. 7 within the term. Despite this robust defence, Love evidently decided against making a career of teaching, and left Oxford shortly afterwards. He became a curate in Bristol, where he died the following year.

The study of Euclid at Oxford could reach rather greater heights than in poor Love's hands. James Williamson, Fellow of Hertford College, was a Scot, born in Elgin and a student at Aberdeen. He offered a lecture course in mathematics, which according to a contemporary letter cost two guineas for the first course, one guinea for the second course, and was gratis thereafter. He also produced an English edition of Euclid's *Elements* with lengthy essays on aspects of the work; commenting on the high standards of this edition, Augustus De Morgan remarked that 'Williamson was a real disciple of Euclid; and he translated so closely, that such words as, not being in the Greek, English idiom renders necessary, are put in Italics'. As with so many eighteenth-century dons, Williamson moved away from Oxford, to become a rural vicar at Plumtree, near Nottingham.

Interest in history: Robertson and Rigaud

The first volume of Williamson's edition of Euclid's *Elements* was published by the University Press in 1781. By this stage of the century, mathematical books were produced only rather sporadically, and were almost always of classical texts. Not since Stirling's work on Newton's classification of cubics in 1717 had a volume of contemporary research mathematics been published in Oxford; the previous two mathematical works from the Press, for instance, had been Halley's edition of the *Spherics* of Menelaus, published eventually in 1758, and Samuel Horsley's 1770 restitution of *Inclinations*, a minor work of Apollonius. Towards the end of the century the opportunity arose for the Press to continue its tradition of restoring the scholarship of the past by

M E N E L A I

S P H Æ R I C O R U M

L I B R I   III.

Quos olim, collatis MSS: *Hebræis* &
Arabicis, Typis exprimendos curavit
Vir Cl. Ed. H A L L E I U S L. L. D.
R. S. S. & Geometriæ Profeſſor Savil.
Oxonienſis.

Præfationem addidit
G. C O S T A R D, A. M.

O X O N I I,
S U M P T I B U S   A C A D E M I C I S,
M DCC LVIII.

The publication history of Halley's Latin
translation of Menelaus' *Spherics* illustrates
eighteenth-century publishing practices.
Printed under Halley's supervision in 1713–14
as pp. 7–70 of a book, it was not published
until 1758, seen through the press by George
Costard, at a cost of 1s 2d, in an edition of 250
of which 188 copies were still unsold 40 years
later.

publishing the papers of Thomas Harriot (see Chapter 3), perhaps the greatest mathematician to have been educated at Oxford.

In 1784 Harriot's manuscripts were rediscovered among stable accounts in Petworth House, Sussex. An ambitious young Austrian, Franz Xaver Zach, who was awarded an honorary doctorate by the University of Oxford in 1786 on the recommendation of the Savilian Professor of Astronomy, Thomas Hornsby, proposed to the University Press an edition of the more significant of Harriot's papers, together with a biographical study. On the Press's behalf the papers were examined by Abraham Robertson, deputy to the absentee Savilian Professor of Geometry, John Smith, whom he succeeded a few years later. It was widely believed that a momentous edition was about to appear. Charles Hutton wrote in his *Mathematical and philosophical dictionary* (1797):

As to his manuscripts lately discovered by Dr Zach, as above mentioned, it is with pleasure I can announce, that they are in a fair train to be published: they have been presented to the university of Oxford, on condition of their printing them; with a view to which, they have been lately put into the hands of an ingenious member of that learned body, to arrange and prepare them for the press.

But alas! Robertson reported back that the papers were in no state for publication and would not contribute to the advancement of science. Zach had in any case moved on to other interests, and so the moment passed. (It is only fair to add that the present generation has not found it any easier to prepare Harriot's papers for publication.)

By the time the papers were returned to Petworth by the Press in 1799, the whole collection was in an even more chaotic condition than Harriot's executors had left them. Furthermore, the transfer was done so quietly that rumours abounded for years that Oxford was sitting on the papers; given the somewhat stately progress of some manuscripts through the Press in the eighteenth century, this was not *a priori* implausible. In order to lay these rumours to rest, the papers were examined again, by the most historically sensitive mathematician to have studied them since Harriot's time, the energetic and versatile Stephen Rigaud. Rigaud was Robertson's successor in the Savilian Chair, first of geometry (1810) and then of astronomy (1827), as well as one of the Delegates of the Press, and as he studied Harriot's papers he grew more and more interested in trying to take on the editorial and biographical role that Zach had abandoned nearly 40 years before. In 1832 he addressed a meeting of the British Association, held at Oxford, on the subject of Harriot's astronomical observations and Zach's misunderstandings of them. But before Rigaud could bring this work to fruition he died, in 1839, leaving the notebooks with all his Harriot researches to the Bodleian Library, where they are still to be found.

Aristippus Philosophus Socraticus, naufragio cum ejectus ad Rhodiensium littus animadvertisset Geometrica schemata descripta, exclamavisse ad comites ita dicitur, Bene sperimus, Hominum enim vestigia video.
Vitruv. Archit. lib 6 Praef.

The continuity of the Oxford classical mathematical texts tradition was emphasized by adapting, for the 1792 Torelli edition of Archimedes, the frontispiece of the shipwreck of Aristippus previously used for Gregory's *Euclid* and Halley's *Apollonius* (see pages 53 and 129). The legend had been introduced to Oxford thinking by Sir Henry Savile two centuries before.

Rigaud made several further contributions to the advancement of scholarship. He was one of the active participants in the growth of Newtonian historical scholarship in the 1830s: his *Historical essay on the first publication of Sir Isaac Newton's 'Principia'* (1838) was a major contribution to the development of modern understanding of Newton as a historical figure, and his *Correspondence of scientific men of the seventeenth century* (1841) is still of great value for its rich selection of letters from or to Henry Briggs, William Oughtred, John Wallis, Edward Bernard, John Keill, Edmond Halley, and others, many associated with Oxford's scientific past.

Perhaps the most far-reaching way in which Rigaud contributed to Oxford scholarship was through his founding of the Ashmolean Society in 1828, together with his friend the geologist William Buckland; this was a club which, in Robert Gunther's words, 'undoubtedly went far to prepare a way for the scientific renaissance of Oxford'. For much of the nineteenth century this was the place for Oxford scientists to meet, to hear papers on current researches and hold discussions on a wide variety of matters. Many of the papers were published by the Society as pamphlets; the leading mathematicians of early and mid-century Oxford contributed, as well as many others across the natural and physical sciences. Until around 1860 the Society met in the Old Ashmolean building (now the Museum of the History of Science); by the time of its amalgamation with the Oxfordshire Natural History Society towards the end of the century, the Oxford Mathematical Society had been founded (see Chapter 13), and the Ashmolean Society had less of a crucial nurturing role to fulfil for Oxford's mathematical community.

Examinations

In 1800 a new system of University-wide examinations came into being. This created the honours degree and made the examination public and a function of the University as a whole, taking responsibility away from the individual colleges. It was still an oral examination, though, for some years to come. We have an account of one of the first examinations under the new system, that of Daniel Wilson (later to be Bishop of Calcutta) on 30 June 1802, in the presence of an audience of about one hundred. It included questions on Greek authors and the Bible, in Hebrew and Greek, but also questions on science:

Whilst sitting apart, the junior examiner, as if casually, asked whether Wilson had read Physics, and then put certain questions, such as, "Whether the angle of refraction was equal to the angle of incidence?" "Whether a ray of light passing from a thin into a denser medium would be deflected from the perpendicular?" &c: all of which were of course answered. Mathematics, logic, and metaphysics were passed by: one of the sciences only being required by the statute.

Stephen Rigaud was a major contributor to the flowering of Newtonian historical scholarship that took place in the 1830s.

ELEMENTS

OF

CONIC SECTIONS

DEDUCED FROM THE CONE,

AND DESIGNED AS

AN INTRODUCTION

TO THE

NEWTONIAN PHILOSOPHY.

BY THE

REV. A. ROBERTSON, D.D. F.R.S.

SAVILIAN PROFESSOR OF ASTRONOMY IN THE UNIVERSITY
OF OXFORD.

OXFORD,

AT THE CLARENDON PRESS.

MDCCCXVIII.

The main geometry textbook in the early 1800s. One of the most successful students of the period was Robert Peel, whose tutor used to recount of his final examination that 'when Peel was questioned on Robertson's *Conic sections*, the way in which he answered called forth the admiration of all that heard him'.

Cyril Jackson, Dean of Christ Church, was a leading promoter of mathematical studies at Oxford, both at institutional level and as a supporter and patron of the mathematicians Abraham Robertson and Stephen Rigaud.

It is gratifying to learn that Wilson passed this test with the highest honour, although six students found the standards set by the examiners too taxing and were failed.

It was in 1807 that the final examinations were first classified (Cambridge had been doing this since 1748) and also separated into two honours schools, the School of Mathematics and Physics, and the School of *Literae Humaniores* (Greek and Latin, with some logic, rhetoric, and moral philosophy). In 1808 the Christ Church undergraduate Robert Peel was the first 'double first', placed in the first class in both classics and mathematics. At this time the examinations were still oral, and still held in public. Contemporary accounts of Peel's examination make clear what a strong impression this procedure could leave. In 1809 he wrote to his tutor, Charles Lloyd, saying he was eager to continue their friendly intercourse 'except as regards conic sections and matters of that kind'. Despite his mathematical achievement, Peel later failed what we might think of as a symbolic test of his scientific understanding: as Home Secretary he dismissed proposals for funding Charles Babbage's difference engine with the words: 'I should like a little previous consideration before I move in a thin house of country gentlemen, a large vote for the creation of a wooden man to calculate tables from the formula $x^2 + x + 41$'.

Cyril Jackson, the powerful and energetic Dean of Christ Church from 1783 to 1809, was a foremost proponent of the changes in Oxford's examination culture, to such an extent that jokes were circulating in London that 'all the undergraduates at Christ Church read the *"Principia"* '. Some of them certainly studied Newton's great text. We have a good record of the mathematics training provided at Christ Church in the early nineteenth century from the letters of George Chinnery, an undergraduate from 1808, the year Peel graduated, to 1811. The extracts in the box (opposite) show how far Chinnery's mathematical skills were developed by the Christ Church system. In the course of four years, from the basic geometry and algebra he arrived with, Chinnery attained a good understanding, if his examiners are to be believed, of conic sections, fluxions (Newtonian calculus), and Newton's *Principia*.

Anthony Story, father of the pioneer photographer and mineralogist Nevil Story Maskelyne, took a double first in classics and mathematics in 1810, two years after Peel. When his son was sitting his Oxford finals in 1845, Anthony Story was bombarding the young man with advice to work harder, and recalled his own final examinations:

I had but 38 hours between my first and second examinations, those in mathematics. Everybody said "Refresh yourself" but I worked 22 hours out of the number, going through every formula and puzzling myself in all the diagrams in the books as much as possible. I was describing the last of the trajectories of Sir I. Newton's *Principia* as I was passing under the gateway into the schools.

A mathematics student in early nineteenth-century Oxford

George Chinnery studied at Christ Church from 1808 to 1811, and wrote to his mother nearly every day throughout his time at Oxford. This correspondence, now in the library at Christ Church, gives us an extraordinarily detailed insight into undergraduate study at the beginning of the nineteenth century.

15 January 1808 Chinnery's first interview with the Dean of Christ Church, Cyril Jackson:

Have you studied Mathematics?—I have been through Simpson's Euclid, Sir—Have you studied all the Books—I have, Sir—and well—I hope I am pretty well master of them. Leave that to us, we shall find it out … he then asked me whether I had studied Algebra; I said I had been through Butler's Algebra; he said he did not know the book and desired to see it as soon as possible.

27 January 1808 Chinnery spends $1\frac{3}{4}$ hours with the Dean (9.30–11.15):

We conversed on Algebra, and on the properties of Figures, on the nature and scientifical meaning of numbers, on the origin, signification & modification of the unit, and the advantages of decimals, on the points of similarity & diference between vulgar & decimal fractions, on the practical uses of both & what deficiencies in calculation brought on the invention of the latter. He really gave me a lecture …

5 December 1809 *I am very much inclined to think that in this learned University, we, reading men, get crammed rather than wholesomely nourished … and I am convinced beyond a doubt that every first class man must ruminate for a whole year after he has left College—or the books which he has read will remain an undigested food upon his head.*

4 February 1810 *I rejoice most amazingly in being able to tell you that the time for the opening of Robertson's Newtonic lecture is fixed; it will be on the 15th. I had a long conversation with him yesterday morning at his house on the subject, and he read the prospectus of the course of lectures to me; among other things, he advised me before the 15th to study the beginning of the Principia (which is that part of Newton that forms the object of our attention) by myself; and this I mean to do. One of the principal difficulties to be overcome is it's being written in latin, and not in latin similar to the classics, but, as may be easily conceived, in a style quite particular to the subject, concise & replete with technical words hardly known by a Horace, a Virgil, or a Juvenal.*

22 February 1810 *Robertson's 4th lecture took place today; there is one difficulty attending them, from their being public, that if in a demonstration (which he does upon a slate before us) there should be any point not immediately clear, one cannot stop him & beg that he will begin over again; this however signifies but little; it obliges to unremitting attention during the hour & a half which is the time allotted to the lecture, and complete really undivided attention will prevent any part of his demonstration being unintel-*

ligible; he is a delightful beautifully slow, beautifully clear lecturer, 1000000000 times better than Lloyd. I like him much.

23 February 1810 *I was most deeply engaged this morning indeed for 4 hours, ie till 12 o'clock in writing down yesterday's lecture, I mean the mathematical lecture; when I am at the lecture I make a few notes on a bit of paper of the subjects treated & discussed; and from these notes I oblige myself in my own room to recollect the whole and make out the demonstrations; this is a very laborious but a very useful & necessary exercise; it is quite upon your plan …*

16 September 1811 *I still continue in all the vigour of study, & much as I study <u>fluxions</u> and <u>fluents</u> I do not find that my spirits <u>flow away</u>. I almost think that fluxions is too tedious a study for any person that attends also to classics (I speak with reference to degrees & examinations): it takes up such an immense portion of time together with other branches of mathematics, that little remains for anything else.*

30 September 1811 *I am convinced that the Mathematics, which I am now reading eight or nine hours a day, have already strengthened my mind & though they cannot give it powers of invention they have certainly given it powers of tenacity. It is impossible that the mind should not be improved in that way if it becomes able, which I mean to make it, to take up well the body of Mathematics in which I shall be examined at the School.*

24 October 1811 *I am not yet altered in opinion with regard to the mathematics: they are singularly inconvenient,—they are destructive of every thing: above all that odious & abominably tedious book called the Conic Sections, which forms an unaccountable degree of foolish infatuation is forced upon us merely because the author of the treatise is alive, & resident in the University: they all agree that at his death it will fall to the ground. The absurd & intolerable part of the business is that it is not left at our option to read the book or not: if a man reads for the first Class, it is a sine qua non I would willingly read any two treatises to this one; but this one is beyond all expression execrable & deserving of being mathematically anathematised … Sophocles from $6\frac{1}{2}$–8. Breakfast. Conic Sections from 9–3, dinner. Conic Sects from $4\frac{1}{2}$–6.*

17 November 1811 *I was three successive and entire days in the Schools, … this indeed is a completely new system, & will in my opinion try the courage of the weak-hearted. The first day was devoted to the solving of logical questions & to translations from English into Latin and from Latin & Greek into English: the second day in examination viva voce in Ethics Rhetoric & the Classics: the third day in mathematical examinations: which three days were Thursday Friday & Saturday. Days of toil they doubtless were … I have this tolerably certain proof that the Examining Masters were fully satisfied with my Mathematics, that when I went up to Mr Dixon the head master to ask him whether I should return to the Schools on Monday for the purpose of working out Newton's Corollaries, he said "No we will not trouble you any further Mr Chinnery, for we are perfectly aware that you are completely master of the Newton"—Also after having examined me viva voce in the fluxions he concluded by saying "you seem Mr Chinnery to have paid very particular attention to the branches of mathematics you have read."*

Less successful as an Oxford student at that time was Percy Bysshe Shelley, who came up to University College in October 1810 and was expelled for contumaciousness the following March. His tutor (possibly the Dean, George Rowley) evidently did not possess the knack of gaining his interest in Euclid, or indeed anything else on the syllabus. His friend Thomas Hogg later recalled:

'They are very dull people here', Shelley said to me one evening soon after his arrival ... 'a little man sent for me this morning, and told me in an almost inaudible whisper that I must read: "you must read", he said many times in his small voice. I answered that I had no objection. He persisted; so, to satisfy him, for he did not appear to believe me, I told him I had some books in my pocket, and I began to take them out. He stared at me, and said that was not exactly what he meant: "you must read Prometheus Vinctus, and Demosthenes de Corona, and Euclid." "Must I read Euclid?" I asked sorrowfully. "Yes, certainly; and when you have read the Greek works I have mentioned, you must begin Aristotle's Ethics ..." '

Nonetheless, Shelley was one of the most scientifically minded of English poets, and his *Letter to Maria Gisborne* describes the contents of a study that sounds very like the descriptions we have of his college rooms during the few months he spent at Oxford:

A half-burnt match, an ivory block, three books,
Where conic sections, spherics, logarithms,
To great Laplace, from Saunderson and Sims,
Lie heaped in their harmonious disarray
Of figures,—disentangle them who may.

During the 1820s there were further changes in the examination procedures. In 1825 separate examiners were introduced for the two honours Schools, which had the effect of producing a somewhat more specialized focus. Although the examination questions continued to be mainly oral, answers were increasingly written down. As at Cambridge, where newer mathematics was introduced to Oxford it was through the examination system, and came about as much through the efforts of undergraduates as from the dons. A Balliol undergraduate, Francis Newman, was remembered as a pioneer of student determination of the syllabus:

He gained a Double First in 1826, being the first man who ever offered in the Schools the Higher Mathematics analytically treated. Cooke [Sedleian Professor of Natural Philosophy] pronounced that they could not, according to the Statute, pass beyond the Geometry of Newton; but Walker [later Reader in Experimental Philosophy], who probably of the three alone knew the subject, persuaded his colleagues to let him examine Newman in the work he offered; and the candidate's answers were so brilliant, that the examiners, not content with awarding his First, presented him with finely bound copies of La Place and La Grange.

It was two years later, in 1828, that the question papers were first printed; some questions from this examination are shown in

The reader in geology, William Buckland, lectured in the Ashmolean Museum on 15 February 1823 to an audience of prominent Oxford academics, including Stephen Rigaud (left-hand end of the back row), the Sedleian professor George Cooke (right-hand end of the middle row), Edward Copleston, Provost of Oriel College (fourth from right, front row), and Robert Peel's former tutor Charles Lloyd, by now Regius Professor of Divinity (fifth from left, back row).

Chapter 11. A consequence of this development, compared with the old oral system, was that the same questions were now to be asked of all candidates.

A Christ Church undergraduate of this generation who took a double first was William Ewart Gladstone (later, like Peel, to be Prime Minister). He recorded in his diary the experience of sitting the mathematics examinations, in December 1831:

December 9. In the Schools six hours with two papers of Algebra & Geometry in which I succeeded better than I had any right to expect.
December 10. Differential Calculus & Algebra papers: did the former but ill: the latter as well as any hitherto.
December 11. God be praised for this day of rest.
December 12. Had Differential Calculus & Hydrostatics. Did better than I expected in the latter, and worse in the former.
December 13. Had Mechanics & Optics. Did but ill in the former: wretchedly in the latter.
December 14. Principia in the morning: considerably disappointed in my performances. Astronomy! (which I had crammed for about half an hour,) during a short time in aftn.

Contrasting Gladstone's experience with that of Wilson 30 years earlier gives some pointers to the strides that Oxford mathematics teaching had already made over the period, even before the further changes later in the century.

Reform in the air

These changes in the examination system were developed against a wider background of ongoing concern and controversy about Oxford's academic standards, in mathematics in particular, from the early century onwards. In 1808 John Playfair, professor of natural philosophy at Edinburgh, remarked in the radical Whig journal *The Edinburgh Review* on the inferiority of British mathematics. Playfair argued that mathematical studies in Britain were lagging so far behind the rest of Europe that people were unable to read even the clearest of major European authors, such as Euler and d'Alembert; no more than a dozen people in the kingdom, he hazarded, could read the major scientific work of the age, Laplace's *Celestial mechanics*. This inferiority stemmed chiefly, he argued, from the deficiencies of the universities, particularly Oxford,

where the dictates of Aristotle are still listened to as infallible decrees, and where the infancy of science is mistaken for its maturity, the mathematical sciences have never flourished; and the scholar has no means of advancing beyond the mere elements of geometry.

This provoked lively discussion at Oxford, and a strong response from Edward Copleston, tutor of Oriel College, in which every phrase of this judgement was dissected. It is clear from what we saw of George Chinnery's education that Playfair's judgement on Oxford, even if informed, might arise merely from a different conception of what constitutes a flourishing mathematical education. Copleston mounted a vigorous defence:

What are the mere elements of geometry? Are Plane and Spherical Trigonometry, are the properties of Conic Sections, of Conchoids, Cycloids, the Quadratrix, Spirals, &c &c the mere elements of Geometry? Is the method of Fluxions included under the same appelation? On all these subjects, lectures both public and private are given.

Copleston also somewhat mischievously cited the works and practice of Professor Playfair against the anonymous reviewer—with such relentless and heavy irony that we can presume he knew the author of the review to be Playfair himself.

Although the most influential, Playfair was not the first to attack Oxford mathematics. Three years earlier, in 1805, a young Cambridge graduate by the name of John Toplis wrote a forthright article 'On the decline of mathematical studies' in the *Philosophical Magazine*. He too argued that mathematics in Britain was neglected, that such mathematics as was studied was of the wrong sort—ancient geometry, instead of modern analysis as practised on the continent—and he too blamed Oxford in particular. The examples of Toplis and Playfair indicate that efforts to reform British mathematics from its eighteenth-century torpor, as they saw it, began rather earlier than the Analytical Society in Cambridge (1812) of Charles Babbage, John Herschel, George Peacock, and their friends, which is commonly stated to have initiated this movement.

Playfair's was the most magisterial of a series of attacks on Oxford education in general, and on its mathematics teaching in particular,

Not only were examinations held in public after the reforms of the early nineteenth century, but students in their first year had to observe the public examination of second-year students in 'Moderations'. Charles Dodgson wrote a jocular account of this process, which he underwent in 1852, noting 'A long table, covered with books … two gloomy-browed examiners, and twelve pale-faced youths'.

throughout the nineteenth century. Playfair and Toplis were surely right in their overall perception that British mathematics lagged behind the continent, a legacy in part of the Newton–Leibniz priority dispute a hundred years before. As we saw, it was not until the late 1820s that continental analytical methods reached Oxford. But even at the end of the nineteenth century, Oxford mathematics teachers on the whole saw their responsibilities as providing a general liberal education, rather than training students in advanced mathematical research techniques.

There were various reasons why external critics should attack the state of Oxford mathematics—the controversy was an expression of many different agendas. Nonetheless, that there might be a problem was gradually taken more seriously at Oxford towards the end of the Georgian period, as the changes in examination procedures testify. To promote further reforms in Oxford mathematics, and to encourage science to occupy a more central place in the curriculum, were the prime issues to which Baden Powell, who studied at Oriel College and graduated with a double first in 1817, applied himself on his election to the Savilian Chair of Geometry in 1827—with what success, you can read in Chapter 11.

Further reading

An agreeable account of the social life of students and dons during the period is to be found in Graham Midgley's *University life in eighteenth-century Oxford*, Yale University Press, 1996. A good insight into academic conditions towards the end of the period is given in William J. Baker's *Beyond port and prejudice: Charles Lloyd of Oxford, 1784–1829*, University of Maine at Orono Press, 1981. The development of Oxford's examination system is explored in Sheldon Rothblatt, 'The student sub-culture and the examination system in early 19th century Oxbridge', in *The university in society*, Vol. 1, *Oxford and Cambridge from the 14th to the early 19th century* (edited by Lawrence Stone), Princeton University Press, 1974, pp. 247–303. Overall, the relevant volume of the history of the University is the recognized standard for comprehensiveness and scholarship: *The history of the University of Oxford*, Vol. V, *The eighteenth century* (edited by L. S. Sutherland and L. G. Mitchell), Clarendon Press, Oxford, 1986.

CHAPTER 10

Thomas Hornsby and the Radcliffe Observatory

Allan Chapman

It is not easy for twentieth-century people to imagine scientists of international distinction like Halley and Bradley making major discoveries from the roof-tops or gardens of their houses with privately owned instruments. But even by the middle of the eighteenth century, astronomy had become the most expensive of the sciences as far as capital equipment was concerned, and a serious astronomer had to have hundreds of pounds at his disposal to acquire the instruments that were necessary to pursue fundamental research.

When James Bradley died in 1762, Thomas Hornsby of Corpus Christi College was elected to succeed him in the Savilian astronomical Chair. Like other active teachers of the physical sciences in eighteenth-century Oxford, Hornsby was already acquiring his own kit of instruments, which came to include a 43-inch transit instrument (1760) and a fine quadrant of 32 inches radius (1767), both by John Bird of London, along with other pieces which were used in a poky room in Corpus that could accommodate only three people.

Hornsby formed a natural successor to Bradley, for he was an energetic teacher of astronomy and physics in the University and on close terms with influential people. He had observed the 1761 transit of Venus from Lord Macclesfield's observatory at Shirburn Castle, Oxfordshire, and also advised the Duke of Marlborough about setting up an observatory at Blenheim Palace. Hornsby recognized, however, that to advance astronomical studies at Oxford, it was essential for the University to possess a major observatory equipped with a full set of instruments, and to this end he petitioned the Radcliffe Trustees in 1768.

The Radcliffe Observatory, built in the 1770s at the instigation of Thomas Hornsby, was the first academic establishment in Europe to combine teaching and original research in astronomy with instruments of the highest quality, made by John Bird.

The Radcliffe trust and the design of the observatory

The £40 000 trust bequeathed by Dr John Radcliffe in 1714 had already provided Oxford University with a scientific library, the Radcliffe Camera (1737), and an infirmary (1759), and it was felt by the Trustees

By the end of the eighteenth century the Radcliffe Observatory was the best-equipped astronomical observatory in the world.

that the provision of an astronomical observatory would be well within the spirit of Dr Radcliffe's scientifically inclined bequest. Land was acquired to the north-west of the city from St John's College, so that the observatory and its spacious grounds stood next door to the Radcliffe Infirmary up what is now the Woodstock Road. Plans were drawn up for an elegant stone building, with two wings (for the meridian instruments) flanking a magnificent central tower, modelled on the Tower of the Winds in Athens. Hornsby, moreover, was to be provided with a spacious house that joined on to the east wing of the observatory via a covered passage to the quadrants and the transit room. The observatory was never an institution under the control of Oxford University's governing body, Congregation, but was governed by the Radcliffe Trustees. Thus, while Hornsby's professorship and salary derived from the University, his observatory and house came from an independent trust.

The elegant architecture of the Radcliffe Observatory, initially designed by Henry Keene but completed by the more fashionable James Wyatt, was dictated by the character and uses of the instruments that it would accommodate. Just as the design of modern observatories is dictated by the need for domes in which to house large aperture telescopes, so those of the eighteenth century were governed by two considerations. One was the need for a single-storey building in which

The Radcliffe Observatory's 161-year tradition of meteorological observations (three times every day from 1774 to 1935) is now sustained from instruments on the lawn in front of the observatory.

to mount the quadrants, zenith sector, and transit in the plane of the meridian. As the quadrants and transit had to command a 180° meridianal arc, from the northern to the southern horizon, the building needed to have clear shutter openings in both walls and across the roof. The main axis of the observatory, therefore, had to run east–west. The second design requirement for a major observatory was for a tall polyhedral building with long narrow windows that could be opened for the insertion of the long-focus refracting telescopes then in use. The 'Great Room', or Octagon Room, in Flamsteed House at the Royal Observatory in Greenwich, which had been designed by Sir Christopher Wren in 1675, was the English prototype for such a room as the one which Keene and Wyatt designed for the Radcliffe Observatory.

In consequence, an eighteenth-century observatory looks very different from a modern (or even an early nineteenth-century) observatory. In the time of Halley, Bradley, or Hornsby, the cutting edge of astronomical research lay in the measurement of the positions of the Sun, Moon, stars, and planets in the coordinates of the right ascension (longitude) and declination (latitude), in the hope of detecting small motions that were in accordance with Newtonian mechanics. Halley's proper motions of stars, his comet researches, and Bradley's aberration and nutation had all been classic discoveries of this kind. Astronomers measured the positions of objects in the sky; they did not generally 'look' at the objects in the way that modern astronomers observe the physical characteristics of bodies. When they did have occasion to examine an object in detail, such as Jupiter's belts, satellites, or cometary nuclei, they used long refracting telescopes inserted through the windows of the central tower. These instrumental requirements, therefore, dominated as they were by the demands of meridian astronomy, resulted in a building which to our eyes looks more like a museum or a mansion than an observatory. Indeed, in 1979, the Radcliffe Observatory was converted into the principal building of Oxford University's new academic foundation, Green College.

Observatory architecture and instrumental needs

The Radcliffe was the last major observatory to be based upon a tall 'Great Room' design. By 1772, John Dollond's newly invented achromatic object glasses were making it possible for powerful refracting telescopes to be made with tubes that were four or five feet long, rather than the 20 or 30 feet of the mid-century, and these no longer demanded the tall windows and ladders of Greenwich. Had the Radcliffe Observatory been designed 30 years later, by which time the long refracting telescope with its single-element object glass had been ousted from serious astronomical research, its central building and main design feature could well have been a dome housing an achromatic refractor rather than an elegant tower. When Cambridge University built its first observatory in 1823, a central dome on a Greek portico dominated its design; when the changing priorities of research instrumentation obliged the Radcliffe Observatory to acquire large equatorially mounted refractors in the nineteenth century, these instruments were housed in separate domed buildings in the grounds.

Being a man of enterprise Thomas Hornsby did not let matters rest, for even while the funding, land, and designs were being secured, he set about commissioning the instruments. John Bird, the greatest instrument maker of the age, was approached. He had equipped the Royal Observatory for James Bradley 20 years before and could produce the most precise graduated angular scales in the world. For the sum of £1260 (finally rounded up to £1300), Bird agreed to construct the five principal instruments, consisting of a pair of eight-foot-radius mural quadrants in brass for £760, an eight-foot-long transit instrument for £150, a twelve-foot-long zenith sector also at £150, and an equatorial sector costing £200. Moreover, the Radcliffe instruments were unique, in so far as Hornsby persuaded Bird to equip each of these pieces with Dollond achromatic object glasses to overcome the old master's dislike of fitting heavier lenses to his instruments. Further orders were also placed with Bird and other specialist makers for the additional instruments, which included a precision siderial regulator clock by John Shelton and several fine achromatic refracting telescopes by Dollond, along with barometers, thermometers, and levels. The entire set of instruments was estimated at £2500—a colossal sum by eighteenth-century standards, and two and a half times the amount that Bradley had received to replace the principal instruments at Greenwich 20 years previously.

With Bird's commission, however, there was a definite touch of urgency, for not only did the master craftsman have full order books, but he was known to be suffering from indifferent health which allowed him to work only at intervals. From the correspondence, one might at first suspect that Bird wanted the Radcliffe commission for the handsome profit which it offered, though since he was a relatively rich, elderly, and ailing bachelor, this was probably a secondary consideration. More important was the prestige of making the instruments for what was clearly going to be a very major observatory, which the visiting Danish scientist Thomas Bugge was to describe in 1777 as 'no doubt the best in Europe, both as regards the arrangement and the instruments'. While Hornsby was busy arranging the bridging finance to translate the Radcliffe Trustees' promises into initial sums of hard cash with which to pay the necessary advances to secure the commission, Bird reminded him that Dr Maskelyne, the Astronomer Royal, was also planning a major order for Greenwich, and if Oxford delayed

John Bird in 1776, at around the time that he constructed instruments for the Radcliffe Observatory.

One of the quadrants made by John Bird for the Radcliffe Observatory. Its scale enabled measurements of the highest precision to be made.

too long, 'the Astro. Roy. will be before you'. Perhaps a touch of high-pressure salesmanship was being applied here; in any case it worked. Hornsby beat Maskelyne with his contract and the Radcliffe Observatory ended up with the finest complete set of astronomical instruments in Europe.

John Bird and the Radcliffe quadrants

Bird's prestige as an instrument maker resided not only in the homogeneity of his beam compass-drawn quadrations, which made it possible to measure declination angles to almost a single arc second, but also in his careful study of the way in which instruments behave when temperatures change. Because the graduated limbs, frame, screws, and wall mounts were all made of brass, Bird knew that when everything expanded in hot weather, the expansion would be homogeneous through the whole structure and cause no thermal stress. He realized that the composite brass and iron instrument that George Graham built for Edmond Halley at Greenwich in the 1720s had been prone to errors that result from the different expansion coefficients of the two metals, and he consequently designed his 1749 quadrant for Bradley entirely of brass. The lessons in precision that he learned from his 1749 Greenwich instrument lay at the heart of the improved design of the Radcliffe instruments. Indeed, Hornsby's quadrants were regarded as the finest that Bird ever made.

Both Oxford quadrants were virtually identical in design, and were mounted on a stone-built meridian wall inside the observatory in the east wing, with full access to the sky through roof and wall shutters. One quadrant commanded the 90° arc from the zenith to the southern horizon while the other looked north, so that all 180° of the meridian were visible. To guarantee maximum rigidity, the eight-foot-radius graduated limb of each quadrant was mounted on a screwed and bolted lattice of heavy brass bars that had to be imported from Holland. The quadrant scales were divided with beam compasses, using a technique which Bird had originally perfected in the 1740s. Each instrument had a *pair* of scales engraved on its eight-foot-radius limbs. One scale was divided into ninety degrees by a series of pre-calculated beam-compass openings which Bird carefully read off from a 'scale of equal parts' to form chords upon the engraved arc of the scale.

It is impossible to divide a scale into ninety equal parts by repeated bisection, as geometers have known since antiquity. In order to side-step this awkward fact, George Graham in 1725 conceived the idea of using a 96-part scale that could be engraved parallel to the conventional one of 90° to act as a cross check upon it, and Bird incorporated these 96-part check-scales into his own quadrants. The north and south quadrants which Bird built for Hornsby each carried a 90-part and 96-part scale, engraved on two slightly different radii about an inch apart, and read with a double set of verniers engraved on the opposite edges of the same 'fiducial' pointer.

To make an observation with one of the quadrants, Hornsby looked through the eight-foot sighting telescope and brought the star image to the intersection of the cross-wires by means of a micrometer screw. He then read off the angle from the vernier on the 90° scale, and repeated the process with that of the 96-part scale. A pre-calculated 'ready reckoner' table enabled him to convert the full and fraction units on the 96-part scale into conventional degrees, minutes, and seconds, and if both scales gave exactly the same angle, then he could enter it into the observing log with confidence. If the two scales did not give an identical figure, then the observer knew that he needed to search for an error. Either way, he knew that this process of geometrical cross-checking facilitated high standards of angular measurement.

Bird's transit and zenith sector

The two quadrants made it possible to measure the declination, or angular distance, between the celestial equator and the pole of any given object in the heavens to within one second of arc. They were used in conjunction with the eight-foot transit instrument which Bird also made for the Radcliffe collection. This consisted of a telescope with a four-inch achromatic object glass of eight feet focal length mounted on a pair of precision trunnions, rather like a cannon, and balanced with counterweights. The trunnions were supported in a pair of 'V bearings' set in massive east–west-lying stone piers that were buried deep in the earth, so that when the telescope of the transit was rotated, it described a circle in the plane of the meridian. Because it was possible to adjust the exact horizontal plane of the V bearings with a precision spirit level, it was easy to use the transit as the definitive meridianal plane of the observatory.

Once this plane had been defined by the transits it could easily be used as a cross-check upon the planes of the flat, vertical quadrants by seeing if stars came to the meridian at exactly the same instant in the telescopes of both instruments when operating at a variety of elevations. On a daily basis, the transit was used to take the precise noonday passage of the Sun, check the observatory clocks, and observe the daily passage of the vernal equinox or first point of Aries, from which the right ascension (or east–west longitude) angle of any object in the sky could be measured.

The twelve-foot zenith sector which Bird built for Hornsby was very similar to the one used by Bradley to discover the aberration and nutation, and consisted of a twelve- foot telescope suspended in the zenith meridian on a pair of transit-like trunnions near the object glass. Apart

John Bird's eight-foot transit (1773) for the Radcliffe Observatory was remounted in 1843.

The lower portion of John Bird's twelve-foot zenith sector (1773) for the Radcliffe Observatory is seen with the observer's chair.

from monitoring the aberration and nutation, the Bird zenith sector was also used to check the vertical alignment of the two quadrants against zenith stars such as γ and β Draconis.

Bird's Radcliffe equatorial instrument

Although the quadrants, transit, and zenith sector represented 'mature' instruments from the standpoint of design by 1770, the Radcliffe commission to Bird contained an order for a piece which was almost experimental in character. This was the five-foot equatorial sector which Bird signed and dated 1775. English craftsmen had experimented with the idea of mounting a pair of graduated circles on to a polar axis ever since Tycho Brahe had described his 'great equatorial armillary' in 1598. The appeal of the equatorial lay in its ability to obtain two coordinates, right ascension and declination, from a single observation, by setting a sturdy axis parallel to the Earth's pole and mounting upon it a vertical (declination) and horizontal (right ascension) circle.

The equatorial sector which Bird designed for Hornsby had a wrought-iron polar axis that was nearly seven feet long; upon this was mounted a declination circle three feet in diameter, a five-foot sector of 28°, and a telescope. The bright comets of 1769 and 1770 almost certainly provided the immediate stimuli that led to the commissioning of the sector, for the instrument was intended for the measurement of declination angles out of the meridian. On successive nights, a comet's coordinates could be traced across the sky, and from these positions an astronomer could compute its orbital characteristics in accordance with Newtonian criteria. One might say that the equatorial sector was a very appropriate instrument for an Oxford observatory, for Edmond Halley had first developed the techniques for cometary orbit calculation 70 years before, and his successor, James Bradley, had further perfected them.

In October 1777, the equatorial sector was examined and sketched by the visiting Dane, Thomas Bugge, who gave a favourable account of it. But Hornsby seems to have made relatively few observations with the instrument and never even kept an individual log book for its observations as he did for his other Radcliffe instruments. Equatorial sector observations were simply recorded on odd bits of paper, for this is how they survive amongst Hornsby's papers. On 25 September 1781, however, Thomas Hornsby used the equatorial sector to measure the positions of William Herschel's new 'comet' (Uranus) among the stars of Gemini, while the instrument was described in the astronomical manuscripts of the Duke of Marlborough in 1786.

Hornsby's working procedures at the Radcliffe Observatory

The broader observing procedures that came to be used at the Radcliffe Observatory had been initiated at Greenwich by John Flamsteed and Edmond Halley, and polished to perfection there by James Bradley. Even so, Hornsby produced several innovations of his own, necessitated in part by the funding situation at the observatory. In most observatories the basic process of observing the heavens would have involved at least two men, one at the transit and another at the quadrant, and perhaps a third to 'keep the clock' to check the timings of the stars as their culminations to the meridian were called out by the observers. But a peculiarity of the Radcliffe Observatory was the absence of funding for staff salaries. The priorities of the Radcliffe Trustees seem to have been to disburse a total sum of £31 661 on the observatory fabric and instruments while omitting to provide any endowment to ensure their efficient use.

Thomas Hornsby himself never expected to be paid for his duties at the observatory, and his accumulated incomes from the Savilian professorship of astronomy, student fees, and (after 1782) the additionally held posts of Sedleian Professor of Natural Philosophy and Radcliffe Librarian left him personally well off. The observatory house, moreover, provided him with a spacious and elegant residence free of charge. But the real problem came from the lack of funds to provide for an assistant. In response, perhaps, Hornsby developed an observing procedure that made him entirely self-sufficient and even led him to turn down offers of assistance.

John Bird's transit and quadrant each had five vertical wires (two on each side of the meridian wire) and one horizontal wire in their respective fields. Hornsby would often observe with the transit as an object the first two vertical wires in the field, and would then leave the transit to catch the actual meridian passage across the centre wire of the quadrant (15 feet away) to secure the declination, after which he would rush back to the transit to watch the object exit across the last two left-hand wires in the field. As he had already established the arc seconds separation between the wires visible in the field of view (corresponding to some 40 or 50 seconds of time), he could easily compute the time of the meridian passage and derive the right ascension. Returning to the quadrant, which he had locked immediately after securing the centre-wire observations, he could then read off the scales and the micrometer to obtain the declination.

Hornsby was a daytime observer of the heavens, and rarely logged anything between midnight and morning (with a few exceptions, such as during his search for Herschel's 'comet' in 1781). Unlike modern observers who need black skies to see the feeble light sources upon

A sundial in Green College commemorates its association with the Radcliffe Observatory. The figure-of-eight shows the position of the Sun's rays (through the sun-shaped gnomon at the top) at 1200 GMT over the course of the year. The vertical line marked with a cross shows the position of local noon at Oxford, and the thinner vertical line to its left is the meridian line indicating noon at Greenwich.

which modern astrophysical research hinges, Hornsby was concerned with accurately measuring the coordinates of a relatively small number of bright objects. Since Flamsteed at Greenwich a century earlier had shown that one could easily see the stars in the daytime with the right kind of telescope, daylight observations of star positions had been relatively common. As the Sun moves through the zodiac in the course of a year, every zone of right ascension passes across the daytime or evening sky at one season or another, so that if one were observing regularly for 30 years, like Hornsby, one could measure the positions of the principal stars several times over and never miss a night's sleep.

The first thing that any eighteenth-century astronomer had to do when commencing operations on a new site, after measuring his initial latitude, was to lay down a meridian line. This was done with the transit instrument and the sidereal (star time) regulator clock. To form a convenient daily check on the transit's position for future reference, a vertical meridian marker was placed in the wall of Worcester College, which was then easily visible 680 yards due south of the observatory, while a northern meridian mark was erected on to the roof of a house outside the observatory grounds. Thomas Bugge describes this northern marker, which consisted of a metal plate drilled through with a hole which was exactly on the meridian and through which daylight was visible. Hornsby knew that his transit was in correct adjustment if the central wire in the field of view fell precisely across the pin-point of daylight that shone through the hole. From this accurately surveyed meridian, extending across 180° of sky and fixed with reference to distant ground-based markers, it was next possible to align the planes of the north-pointing and south-pointing quadrants.

The quadrants were secured to their walls with screw-adjusters, so that their planes could be brought parallel to the transit's meridian against the due north and due south passage of stars. The exact vertical situation of the quadrants was further confirmed by simultaneous observations of zenith stars as seen in the zenith sector. The quadrants, once secured to their walls, could be used to establish the exact latitude of the observatory. All of the instruments, therefore, cross-checked each other as their different principles of mechanical and geometrical operation worked in unison to delineate the cardinal points of the heavens. In this way, Hornsby could be thorough in his elimination of errors.

Another source of error in astronomical observations is meteorological: air temperature and barometric pressure affect the refractive index of the atmosphere and result in a variation of angular displacement in the images of stars; this was recognized by Bradley at Greenwich in the 1740s. Hornsby made full atmospheric corrections to his Oxford observations, using barometers and thermometers supplied by Bird and the finest London makers.

Thus the superb mechanical quality of his astronomical instruments was not only completed by meticulous cross-checking, but also monitored for potential meteorological errors when in use. These fastidious procedures bore fruit in the 78 000 transit and 20 000 quadrant observations that Hornsby made at the Radcliffe Observatory between 1774 and 1803. Subsequent analysis has found his right ascensions to be accurate to 0.0834 seconds of time and his declinations to 1.62 arc seconds. Harold Knox-Shaw and his colleagues, who reduced and published many of the old Radcliffe observations in the 1930s, were led to conclude that in some respects Hornsby was an even better observer than James Bradley. Hornsby had the advantage, of course, of using instruments of a slightly more mature design, with large-aperture achromatic telescopes, than those available to Bradley at Greenwich in 1750.

Hornsby as a teacher of astronomy and the physical sciences

The Radcliffe Observatory was not only a monument to Thomas Hornsby as an individual; it was also an indicator of the vigour of, and

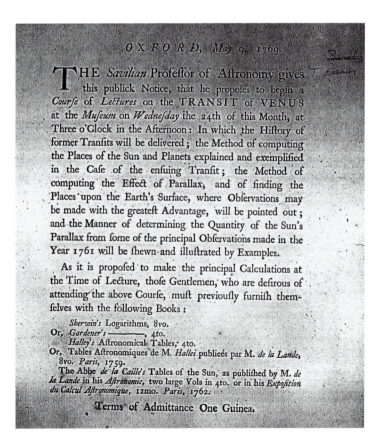

OXFORD, May 6, 1769.

THE *Savilian* Professor of Astronomy gives this publick Notice, that he proposes to begin a *Course* of *Lectures* on the TRANSIT of VENUS at the *Museum* on *Wednesday* the 24th of this Month, at Three o'Clock in the Afternoon: In which the History of former Transits will be delivered; the Method of computing the Places of the Sun and Planets explained and exemplified in the Case of the ensuing Transit; the Method of computing the Effect of Parallax, and of finding the Places upon the Earth's Surface, where Observations may be made with the greatest Advantage, will be pointed out; and the Manner of determining the Quantity of the Sun's Parallax from some of the principal Observations made in the Year 1761 will be shewn and illustrated by Examples.

As it is proposed to make the principal Calculations at the Time of Lecture, those Gentlemen, who are desirous of attending the above Course, must previously furnish themselves with the following Books:

Sherwin's Logarithms, 8vo.
Or, *Gardener's* ———, 4to.
Halley's Astronomical Tables, 4to.
Or, Tables Astronomiques de M. *Hallei* publieés par M. *de la Lande,* 8vo. *Paris,* 1759.
The Abbe *de la Caille's* Tables of the Sun, as published by M. *de la Lande* in his *Astronomie,* two large Vols in 4to. or in his *Exposition du Calcul Astronomique,* 12mo. *Paris,* 1762.

Terms of Admittance One Guinea.

A 1769 lecture notice of Thomas Hornsby indicates a demanding teaching level and his expectations that students would come equipped with logarithmic and astronomical tables.

commitment to, the physical sciences in eighteenth-century Oxford, as well as being part of a continuing astronomical enterprise initiated by Edmond Halley that had previously been embodied in the re-fitting of the Royal Observatory in Greenwich. In addition to pure research, Thomas Hornsby, like Edmond Halley and James Bradley before him, was an enthusiastic scientific teacher.

On taking up his Savilian Chair in 1763, Hornsby set about the business of lecturing 'to the younger part of the University', while part of his petition to the Radcliffe Trustees in 1768 was the request for facilities where he might 'read a course of lectures in Practical Astronomy [observational astronomy]; which I was the rather disposed to undertake as it had never been publicly attempted at any University'. Although Bradley, Halley, and earlier Savilian professors had delivered astronomical lectures at Oxford, they had lacked a first-class establishment in which to do it. Considering the unrivalled quality of the instruments and facilities which the Trustees placed at his disposal, Hornsby's initiative made the Radcliffe Observatory the first British, if not European, academic establishment to combine teaching and original research in astronomy. He also suggested that the members of Oxford colleges might receive practical astronomical instruction at the observatory free of charge, as a way of fostering skills in a subject that was of such importance to a maritime nation.

In addition to astronomy, Hornsby inherited the course in experimental philosophy or physics from Bradley and Bliss, and mention was made at the end of Chapter 8 of the special lecture to ladies which he delivered under Mrs Bliss's sponsorship in 1765. Several prospectuses survive for courses which Hornsby gave in the University after he became Savilian professor, and one sees how wide was the range of his teaching. In addition to many aspects of astronomy, Hornsby lectured on electrostatistics, magnetism, the air pump, and optics, using what Thomas Bugge called 'a beautiful collection of physical instruments of his own'. The nucleus of this collection seems to have been purchased from the widows of Dr Bradley and Dr Bliss, although Hornsby augmented the set so that new experiments could be added to the course. In 1790 the eminent instrument maker Edward Nairne valued Hornsby's private collection of physics apparatus at £375 14s 6d, and it was purchased for the University through a benefaction of Lord Leigh who, like Hornsby, had been a pupil of Bradley.

When Thomas Bugge visited Oxford in October 1777, Hornsby's astronomical lecture room at the observatory had not yet been finished, and teaching still took place in 'the Museum' in Broad Street, which had been the venue for Oxford science lectures for almost a century. Bugge had not only been impressed by the observatory itself, but also by its director and the energy with which Dr Hornsby pursued his work. By 1785, Hornsby conscientiously discharged the duties of two University professorships and the readership in experimental philosophy, administered the Radcliffe Library of scientific books, and was in the midst of a programme of intensive astronomical observations.

In November 1781, Hornsby outlined his teaching schedule to the Duke of Marlborough, and his letter further implies that he used his electrical demonstration equipment for medical purposes:

As soon as I have breakfasted I must go to the Museum. I have sometime an hour and a halfs work or even more, sometimes not so much. I must be home by eleven O'clock, because from that time to the ☉ [Sun's] Passage I can only be able to Electrify the poor unhappy Lady whose Eyes the light of the Sun will never again revisit. They roll in vain to find the piercing ray, & find no dawn. As soon as I have obs.^d the Sun, & perhaps in 5 minutes dressed myself I must set out for [my] Lecture which is not over till a little before two. I then have my things to put away, & by a quarter before 3 I am in general at home again. After that time I am at your Grace's service.

And when this was over, he had his astronomical observations to get on with.

Hornsby's achievement

It is a pity, therefore, that Thomas Hornsby is not better known today outside the more specialist studies of the history of Oxford University. In comparison with Halley and Bradley, of course, Hornsby made no fundamental discoveries, and despite his obvious energies probably spread himself rather too widely. As with many busy and conscientious members of university communities, it was upon his students and colleagues rather than posterity that he had his greatest impact.

It may also be argued that he lived somewhat in Bradley's shadow, for Hornsby's truly enduring scholarly achievement, other than effectively establishing the Radcliffe Observatory, was the publication of James Bradley's *Observations* in two massive folio volumes in 1798 and 1805. There must have been something disarmingly modest in the character of Dr Hornsby. Few scientists with the drive and persuasiveness to raise up the finest and best-equipped observatory in Europe from the green fields, and who had compounded themselves into four senior University appointments, would have laid aside their own tens of thousands of superb observations to publish the works of an illustrious predecessor. Another measure of his modesty is that no portrait or visual likeness of Hornsby is known to survive.

Hornsby's observations were examined by astronomers from the late nineteenth century onwards, and the quality was found to be excellent. His pre-Radcliffe Observatory work on the determination of the solar parallax from the 1769 transit of Venus was very accurate, and produced the value of 8.78 arc seconds which is very close to the value of 8.79415 accepted today. Arthur Rambaut, the Radcliffe Observer from 1897 to 1923, compared the observations of Hornsby and the positions of the same stars when observed by Bradley and analysed by F. W. Bessel in *Fundamenta astronomia* (1818), and drew the most favourable conclusions: Hornsby's right ascensions, in particular, were found to be more accurate than Bradley's.

In spite of Rambaut's efforts to persuade the Royal Society to publish them, the vast bulk of Hornsby's tens of thousands of observations remained unpublished over a century after his death. It was not until 1932 that Rambaut's successor Harold Knox-Shaw obtained funding to publish a reduction of 54 000 of Hornsby's observations, made between 1774 and 1798. In addition to analysing the slight changes that had taken place to the quadrant and which were apparent in the observations, Knox-Shaw concluded that, while Hornsby's declinations (made with a quadrant) were slightly less regular than Bradley's observations with the eight-foot Bird quadrant at Greenwich, his transit observations—which formed by far the greatest class of observations made at the Radcliffe Observatory—were superior to Bradley's. The greater part of Hornsby's observations still await analysis and publication.

The establishment of the Radcliffe Observatory was, in many ways, the crowning achievement of the Oxford tradition in practical astronomy, commenced by the first Savilian professors and developed by Edmond Halley, which emphasized the close relationship existing between front-rank research and sound teaching. Yet, whilst in 1780 the University possessed the finest astronomical observatory in the world, it had in some respects invested its money in what was soon to become an outmoded technology. As already noted, had the Radcliffe Observatory been founded only a few decades later, it might have had a domed refracting telescope instead of a tower as its main architectural feature; nor would it have had quadrants and a zenith sector. The Radcliffe, in fact, was the last major observatory to use quadrants for the measurement of declinations; even before Hornsby had got the beautiful Bird quadrants into their final adjustment, Jesse Ramsden and John Troughton (and soon after, Edward Troughton) in London were experimenting with the more versatile and less cumbersome astronomical circle. By 1810, when Hornsby died, the Radcliffe might still have been the most beautiful observatory in Europe, but its now rather antiquated quadrants meant that it was no longer the most technically advanced.

Hornsby's successors

Hornsby was followed into both the Savilian Chair and Radcliffe observership by Abraham Robertson, a remarkable man who had come to Oxford from Berwick-on-Tweed as a self-educated domestic servant.

After being reprimanded by his master for not paying attention when waiting at table, Robertson admitted that he had been engrossed in a mathematical problem being discussed between his master and a guest. He had the temerity to point out that both gentlemen had been wrong and proceeded to furnish the correct proof. Instead of throwing him out for insolence, however, his master, an Oxford apothecary named John Ireland, realized that his servant's talents were being wasted, and secured for him a servitorship (or poor student's place) at Christ Church. From there, Robertson went from strength to strength, becoming Savilian Professor of Geometry in 1797 and finally succeeding Dr Hornsby. Because he lacked Hornsby's diverse sources of income beyond his Chair, Robertson's income was augmented to £300 per annum by the Radcliffe Trustees in 1812, thereby commencing the wise tradition of paying a proper salary to the observatory director. It is unlikely that Robertson needed the additional money, however, for as a deeply religious widower of simple tastes and charitable disposition, he seems to have given much of it away to maintain the poor of his native parish and to similar humane enterprises.

When Robertson died in 1826, he was followed as Radcliffe Observer by Stephen Peter Rigaud, and he in turn by Manuel Johnson in 1839. But Johnson was never made Savilian Professor of Astronomy as his predecessors had been, probably because of his personal association with John Henry Newman and the high church tractarian, or 'Oxford movement', which aroused great opposition in the University, especially from influential Evangelicals. Thus, from 1839 onwards, the Savilian Chair and the Radcliffe observership became formally separated. It is perhaps worthy of note that, after resigning his Oriel fellowship in 1845, Newman spent his last night in Oxford as Johnson's guest at the Observer's house at the Radcliffe Observatory, before leaving for Italy to become a Roman Catholic priest.

Over the first four decades of the nineteenth century, the failure of the Trustees to keep the instrumentation up to date meant that the Radcliffe Observatory became an increasingly antiquated institution. No major observatory was still using quadrants by 1830, although it was not until 1836, when Rigaud obtained the new six-foot mural circle from Thomas Jones of London, that the meridian work of the observatory was brought up to date once more. In 1843, William Simms was also engaged to build a new transit for the observatory, although still using the same four-inch Dollond object glass that Bird had supplied with the original transit in the early 1770s.

Similarly, the $4\frac{1}{2}$-inch Dollond refractor of ten feet focal length, which had been a powerful and advanced instrument to Hornsby in 1774, was frankly feeble five decades later. Although an eight-inch reflector of ten feet focal length had been obtained from Sir William Herschel in 1812 (itself an instrument of modest aperture for that

The Merz-Repsold heliometer equipped the Radcliffe Observatory for high-precision astrometric observations in the mid-nineteenth century.

From his appointment in 1870, Charles Pritchard brought an energy to the Savilian Chair of Astronomy which had not been seen for many years. As well as being a superb teacher, he designed and supervised the building of the new University Observatory in the University Parks.

date), it is not easy to deduce when the Radcliffe Observatory obtained its first 'modern' equatorially mounted refractor.

The Radcliffe Observatory's commitment to bringing its instrumentation up to the standard required to do front-rank research was made very clear in 1841 when the Trustees commissioned the magnificent $7\frac{1}{2}$-inch Merz-Repsold heliometer from Hamburg. When the instrument eventually became operational in 1848, it made the Radcliffe one of the best equipped observatories in the world for undertaking precision astrometric work. The new Repsold heliometer also saw a third man added to the Radcliffe staff, when Norman Pogson joined John Lucas as second assistant to Johnson. It now seems incredible that a major observatory could have lacked a salaried director during the first 40 years of its existence, and that not until 1840, when Lucas was appointed, did that director possess any assistance beyond the services of a handyman.

Meanwhile, the effect of the separation of the posts of Radcliffe Observer and Savilian Professor of Astronomy began to be noticeable. Rigaud's successor in the Savilian Chair, George Johnson, had no strong interest in astronomy, but was at any rate a sound churchman—he moved on shortly to be professor of moral philosophy and ended his career as Dean of Wells; an unsuccessful attempt had been made to interest the astronomer John Herschel in the Savilian position after Rigaud's death, which would have led to a rather different outcome. William Donkin, who succeeded Johnson in 1842 when still in his late 20s, was a far better mathematician, but pursued wide interests and did not devote much of his creative energies to astronomy. Donkin was followed in 1870 by a highly energetic 62-year-old, Charles Pritchard, who applied himself with zeal to promoting Oxford astronomy. Now that the Radcliffe Observatory was no longer part of the Savilian empire, he secured money for building a new observatory in the University Parks, which was constructed according to his design and opened in 1875 as the 'New Savilian Observatory for Astronomical Physics'. Here he pioneered photographic measurement of stellar parallaxes, and was also a highly effective and widely admired teacher, with as many as 15 students at a time receiving practical astronomical instruction from him and his assistants.

By the end of the nineteenth century, then, there were two observatories in Oxford, with separate staffing. The Radcliffe Observatory was the better equipped, receiving greater continuing support from the Radcliffe Trustees than the University Observatory did from the University. By the 1920s, the murky air of Oxford became ever less suitable for observing the heavens, and in 1924 the Radcliffe Observer, Harold Knox-Shaw, urged that the Radcliffe Observatory be relocated, at first just outside Oxford. By the end of the 1920s the plan was to move the observatory abroad, in which Knox-Shaw was supported by

The New Savilian Observatory for Astronomical Physics was constructed in the University Parks to Charles Pritchard's design in 1875. It was closed 110 years later.

the Savilian Professor of Astronomy, Pritchard's successor Herbert Hall Turner. After a lengthy legal battle in which the University took the Radcliffe Trustees to court in a forlorn preventative attempt, the Radcliffe Observatory was finally moved to Pretoria, in South Africa, in 1935. Its 74-inch reflecting telescope was the largest in the southern hemisphere when it was installed after the Second World War.

Thereafter the University Observatory was the only working observatory in Oxford. Under Turner's successor, the Canadian astrophysicist Harry Hemley Plaskett, the astronomy department developed postgraduate research activities, particularly into solar physics. Although deteriorating visibility conditions forced the University Observatory to be closed in 1985, Oxford astronomy and astrophysics nonetheless enjoy a position of international eminence today. Now, computer link-ups with the world's greatest telescopes in Australia, Hawaii, and the Canary Islands bring fundamental research data to the desks of Oxford astronomers in a way that would have seemed incredible to Halley, Bradley, and Hornsby.

Further reading

The scientific records of the Radcliffe Observatory are now preserved in the libraries of the Museum of the History of Science, Oxford, and the Royal Astronomical Society, London, while the Bodleian Library in Oxford holds the administrative records of the Radcliffe Trust. Most of the Radcliffe Observatory instruments are held by the Museum of the History of Science, although the Science Museum in South Kensington has the south-facing quadrant of 1772 by Bird. This instrument, alas, has been 'in store' since the Science Museum closed its astronomy gallery in the late 1980s. For an account of Bird's equatorial sector, see Allan Chapman, 'Out of the meridian: John Bird's equatorial sector and the new technology of astronomical measurement', *Annals of Science* **52** (1995), 431–63.

One of the first published accounts of the Radcliffe Observatory appeared in James Ingram's *Memorials of Oxford*, Oxford, 1832. The best modern studies are in Ivor Guest's *Dr. John Radcliffe and his Trust*, Radcliffe Trust, 1991, and in A. D. Thackeray's *The Radcliffe Observatory 1772–1972*, Radcliffe Trust, 1972. Gerard L'E. Turner discusses the observatory's role in his essay 'The physical sciences' in *The history of the University of Oxford*, Vol. V, *The eighteenth century* (edited by L. S. Sutherland and L. G. Mitchell), Clarendon Press, Oxford, 1986, pp. 659–81. A. V. Simcock looks at Hornsby's activities as a physical sciences teacher in *The Ashmolean Museum and Oxford science, 1683–1983*, Museum of the History of Science, Oxford, 1984; see also Vols. II and XI of Robert T. Gunther, *Early science in Oxford*, Oxford, 1923, 1937.

A reduction of Hornsby's observations was published by Harold Knox-Shaw, J. Jackson, and W. H. Robinson, *The observations of the Reverend Thomas Hornsby, D. D. ... made at the Radcliffe Observatory, 1774–1798*, Oxford University Press, 1932.

CHAPTER II

The mid-nineteenth century

Keith Hannabuss

The nineteenth century was a period of major change in the University as it recovered after a century of relative torpor. At the start of the century the University was something of a cross between a seminary for the Anglican clergy and a finishing school for the gentry. Most of the dons were in holy orders and were obliged to relinquish their fellowships upon marriage. This had the virtue of providing a steady supply of new appointments, but made it difficult to sustain any kind of academic career, even though professors and Heads of Houses were generally exempt from the celibacy requirement.

In the course of the century Oxford was gradually transformed into some semblance of a modern university, in which both teaching and research were taken seriously. These changes were helped along their way by various commissions into the University, and Baden Powell and Henry Smith, successive Savilian professors of geometry, served on the 1850–2 and 1877 commissions, respectively. Not surprisingly, this transition was accompanied by considerable internal tensions, as different visions of the future competed with one another.

The first reform came in 1800 with the institution of examinations for the honours Schools of Classics and Mathematics; previously, there had been only a perfunctory oral examination to determine whether the candidate could graduate, and there was no classification of degrees. The colleges varied in the speed with which this reform was implemented, with Christ Church, Oriel, and Balliol amongst the earliest colleges to give full support to the new examination system. Christ Church later faltered, and Oriel, the leading college intellectually in the 1820s, was riven by the tractarian controversies of the 1830s. Balliol College was more fortunate and, owing to the wisdom and foresight of two of its Masters, came to occupy the foremost position by the middle of the century; this was despite the enforced resignation of its mathematics lecturer, William Ward, who had been publicly degraded in the Sheldonian Theatre following condemnation of his Tract 90. About half of the mathematicians whose work we discuss were educated at Balliol.

In 1849 a new School of Natural Science was approved, surprisingly the least controversial of three proposals considered—Theology was

The University Museum, constructed in the late 1850s, realized in brick and iron Oxford's mid-century aspirations to improve the facilities for teaching mathematics and science in terms that recall the soaring confidence of medieval buildings such as the Divinity School (see page 30).

Sample questions from early examination papers. The papers of Easter 1828 comprised the first published examination.

Honour School of Mathematics, Easter 1828

Paper I

1. What decimal of a week is 1 hour 7 minutes and 14 seconds?

4. Find $\tan\left(\frac{\pi}{4} - \frac{\theta}{2}\right)$ in terms of $\tan\theta$.

Paper III

2. Show that impossible roots enter equations by pairs.

Honour School of Mathematics, Trinity 1828

Paper I

8. A dealer sells a horse for 56l and gains as much percent as it cost. What did it cost?

Paper IX

1. The altitude of the sun was observed to be half of his declination at six o'clock. Prove that twice the sine of the latitude of the place = the secant of the sun's latitude.

8. Explain the principal methods of finding longitude at sea.

Honour School of Mathematics, Easter 1829

Paper I

5. A box in the form of a parallelepiped is filled with spherical balls. What portion of the space is unoccupied?

Paper IV

1. Differentiate

$$u = \frac{\sqrt{1+x^2} - x}{\sqrt{1+x^2} + x}.$$

2. Apply Maclaurin's Theorem to find the tangent in terms of the arc.

Honour School of Mathematics, Michaelmas 1849

Paper I

3. Extract the square root of .0313487, the cube root of 8242408 and the fourth root of $x^4 - 2x^3 + \frac{3}{2}x^2 - \frac{x}{2} + \frac{1}{16}$ explaining the processes; and show that if the square root of a whole number has any decimal figures they neither terminate nor recur.

Paper II

5. Find the number of permutations of n things taken all together, when p are of one sort, q of another, r of a third and so on.

6. Prove the formula $4\tan^{-1}\frac{1}{5} - \tan^{-1}\frac{1}{239} = \frac{\pi}{4}$ and state its use.

12. Under what conditions will the diametral planes of surfaces of the second order intersect in a point?

rejected, and History was accepted only as a joint School with Jurisprudence. This change affected mathematics too, as shortly afterwards the University Museum, built as a centre for both science and mathematics, enhanced the status of the subject by providing lecture rooms, offices for the professors, and a lending library. The building of the museum was started in 1855, and by 1860 it was complete enough to host the British Association meeting at which T. H. Huxley and Bishop Samuel Wilberforce clashed in a debate on Darwin's theory of evolution.

Baden Powell

Lest any confusion arise, we mention at the outset that Robert Stevenson Baden-Powell, the founder of the Boy Scout movement, was the sixth son of Baden Powell by his third wife; the Christian name Baden was incorporated into the surname only in that generation.

Although Baden Powell wrote a number of useful textbooks on calculus and geometry, and had good contacts with the mathematicians who had founded the Analytical Society at Cambridge and with Augustus De Morgan in London, he was primarily a physicist rather than a mathematician. His main scientific contributions were in optics, but he was also a theologian and popularizer of science.

After graduating from Oriel College in 1817 with first class honours in classics and mathematics, he was ordained, and it was whilst working in a parish that he started the research into the infra-red end of the solar spectrum on which his early reputation was based. Powell was among the first British scientists to embrace the wave theory of light, which he expounded in various survey articles and books. He maintained a lively correspondence with George Airy and William Rowan Hamilton, and later with George Gabriel Stokes. Indeed, Hamilton's work on dispersion was stimulated by an enquiry of Powell's, and was first published (with Hamilton's blessing) in a survey article by Powell.

When the astronomer and historian of mathematics, Stephen Rigaud, migrated to the Chair of astronomy in 1827, Powell was elected as his successor. According to Oxford gossip, Charles Babbage, the computer pioneer, had tried hard to get elected, although this is not borne out by Babbage's correspondence. In any case, Powell's church and University connections ensured his election, and Babbage became Lucasian Professor at Cambridge in the following year. The two professors became firm friends, sharing a keen interest in the relationship between science and religion and, in particular, the place of miracles in the natural order.

The new Savilian professor was shocked and dismayed by the low esteem accorded to mathematics in the University. He had been advised not to give an inaugural lecture on arrival, as he would almost surely not attract an audience. A lecture delivered a few years later in 1832, *On the present state and future prospects of mathematics at the University of Oxford*, gives a good idea of the problems he faced and his strategy for dealing with them. The introduction of the honours Schools in 1800,

Baden Powell began his tenure of the Savilian Chair of Geometry with high hopes for the part that he could play in raising the standards of mathematical education in Oxford.

THE PRESENT STATE

AND

FUTURE PROSPECTS

OF

MATHEMATICAL AND PHYSICAL

STUDIES

IN THE UNIVERSITY OF OXFORD,

CONSIDERED

IN A

PUBLIC LECTURE,

INTRODUCTORY TO HIS USUAL COURSE,

IN EASTER TERM, MDCCCXXXII,

BY THE

REV. BADEN POWELL, M.A. F.R.S.

OF ORIEL COLLEGE,

SAVILIAN PROFESSOR OF GEOMETRY.

OXFORD,

PRINTED BY W. BAXTER FOR THE AUTHOR.

SOLD BY J. H. PARKER;

AND BY J. G. AND F. RIVINGTON, LONDON.

1832.

admirable though it may have been in providing the students with a challenge, had apparently narrowed their sights to the subjects of the examinations and away from the wider-ranging lectures of the science professors. Baden Powell noted that not only had the number of students at science lectures actually declined, but that of those candidates who did study mathematics 'though a certain portion had "got up" the four books of Euclid, not more than two or three could add Vulgar Fractions or tell the cause of day and night or the principle of a pump'. His complaint is corroborated by another examiner's story of a student who reproduced a proof from Euclid perfectly, but in his diagrams drew all the triangles as circles.

Baden Powell argued that the future was likely to be dominated by science and technology, and that the ignorance and hostility displayed by the Establishment towards science was likely to endanger its own position. In particular, if the University persisted in neglecting the sciences, then future leaders of the nation would go elsewhere for their education.

Although Powell set about changing things with great energy, he unfortunately lacked the political skills to carry through more than a few modest changes, such as printing the examination papers so that students could hone their skills. Of the whole package of reforms that he had proposed to raise the status of mathematics and science on taking up his Chair in 1827, most were rejected by the University.

More successful was his proposal to create scholarships for the promotion of mathematical studies. These were founded in 1831 and a fund for their endowment was raised by contributions from many of the colleges and from individual members of the University. There were at first three scholarships with a stipend of £50, each tenable for three years, but this number was later changed to four scholarships of £30, each tenable for two years. Before long, the mathematical scholarships were providing the career opportunities that Powell had envisaged, and they are a fitting tribute to his foresight. Early holders of these scholarships included Robert Anstice (a neglected figure who did interesting later work on combinatorial designs), William Donkin (later Savilian Professor of Astronomy), Bartholomew Price (Sedleian Professor of Natural Philosophy), and William Spottiswoode (president of the Royal Society).

In general, however, Powell decided to leave the reforms to his colleagues and he set out instead on a crusade to educate people through his popular lectures and writings. He was an accomplished public speaker, renowned for his illuminating lecture demonstrations. When the British Association for the Advancement of Science (BAAS) was founded in 1831, he became an enthusiastic supporter.

There was, in fact, an almost symbiotic relationship between the British Association and Oxford scientists. The visits of the BAAS gave

Sample questions from the Senior Mathematical Scholarship 1851. This is the examination in which Henry Smith (see Chapter 12) was the successful candidate.

Paper I

3. If $f(x)$ be a function of x finite and continuous for all values of x between a and $a + h$, and if all the derived functions of $f(x)$ up to the nth vanish when $x = a$ and are finite and continuous for all values of x between a and $a + h$, but the nth does not vanish then

$$f(a+h) - f(a) = \frac{h^n}{1 \cdot 2 \cdot 3 \ldots n} f^{(n)}(a + \theta h)$$

where θ is a proper primitive fraction. How does the above proportion enable us to determine orders of infinitesimals? Considering x to be the infinitesimal base determine the order of $\log_e(1 + x)$, $\sin x - \tan^{-1} x$, $e^{-\frac{1}{x}}$.

Paper II

7. If V be the potential of attraction prove that

$$\frac{d^2 V}{dx^2} + \frac{d^2 V}{dy^2} + \frac{d^2 V}{dz^2} = 0 \text{ or } = -4\pi\rho.$$

In what cases is the latter value to be used and why?

Paper III

5. Find the value of the triple integral

$$\iiint e^{ax + by + cz} dx\, dy\, dz$$

the limits being given by the condition $x^2 + y^2 + z^2 \overline{\overline{<}} h^2$.

Paper IV

5. Explain the physical meaning of the term $k \sin(g\theta - \gamma)$ in the expression for the Moon's orbit to the Equator: and of the terms

$$\left\{ 1 + \varepsilon \cos(\theta - \alpha) + \frac{15}{8} m\varepsilon \cos(\theta - 2\beta + \alpha) \right\}$$

in the expression of the Moon's parallax.

Paper V

11. The height of a mountain has been measured by equally skilful observers and by the same method. Their results are respectively 1000, 1003 and 1005 feet; the first being the mean of 3 observations the second of 2 and the third of 4. Find the most probable height and estimate the average error of a single observation.

12. Determine the differential equation of vibrating chords, and integrate it.

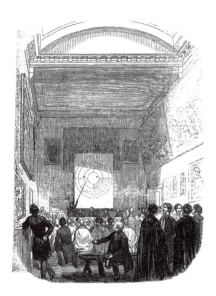

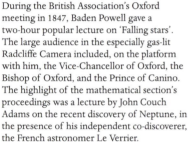

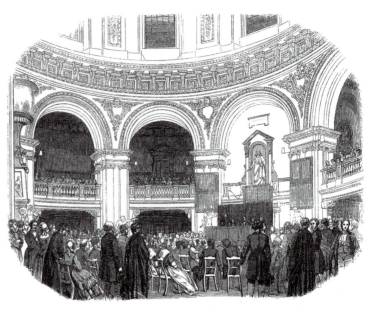

During the British Association's Oxford meeting in 1847, Baden Powell gave a two-hour popular lecture on 'Falling stars'. The large audience in the especially gas-lit Radcliffe Camera included, on the platform with him, the Vice-Chancellor of Oxford, the Bishop of Oxford, and the Prince of Canino. The highlight of the mathematical section's proceedings was a lecture by John Couch Adams on the recent discovery of Neptune, in the presence of his independent co-discoverer, the French astronomer Le Verrier.

science a higher profile at Oxford, whilst invitations to the nation's oldest university lent the infant Association respectability and prestige. Not only did the Association hold its second meeting in Oxford in 1832, but it returned there in 1847 and 1860; indeed, it was in Baden Powell's house at the 1847 meeting that John Couch Adams and Urbain Le Verrier, whose independent calculations had led to the discovery of Neptune, met for the first time. The relationship with the BAAS turned sour only in 1882 when a group of Oxford professors, from which Henry Smith was conspicuously absent, objected to the choice of Arthur Cayley as president, instead of Lord Salisbury. The 1883 meeting was moved from Oxford to Southport, and the Association did not return to Oxford for another 12 years.

It was through his clearly written popular books and articles that Powell had the greatest impact. He by no means restricted himself to mathematics and physics, and Charles Darwin later acknowledged that he had been strongly influenced by Powell's ideas when he was working on his theory of evolution. Powell returned this compliment by becoming the first prominent churchman to embrace Darwin's views on *The origin of species* in his contribution to the volume of *Essays and reviews*, published just before his death in 1860.

For many years, Powell's colleague in the other Savilian Chair was William Donkin, who had graduated from University College. A modest unassuming north-countryman, when asked for his requirements in the new University Museum, he replied that a hut in the grounds would meet his needs perfectly.

Early photographs of the two Savilian professors, Baden Powell (above) and William Donkin (below), were taken by Nevil Story Maskelyne, probably in the 1850s.

Donkin published only 16 papers in his life, but some of these were of greater importance than his current neglect might suggest. His early papers on the least squares method have some importance in the history of statistics; an 1854 paper extended Hamiltonian mechanics to cover the case of time-dependent systems, a generalization made independently at about the same time by M. V. Ostrogradsky; an accomplished musician, he also wrote several papers on early Greek music. His main interest was acoustics, but unfortunately his major treatise on the subject remained unfinished at the time of his death in 1869.

Statistics in nineteenth-century Oxford

William Donkin was not the only Oxford mathematician to contribute to the development of statistics; Oxford might even have developed a great statistical school had an unlikely scheme of Florence Nightingale and Benjamin Jowett come to fruition.

The 1860 volume of *Essays and reviews*, in which Baden Powell's essay appeared, caused a storm of protest in ecclesiastical circles, not for Powell's support of Darwin, but because of his critique of miracles. By the time the storm broke, however, Powell was dead and it was another essayist, the Regius professor of Greek, Benjamin Jowett, who inherited the brunt of the attacks within the University. Jowett weathered this and many other controversies, and went on to become Master of Balliol College from 1870 to 1893 and Vice-Chancellor of the University from 1882 to 1886.

Around the time that his controversial essay was in print, Jowett was introduced by his former pupil, the poet Arthur Hugh Clough, to Florence Nightingale. After two years' correspondence, Miss Nightingale asked Jowett to come and administer holy communion to her once a month, in the house where she had lain as an invalid since returning from the Crimean War. Florence Nightingale was interested in using statistical methods to analyse the mortality data that she had collected in the Crimea, and she had made a careful study of Adolphe Quetelet's pioneering works on statistics. She had been guided in her study by William Farr of the General Registry, and it was he who suggested to her the idea of a Chair in statistics. The idea assumed a new importance for her when, in February 1874, she heard of the death of Quetelet: 'The only fitting memorial to Q. [is] to introduce *his science* in the studies of Oxford', she noted.

Jowett was by no means a mathematician, and generally preferred things not to be 'overloaded ... with mathematics', but he also believed that statistical methods would revolutionize medicine and public health care. He leapt on the idea, and for almost 20 years it was to form a recurring theme of their correspondence. Jowett was well skilled in University politics and knew well how to turn such ideas into reality,

A portrait of Florence Nightingale by Parthenope, her sister.

Nightingale's interest in statistics arose from her studies of mortality in the Crimean campaign of the early 1850s. This shows one of her polar diagrams of mortality in British military hospitals, a mode of representation that she invented for statistical data.

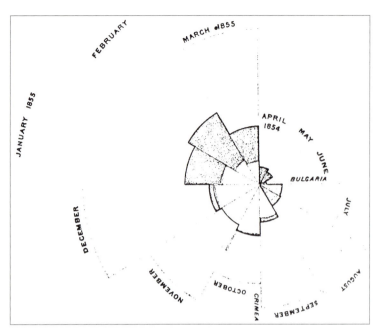

Benjamin Jowett, drawn in 1871.

but unfortunately the more practical his suggestions became, the more Miss Nightingale hesitated. She had been alarmed at Francis Galton's advice that any professor of statistics must be active in research as well as teaching, and feared that her gift would 'only end in endowing some bacillus or microbe and I do not wish that'. With less justification, Galton also advised that lacking a final honours School and students in statistics, Oxford could not hope to attract an able man.

In October 1890, Jowett made one last attempt to persuade Miss Nightingale, proposing that they each 'give or bequeath 2000£ for the endowment … and then we might go about begging of the rich people of the world'. Finally, only a month before his death in 1893, Jowett admitted defeat and altered his will to withdraw the legacy of £2000 to Miss Nightingale 'as I fear that there is no possibility of realizing the scheme to which it was originally to have been applied'.

In the event, Galton's advice was unduly pessimistic, since the University was able to appoint a statistician of the highest calibre to another Chair. Francis Edgeworth had been a pupil of Jowett at Balliol College in the late 1860s, and in 1891 returned to Oxford as Drummond Professor of Political Economy. Although Edgeworth had actually studied classics, Jowett apparently liked to teach his pupils some economics as well and, according to John Maynard Keynes, it was apparently this which stimulated his interest in the 'moral sciences'. It is often said that modern mathematical statistics was founded by Francis Galton, Karl Pearson, and Francis Edgeworth.

Little is known about Edgeworth's career during the first few years after he left Oxford, but he was called to the Bar in 1877. However, he

Francis Edgeworth is regarded as one of the founders of modern mathematical statistics.

William Spottiswoode, a Balliol student in the 1840s, combined his mathematical and scientific interests with a career in the family printing house.

must have done some extensive mathematical reading at the same time, for his *New and old methods of ethics*, published in the same year, already shows a confident use of mathematical techniques such as the calculus of variations. Over the next few years, he extended this still further, and in 1881 published *Mathematical psychics: an essay on the application of mathematics to the moral sciences*.

Edgeworth's main contributions to statistics were in the investigation of multi-dimensional normal distributions and the analysis of variations, but he is equally remembered for his contributions to mathematical economics. Towards the end of his life he numbered amongst his friends T. E. Lawrence (Lawrence of Arabia) and the poet Robert Graves. Edgeworth and Graves at one time planned to write a parody of the government Blue Books, and Edgeworth went to buy one to use as a model. When asked which volume he wanted, he replied that it did not matter, whereupon the clerk, mistaking indifference for embarrassment, produced one on venereal diseases.

Francis Galton acknowledged that his interest in statistics was strongly influenced by another of Jowett's pupils. William Spottiswoode had been a student at Balliol College in the 1840s, and continued to publish a steady stream of papers on geometry and mathematical physics, whilst running the family printing business. In an 1861 paper, he attempted to analyse statistically whether the mountain ranges in Asia had been formed by a single cause or by many. Galton said that it was this paper which stimulated him to apply statistics in the social sciences. Spottiswoode was elected president of the Royal Society in 1878, and held that office until his sudden death from typhoid in 1883. He is buried in Westminster Abbey.

A final example of Oxford's role in the development of mathematical statistics is William Sealy Gossett, alias 'Student' of the *t*-distribution. Gossett was also an Oxford graduate, and studied mathematics and natural science at New College in the closing years of the century.

Charles Lutwidge Dodgson

Yet what are all these gaieties to me
Whose thoughts are full of indices and surds?
$$x^2 + 7x + 53$$
$$= \tfrac{11}{3}.$$

This self-portrait by Oxford's most famous mathematician, as the man locked away with his books whilst others enjoy themselves at a ball, is clearly not meant to be taken too seriously. Charles Lutwidge Dodgson, better known as Lewis Carroll, could see the funny side of everything. His humorous pamphlets on University matters display the

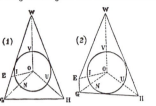

Prop. VI. Pr.

To remove a given Tangent from a given Circle, and to bring another given Line into contact with it.

Let UNIV be a Large Circle, whose centre is O (V being, of course, placed at the top), and let WGH be a triangle, two of whose sides, WEG and WH, are in contact with the circle, while GH (called 'the base' by liberal mathematicians,) is not in contact with it. (See Fig. 1.) It is required to destroy the contact of WEG, and to bring GH into contact instead.

Let I be the point of maximum illumination of the circle, and therefore E the point of maximum enlightenment of the triangle. (E of course varying perversely as the square of the distance from O).

In *The dynamics of a parti-cle*, Dodgson satirized in Euclidean style the 1865 election for the Oxford University parliamentary seat, contested by William Ewart Gladstone (WEG), Sir William Heathcote (WH), and Gathorne Gathorne-Hardy (GH).

same ability to parody arguments and drive them well beyond their logical conclusions that is so familiar from his children's books.

The new method of evaluation, as applied to π takes a light-hearted look at a controversy surrounding Benjamin Jowett's stipend as Regius professor of Greek. *The dynamics of a parti-cle* also casts mathematics in an unfamiliar light, and is surely one of the most amusing election pamphlets ever written, when, as Dodgson had hoped, Gathorne Gathorne-Hardy displaced William Gladstone from the Oxford University parliamentary seat:

Who shall say what germ of romance, hitherto unobserved, may underlie the subject? Who can tell whether the parallelogram, which in our ignorance we have defined and drawn, and the whole of whose properties we profess to know, may not all the while be panting for exterior angles, sympathetic with the interior, or sullenly repining at the fact that it cannot be inscribed in a circle? What mathematician has ever pondered over an hyperbola, mangling the unfortunate curve with lines of intersection here and there, in his efforts to prove some property that perhaps after all is a mere calumny, who has not fancied at last that the ill-used locus was spreading out its asymptotes as a silent rebuke or winking one focus at him in contemptuous pity?

A proposal from Professor Robert Clifton in 1868 for a physics laboratory (the later Clarendon Laboratory) was countered by Dodgson with a proposal for a typically Carrollian Mathematical Institute (see the box opposite).

Charles Dodgson by no means restricted himself to parochial University politics. He was also interested in electoral theory, and in 1884 made a number of suggestions to his friend, Lord Salisbury, shortly to become Prime Minister. Some of these have been adopted, such as the idea that no results should be announced until the polls

Dodgson's sitting-room at Christ Church was part of his ten-room suite on Staircase 7 of Tom Quad. He commissioned the tiles around the fireplace from the potter William De Morgan, son of the mathematician Augustus De Morgan.

'How I look when I'm lecturing'—a self-portrait by Dodgson.

On the needs of mathematics for rooms in the new Museum, 1868

A letter sent by Charles Dodgson

THE OFFER OF THE CLARENDON TRUSTEES

"Accommodated: that is, when a man is, as they say, accommodated: or when a man is—being whereby—he may be thought to be accommodated; which is an excellent thing."

DEAR SENIOR CENSOR:

In a desultory conversation on a point connected with the dinner at our high table, you incidentally remarked to me that lobster sauce, "though a necessary adjunct to turbot, was not entirely wholesome."

It is entirely unwholesome. I never ask for it without reluctance: I never take a second spoonful without a feeling of apprehension on the subject of possible nightmare. This naturally brings me on to the subject of Mathematics, and of the accommodation provided by the University for carrying on the calculations necessary in that important branch of Science.

As Members of Convocation are called upon (whether personally, or, as is less exasperating, by letter) to consider the offer of the Clarendon Trustees, as well as every other subject of human, or inhuman, interest, capable of consideration, it has occurred to me to suggest for your consideration how desirable roofed buildings are for carrying on mathematical calculations; in fact, the variable character of the weather in Oxford renders it highly inexpedient to attempt much occupation, of a sedentary nature, in the open air.

Again, it is often impossible for students to carry on accurate mathematical calculations in close contiguity to one another, owing to their mutual inter-ference, and a tendency to general conversation: consequently these processes require different rooms in which irrepressible conversationalists, who are found to occur in every branch of Society, might be carefully and permanently fixed.

It may be sufficient for the present to enumerate the following requisites: others might be added as funds permitted.

A. A very large room for calculating Greatest Common Measure. To this a small one might be attached for Least Common Multiple: this, however, might he dispensed with.

B. A piece of open ground for keeping roots and practising their extraction: it would be advisable to keep Square Roots by themselves, as their corners are apt to damage others.

C. A room for reducing Fractions to their Lowest Terms. This should be provided with a cellar for keeping the Lowest Terms when found, which might also be avail-able to the general body of undergraduates, for the purpose of "keeping Terms".

D. A large room, which might be darkened, and fitted up with a magic lantern, for the purpose of exhibiting Circulating Decimals in the act of circulation. This might also contain cupboards, fitted with glass-doors, for keeping the various Scales of Notation.

E. A narrow strip of ground, railed off and carefully levelled, for investigating the properties of Asymptotes, and testing practically whether Parallel Lines meet or not; for this purpose it should reach, to use the expressive language of Euclid, "ever so far".

This last process, of "continually producing the Lines", may require centuries or more: but such a period, though long in the life of an individual, is as nothing in the life of the University.

As Photography is now very much employed in recording human expressions, and might possibly be adapted to Algebraical Expressions, a small photographic room would be desirable, both for general use and for representing the various phenomena of Gravity, Disturbance of Equilibrium, Resolution, etc., which affect the features during severe mathematical operations.

May I trust that you will give your immediate attention to this most important subject?

Believe me,
Sincerely yours
MATHEMATICUS

February 6, 1868

have closed, whereas others, such as proportional representation, have not. Some of this work on elections was quite mathematical in spirit and represents one of his own most innovative contributions to mathematics. Unfortunately, little of it was published, and the originality of his ideas was recognized only after their rediscovery by others.

A similar fate overtook his work in logic, where his most important contributions remained unknown until the rediscovery of the galley proofs of the unpublished second volume of his *Symbolic logic*, some seven decades after his death. Whereas Dodgson's first book on symbolic logic and his puzzles and games were aimed at a more general readership, this monograph shows that he had anticipated a number of subsequent developments.

Among Charles Dodgson's published mathematical works his more advanced *An elementary treatise on determinants* stands out; this is the book which, according to gossip, Dodgson presented to Queen Victoria as his next publication after *Alice's adventures in Wonderland*, a rumour that he firmly denied. A clear concise exposition of the subject, it might have become more popular had he resisted the temptation to introduce non-standard notation and terminology, such as 'block' for 'matrix'. It includes the first precise statement of the theorem linking the solutions of simultaneous linear equations to the rank of the coefficient matrix (the so-called 'Kronecker–Capelli theorem'). It also contained Dodgson's own new 'condensation algorithm' for evaluating determinants, which had already appeared in the *Proceedings of the Royal Society* in 1866; this algorithm is almost as efficient as that of triangularizing the matrix, but has the advantage of involving only local calculations. Perhaps with the advent of parallel processing this may be of interest once more; it has recently excited the attention of combinatorialists.

However, the majority of Dodgson's 20 books are on mathematics or logic and many reflect his interest in mathematical education. Although he was mathematical lecturer at Christ Church, much of his teaching was not for the final honours School of Mathematics but for Responsions, the test of basic skills that served as a sort of equivalent of an O-level or GCSE in mathematics. Most of his books treat the elements of the subjects that he taught, such as Euclidean geometry. In particular, *Euclid and his modern rivals* was more concerned with the current controversies concerning the teaching of Euclidean geometry than with the geometry itself. His last publication was a letter in *Nature* describing a new algorithm for long division.

For published clues as to Dodgson's more advanced mathematical thinking we must return to Lewis Carroll's children's books, and, in particular, to his two *Sylvie and Bruno* books. There one finds a description of a Möbius band, and how to make a model of a projective plane out of three handkerchiefs. Even more tantalizing is his discussion of a

AN

ELEMENTARY TREATISE

ON

DETERMINANTS

WITH THEIR APPLICATION TO

SIMULTANEOUS LINEAR EQUATIONS
AND ALGEBRAICAL GEOMETRY.

BY

CHARLES L. DODGSON, M.A.

STUDENT AND MATHEMATICAL LECTURER OF CHRIST CHURCH, OXFORD.

London:
MACMILLAN AND CO.
1867.

In *An elementary treatise on determinants* Dodgson introduced his new method of evaluating determinants by 'condensation'.

Constructing the projective plane out of handkerchiefs, an illustration in *Sylvie and Bruno concluded* for which Dodgson gave very precise instructions.

Bartholomew ('Bat') Price, a leading figure in the development of Oxford University Press, photographed by his friend Charles Dodgson.

tea party in a freely falling room and in a room accelerated by means of an attached cord. This predates by some 15 years Einstein's famous discussion of the equivalence principle, a cornerstone of the general theory of relativity.

It is thought that Lewis Carroll's parody of *Twinkle, twinkle little star* ('Twinkle, twinkle, little bat') in *Alice's adventures in Wonderland* refers to his former tutor, Bartholomew Price of Pembroke College, who was popularly known as Bat Price. From 1853 until 1898, Price was the Sedleian Professor of Natural Philosophy. His influence was felt less in research than in teaching, where he quickly established a reputation as one of the leading mathematical tutors. His reading parties were famous, and it was at the 1854 party in Whitby that Charles Dodgson was groomed for his finals examination. However, Price's most enduring contribution to the University came through his long period as Secretary to the Delegates to the University Press. His sound judgement and reforms, starting with his proposal for the new 'Clarendon Press' series, transformed the Press into a modern academic publishing house.

Charles Dodgson was by no means the only Oxford mathematician to write stories of fantasy. The eccentric geometer Charles Hinton, a student at Balliol College in the 1870s, wrote articles and romances on life in higher and lower dimensions which are believed to have influenced Edwin Abbott's more famous *Flatland*. Hinton developed a remarkable geometrical intuition for the geometry of three and higher dimensions, and it was he who introduced the word *tesseract* for a four-dimensional hypercube. His suggestion that space might have a

Charles Hinton's mechanical baseball pitcher.

slight but imperceptible extension in dimensions beyond the third is reminiscent of modern compactified Kaluza–Klein theories.

Hinton married Mary Boole, the eldest daughter of George Boole, the pioneer of mathematical logic. He also married Maud Wheldon, and was tried for bigamy at the Old Bailey. After serving a three-day sentence, he fled the country for Japan. Thence he moved to Princeton University, where he invented a gunpowder-powered baseball pitcher with rubber 'fingers' to impart spin. This was used by the college team for batting practice until, intimidated by its ferocity, they refused to face it any longer.

Oxford scholars of the 1880s

Dodgson's pamphlets evidently assumed a deeper mathematical knowledge on the part of readers than Powell's students had displayed some 40 or 50 years earlier. Another indication of the recovery in the quality of the students is the fact that in the 1880s mathematical scholarships were won by Leonard James Rogers, Percy John Heawood, and John Edward Campbell.

Leonard Rogers was the son of Thorold Rogers, Edgeworth's predecessor as Drummond professor, and studied classics, mathematics, and music at Balliol College. After graduating, he became professor of mathematics at the Yorkshire College, later the University of Leeds. He was described by G. H. Hardy as 'a mathematician of great talent, but comparatively little reputation … a fine analyst … but no one paid much attention to anything he did'.

He is mainly remembered for the remarkable Rogers–Ramanujan identities, of which the first is

$$1+\sum_{m=1}^{\infty}\frac{q^{m^2}}{(1-q)(1-q^2)\,\dots\,(1-q^m)}=\prod_{n=0}^{\infty}\left[(1-q^{5n+1})(1-q^{5n+4})\right]^{-1}.$$

Although these identities did not come to prominence until 1913, when they were conjectured by Ramanujan (who was unable to find a proof), they had already appeared as a corollary of a more general theorem published by Rogers in 1894. A generalization of the above identity, published by Selberg in 1936, also proved to be contained already in Rogers' work. The president of the Royal Society noted at his death that he was 'remarkable among our fellows, in that, according to a well-informed biographer, science was to him almost distasteful'.

Percy Heawood was the mathematician who discovered the gap in Alfred Bray Kempe's celebrated 1879 'proof' of the four colour theorem, that every map drawn on the plane can be coloured with just four colours. Heawood heard about it in a geometry lecture given by Henry Smith in 1880, and continued to write papers on this subject until he was 90 years old. After graduating from Exeter College, he was

Leonard Rogers.

appointed a lecturer at the Durham Colleges, now Durham University, where he later became Vice-Chancellor.

The third mathematical scholar of the 1880s was John Campbell, who wrote the first English monograph on Lie groups, and whose name is commemorated in the fundamental Baker–Campbell–Hausdorff formula linking Lie groups and Lie algebras. After graduating, he remained as tutor at Hertford College, a more distinguished figure than some of the professors in the early years of the twentieth century. He was president of the London Mathematical Society from 1918 to 1920.

The success of Rogers, Heawood, and Campbell owed much to the stimulating research environment at Oxford, created by the succeeding Savilian professors of geometry, Henry Smith and James Joseph Sylvester, to whom the next two chapters are devoted.

Percy Heawood (above) and John Campbell.

Further reading

The transformation of Oxford in the nineteenth century is described in W. R. Ward's *Victorian Oxford*, Frank Cass, London, 1965, and in A. J. Engel's *From clergyman to don*, Clarendon Press, Oxford, 1983.

Although primarily devoted to his theological and popular works, P. Corsi's *Religion and science*, Cambridge University Press, 1988, provides a useful account of Baden Powell's career. Florence Nightingale's work on statistics is described in I. B. Cohen's popular article, 'Florence Nightingale', *Scientific American* (March 1984), 98–107, and in M. Diamond and M. Stone's three articles on 'Nightingale on Quetelet' in the *Journal of the Royal Statistical Society* **A144** (1981), 66–79, 176–213, and 332–51. Most of the correspondence between Florence Nightingale and Benjamin Jowett is collected together in *Dear Miss Nightingale* by E. V. Quinn and J. M. Prest, Clarendon Press, Oxford, 1987.

Francine Abeles' book *Mathematical pamphlets of Charles Lutwidge Dodgson and related pieces* was published by the Lewis Carroll Society of North America in 1994. There are also numerous biographies of Lewis Carroll, as well as a collection of his letters edited by M. N. Cohen (Oxford University Press, 1979), and selections from his diaries, edited by R. L. Green (Cassell, 1953), but few of these give more than a passing reference to his mathematical activities. Of the recent biographies, that by Morton Cohen (Macmillan, 1995) probably devotes the most attention to Charles Dodgson the mathematician, whilst that by Donald Thomas (John Murray, 1996) gives more of the Oxford political background. Martin Gardner's *The annotated Alice*, Penguin, 1970, is a useful source for some of the mathematical ideas embedded in the *Alice* books, and his more recent *The universe in a handkerchief*, Springer, New York, 1996, is an invaluable reference book for Dodgson's games and puzzles.

CHAPTER 12

Henry Smith

Keith Hannabuss

Baden Powell and William Donkin put Oxford mathematics back on the national map, but it was Henry Smith who brought it international recognition. A foreign member of the Prussian Academy of Sciences, he carried off major prizes offered by both the Berlin and Paris Academies.

One of the puzzling features when one looks at Henry Smith's work is why his name is so little known, even amongst those who make regular use of the ideas that he introduced. Several historians of mathematics have ranked him with Arthur Cayley and James Joseph Sylvester among the great pure mathematicians of the nineteenth century, and the German *Brockhaus Enzyklopädie* still gives him a longer entry than Sylvester. It seems that his reputation has always been higher on the continent. In an obituary notice to the Accademia dei Lincei, Luigi Cremona commented that Henry Smith's achievements had not received the acclaim which they deserved in England, because of his extreme modesty. The neglect of his contributions was undoubtedly further exacerbated by the fact that he worked in areas such as number theory, which were outside the main focus of English mathematics at the time. There were other factors, though, which we shall mention below.

Henry John Stephen Smith was born on 2 November 1826 in Dublin. His father, a lawyer, died soon afterwards of a liver disease, and the widow moved with her four children to England. Henry Smith was educated first by his mother, and then by a succession of private tutors, before spending three years at Rugby School and then winning the top scholarship to Balliol College. His entry to Oxford was delayed first by smallpox and then malaria, contracted during visits to the continent. He managed, however, to put his convalescence to good use by attending the physics lectures of François Arago in Paris.

Eventually, at the age of 20, Smith arrived in Oxford to resume his undergraduate career. Within two years, he gained first class honours in both classics and mathematics, and became president of the Oxford Union. He also won the major University prizes in both classics (the Ireland Scholarship) and mathematics (the Senior Mathematical Scholarship).

Henry Smith, around the time of his election to a mathematical lectureship at Balliol in 1849. This photograph, taken by his friend and collaborator Nevil Story Maskelyne, captures the ebullience and charm that endeared him to so many.

Frederick Temple, Smith's mathematics tutor at Balliol, was later Archbishop of Canterbury.

The old front of Balliol College in about 1860. Henry Smith had a third-floor room next to the tower.

Smith's classics tutor was Benjamin Jowett, who became a life-long friend. Following William Ward's resignation, mathematics was taught at Balliol by Frederick Temple, who later became chaplain to Queen Victoria, Bishop of London, and Archbishop of Canterbury. Temple's interest was in mathematical puzzles and games, and this continued even during his later ecclesiastical career. He often worked on problems as a diversion during boring episcopal meetings; in the course of one particularly tedious meeting in the 1880s, he produced a 'proof' of the four colour theorem. When Temple resigned his fellowship in 1849, it was almost inevitable that Henry Smith would be chosen as his successor. Nonetheless, Smith himself seems to have been taken slightly by surprise. Reporting the news to his sister, he protested that

they prefer having work ill done by an in-college man to having it well done by an out-college one, and this I think is not wise. I could only say that I felt myself quite incompetent, and I don't remember that I ever spoke with a clearer conscience.

Until this time, Smith seems to have been undecided as to whether he should follow a career in classics or mathematics, but with his election to this lectureship, he seems to have settled on mathematics. His first paper on geometry dates from the next year.

There was, however, still to be one more diversion. When, in 1849, the University decided to create an examination in natural science, there was at first no requirement for practical work to be undertaken. However, Balliol decided that it would nonetheless be a good idea to build and equip a laboratory for this purpose, and Henry Smith was asked to run it. To prepare him for this task, the college sent him to study with Nevil Story Maskelyne in the Old Ashmolean Museum base-

North Oxford in Henry Smith's time
(*c.* 1860s), before the large-scale conversion of
the area into academic housing. The Old
Ashmolean building, where he started
chemistry, is at the front right. The new
University Museum is on the edge of town, at
the top of the picture, with the Keeper's
house to its right, beyond Wadham College.

ment, and with August Hofmann at the Royal College of Chemistry.
Thus it was that Smith ran the first college science laboratory in Oxford
from 1853 until 1860, with a two-year intermission from 1856 to
1858 when Professor Benjamin Brodie used it whilst awaiting the
completion of his new laboratory in the University Museum.

In fact, Smith's most distinguished pupils were scientists, rather than
mathematicians. His very first pupil was Augustus Vernon Harcourt,
who, with the mathematician William Esson (later Savilian Professor of
Geometry), undertook the first detailed studies of chemical kinetics.
Smith himself worked with Brodie on his eclectic atomic theory, and
with Maskelyne on crystallography, although these were always somewhat tangential interests.

Report on the theory of numbers

Like most Oxford scientists, Henry Smith had become an active
member of the British Association for the Advancement of Science.
The Association used to commission reports on various different
branches of science, and in 1858 Smith was asked to prepare a report
on the state of number theory.

Smith's first paper on number theory had been published in 1854.
Unusually for the time, it was written in Latin, perhaps in homage to

Beginning of Henry Smith's *Report on the theory of numbers.*

> *Report on the Theory of Numbers.*—Part I.
> By H. J. STEPHEN SMITH, M.A., *Fellow of Balliol College, Oxford.*
>
> 1. THE 'Disquisitiones Arithmeticæ' of Karl Friedrich Gauss (Lipsiæ, 1801) and the 'Théorie des Nombres' of Adrien Marie Legendre (Paris, 1830, ed. 3) are still the classical works on the Theory of Numbers. Nevertheless, the actual state of this part of mathematical analysis is but

Carl Friedrich Gauss, whose *Disquisitiones arithmeticae* (1801) served as an inspiration. In the end, Smith produced no fewer than six reports over the period from 1859 to 1865, one in each year except 1864. These presented a panoramic view of the whole subject as it existed at that time. The basic contributions of Fermat, Legendre, and Gauss were joined by newer work, such as that of Jacobi on elliptic functions, Kummer's theory of ideals, and the latest ideas of Kronecker and Hermite. The work was well received by continental mathematicians, and Kronecker praised it highly for its lucidity and insight. After Smith's death, the six reports were gathered together into a single volume: the *Report on the theory of numbers.*

It is interesting, in the light of later events, to note a comment made by Hermann Minkowski in the last decade of the century. He and David Hilbert had been asked to prepare a sort of updated German version of the report, the *Zahlbericht.* Minkowski never managed to complete his half, and in 1896 he wrote to Hilbert to tell him that he should go ahead and publish his volume on its own. As one of the excuses for his failure to produce his own portion of the report, Minkowski cited the publication of Smith's report in a single volume, saying that the balance of the literature had now completely changed, and that there was no longer the same need for a volume of the kind he had been asked to prepare.

Savilian Chair

When Baden Powell died in 1860, the Oxford mathematicians petitioned the electors in support of Henry Smith, and he was duly elected to the Savilian Chair. The election was, however, not uncontested. The outside candidate was George Boole, but he was all too aware of the religious controversies that had engulfed Powell and Jowett following their contributions to *Essays and reviews.* (He had written to Augustus De Morgan in November 1860 stating that he considered Jowett's essay 'the *best,* and for the sake of mathematics I am sorry to add Baden Powell's nearly the worst'.) Boole hesitated to move to so intolerant an atmosphere, and eventually compromised by entering his name as a candidate without submitting any testimonials. This seems to have given the impression that his application was not serious. So, as Smith's friend, Charles Pearson, ironically commented,

Eleanor Smith with Henry Acland, the Regius professor of medicine who was the inspiration behind the University Museum.

the electors for the chair chose [Smith], as I have understood without hesitation, and taking the view that as no other Oxford man was candidate, and as Henry Smith was pre-eminently qualified, it was needless to scrutinise the testimonials of outsiders.

Bearing in mind Smith's diffidence when appointed mathematics lecturer, it is not surprising that he himself felt distinctly uncomfortable about the cursory attention paid to a mathematician for whom he had the greatest respect.

Despite his election to the Chair, his financial position forced him to continue his routine teaching as before, in addition to his professorial duties. Smith's precarious finances, the result of an imprudent youthful investment, may also be the reason why he never married. After the death of their mother in 1857, his elder sister Eleanor kept house for him. Eleanor was a remarkable woman in her own right. She founded the first series of women's lectures at Oxford, and later served on the council that established Somerville College.

However strange Smith's election in preference to Boole may have looked at the time, he soon amply justified the electors' confidence. Often, in the course of preparing his *Report on the theory of numbers*, he added investigations of his own. Sometimes these were too substantial to find a place in the report itself, and over the years he published a string of papers taking up themes that he had found in the existing literature.

The Smith form of a matrix

In his paper on linear Diophantine equations, Smith proved the following theorem:

Let $\mathbf{A}$ be a matrix with integer coefficients. Then there exist invertible matrices $\mathbf{P}$ and $\mathbf{Q}$ with integer entries such that

$$\mathbf{PAQ} = \begin{pmatrix} \alpha_1 & 0 & 0 \ldots & 0 \\ 0 & \alpha_2 & 0 \ldots & 0 \\ 0 & 0 & \alpha_3 \ldots & 0 \\ \cdot & & \cdot & \\ \cdot & & \cdot & \\ \cdot & & \cdot & \\ 0 & 0 & 0 \ldots & \alpha_n \end{pmatrix},$$

where $\alpha_1 \mid \alpha_2$, $\alpha_2 \mid \alpha_3$, $\ldots$, $\alpha_{n-1} \mid \alpha_n$. Moreover, the α_j are unique up to sign, and can be taken as

$$\alpha_1 = \nabla_1, \; \alpha_2 = \nabla_2/\nabla_1, \; \ldots, \; \alpha_r = \nabla_r/\nabla_{r-1}, \quad \text{and} \quad \alpha_j = 0 \text{ for } j > r,$$

where $r = \operatorname{rank}(\mathbf{A})$ and ∇_j is the greatest common divisor of the $j \times j$ minors of $\mathbf{A}$.

His method of proof was essentially the same as we would use today: induction on the size of the matrix, coupled with elementary row and column operations.

A system of linear Diophantine equations can be written in matrix form as

$$\mathbf{Ax} = \mathbf{b}.$$

Since $\mathbf{Q}$ is an invertible matrix, we can introduce new variables $\xi = \mathbf{Q}^{-1}x$ to obtain equivalent equations $\mathbf{AQ}\,\xi = \mathbf{b}$, or, by suitable combination,

$$\mathbf{PAQ}\,\xi = \mathbf{Pb}.$$

We already know that $\mathbf{PAQ}$ is the diagonal matrix above, and so, if we write $\beta = \mathbf{Pb}$, we arrive at the simple equations

$$\alpha_j \xi_j = \beta_j, \quad \text{for} \quad j = 1, \ldots, n.$$

In this form, it is easy to see whether the equations admit any solutions and to find all the solutions when they do. Since Smith's method is a constructive one, this provides a practical procedure for handling such equations.

We now recognize that Smith's result is of much wider application than this, for it extends immediately to matrices whose entries lie in principal ideal domains, and provides the backbone to the proof of the cyclic decomposition theorem for finitely generated modules. In his published work at least, Smith does not seem to have taken this next step; that was left to Frobenius and Stickelberger. In his *Theorie der linearen Gleichungen mit ganzen Coefficienten*, published in Crelle's *Journal* in 1879, Frobenius showed how Smith's theorem applied to matrices with entries in a polynomial ring could yield the Jordan canonical form; a joint paper with Stickelberger, *Über Gruppen von vertauschbaren Elementen*, published in the same volume, then derives the cyclic decomposition of a finitely generated abelian group from the same principles.

One of the earliest and most famous of these is his work on linear Diophantine equations (equations whose solutions are integers), which appeared in 1861, within a few months of his election to the Chair. In this paper he proved the existence of what is now sometimes called the *Smith form* of a matrix. As its title *On systems of linear indeterminate equations and congruences* suggests, Smith was interested in this result because of its application to linear Diophantine equations.

Sums of squares

Another investigation that had its origins in the *Report* was his study of the representation of numbers as sums of squares. In the report he had announced his intention of presenting the theory of quadratic forms in six parts, but only the first of these, covering the theory of binary forms, ever appeared. It seems that he judged the more general version of the theory to be the subject for research rather than report, and he worked on it himself over the next few years.

The idea of expressing an integer as a sum of squares is a simple one, whose origin can certainly be traced back as far as the 1621 edition of Diophantus prepared by Bachet de Méziriac. In 1770, Lagrange had completed the proof of the basic result that every positive integer can be expressed as the sum of four squares. For instance, the integers from 5 to 9 can be written as follows; the example of 7 shows that four squares really are needed:

$$5 = 2^2 + 1^2 + 0^2 + 0^2,$$
$$6 = 2^2 + 1^2 + 1^2 + 0^2,$$
$$7 = 2^2 + 1^2 + 1^2 + 1^2,$$
$$8 = 2^2 + 2^2 + 0^2 + 0^2,$$
$$9 = 3^2 + 0^2 + 0^2 + 0^2.$$

This decomposition is not usually unique; for example, we can also write 9 as

$$9 = 2^2 + 2^2 + 1^2 + 0^2, \text{ or as } 9 = 2^2 + 1^2 + 0^2 + (-2)^2.$$

The question now arises: *in how many ways* can a given positive integer be represented as a sum of four squares—or, for that matter, as a sum of any given number of squares?

For two, four, six, and eight squares, as Smith reported, the problem had been solved by the work of Carl Jacobi on the theta function. Much earlier, Gauss had shown how to count representations as a sum of three squares, but the cases of five and seven squares remained. It was to these that Smith now turned his attention. In fact, he turned his back on Jacobi's elliptic function methods and returned to Gauss's more algebraic approach, which he was able to generalize in a very subtle way, using his own earlier work on the diagonalization of integer matrices. The resulting theorem was powerful enough to handle any number of squares, although the application of the general results to

The widespread affection for Henry Smith across the University community can be gleaned from this caricature. As a former student wrote after Smith's death, 'Alas! … that his flowing beard and flying gown will be seen no more, for his place cannot be filled.'

particular cases required enormously detailed calculation. Nonetheless, Smith was able to carry his investigation through to a successful conclusion, and his results were published in the *Proceedings of the Royal Society* in 1867. He included only his formulas for the number of decompositions in terms of five and seven squares, and an outline of the methods he had used. He did not include the detailed proofs because, as he said, it was

practically impossible to test the results by any independent process. The demonstrations are simple in principle, but require attention to a great number of details with respect to which it is very easy to fall into error.

He promised to publish the full details as soon as he had had time to put them into a convenient form.

The insistence on an independent check of his formulas for sums of squares is typical of the high standards he maintained. Where other mathematicians might develop a subject in a whole series of papers, Smith would wait until he had the complete theory. Unfortunately, Smith's time was now much in demand, as tasks and duties accumulated in the wake of his election to the Savilian Chair. These duties were not just the natural concomitants of his office; they came rather as a consequence of his enormous erudition and immense personal charm. Years later, the classical scholar, Evelyn Abbott, wrote:

His genius and accomplishments, his grace and gentleness, his sound judgment, his brilliant conversation, are a tradition in Oxford … he possessed in an extraordinary degree the gift of conciliation, and could say the happy word which quells a rising storm … whenever he was present he was the life of the party, charming in his conversation and possessing an inexhaustible fund of amusing stories, which he told with admirable grace and effect.

The flavour of his wit may be gauged by the story that when Sir Norman Lockyer, the editor of the journal *Nature*, was being criticized for being a little heavy-handed, Smith commented that 'he sometimes forgets that he is only the editor and not the author of *Nature*'. In the middle of a lecture, he is said to have broken off to remark 'It is the peculiar beauty of this method, gentlemen, and the one which endears it to the really scientific mind, that under no circumstances can it be of the smallest possible utility.'

Not only did Smith have a genius for finding the compromise which would command general support, he was also a powerful speaker who could present a case in the most persuasive manner. At a time when many reformers in Oxford were rather abrasive characters, these were skills which were much in demand. It is small wonder, therefore, that Smith was often called upon to cast new schemes into a practical form. However, Smith's qualifications for such a role went well beyond his personal popularity. His expertise in science and classics, as well as in mathematics, earned him the confidence of all sides in tricky inter-

BARCHESTER TOWERS.

BY

ANTHONY TROLLOPE,

AUTHOR OF "THE WARDEN."

IN THREE VOLUMES.

VOL. III.

LONDON:

LONGMAN, BROWN, GREEN, LONGMANS, & ROBERTS.

1857.

Henry Smith's activities in the 1850s attracted attention well beyond Oxford. His participation in religious controversies on the plurality of worlds earned him an oblique reference (as 't'other man') in Anthony Trollope's novel *Barchester Towers* (1857), in which one character asks 'Are you a Whewellite or a Brewsterite or a t'other manite, Mrs Bold?'

disciplinary disputes. The professor of Latin, John Conington, once remarked: 'I do not know what Henry Smith may be at the subjects of which he professes to know something; but I never go to him about a matter of scholarship in a line where he professes to know nothing, without learning more from him than I can get from anyone else.'

Confirmation of this picture may be found in the long obituary notice which appeared in *The Times*:

It is probable that of the thousands of Englishmen who knew Henry Smith, scarcely one in a hundred ever thought of him as a mathematician at all. He was a classical scholar of wide knowledge and exquisite taste, and there were few who talked to him on English, French, German or Italian literature who were not struck by his extensive knowledge, his capacious memory, and his sound critical judgment. French and Italian he spoke fluently and wrote easily, and of German he also had a complete colloquial command.

Soon after completing his paper on sums of squares, Smith was elected to the University's Hebdomadal Council, and during the 1870s he sat on royal commissions into scientific education (the Devonshire Commission) and the universities. He was appointed Chairman of the Meteorological Council, charged with providing the first scientifically based weather forecasts for the country, and in 1874 became Keeper of the University Museum, an office which, according to his sister, he accepted for 'the honour and ... in order to be near the croquet ground'.

The University Museum in 1860. Henry Smith lived with his sister Eleanor in the Keeper's house on the right.

The German Ferdinand Lindemann was among the foreign mathematicians Smith welcomed to England.

International relations

Smith's wide contacts on the continent also gave him a far broader perspective than most of his colleagues. During the election of the Lee's Professor of Experimental Philosophy in 1865, Smith worked hard to attract Hermann von Helmholtz to the Chair from Heidelberg. (Helmholtz mentioned learning of the plan from Smith when they met for lunch with the French mathematician Charles Hermite in Paris, early in April 1865.) Unfortunately the scheme failed, for reasons which are not entirely clear.

Over the years, administrative responsibilities led to increasingly heavy demands on Smith's time, although he himself said that these did not detract from his study of mathematics, on which it was not possible to work in a concentrated way for more than a few hours a day. His research was carried on late into the night, after he had returned from other engagements. Nonetheless, his other commitments may have robbed him of the time he needed to write up his results for publication. This is particularly likely, bearing in mind his reluctance to publish ideas until they had fully matured.

Despite the increasing calls on his time, the 1870s were a very productive period for Smith. He travelled widely on the continent each summer, visiting foreign mathematicians, and he seems frequently to have invited them back to England. In 1873, when Smith was president of the mathematics and physics section of the British Association meeting in Bradford, he invited both Hermite and Klein to the meeting. Hermite talked on the irrationality of e; presumably his newly discovered proof of the transcendence of e would have been too technical for such a general audience. One might also speculate whether, had the meeting been one year earlier, Felix Klein would have presented the mathematical world with a Bradford programme for relating group theory to geometry, instead of the Erlanger Program. In 1876, Smith entertained a then unknown young German geometer, Ferdinand Lindemann, in Oxford. It seems that during his visit Smith spent some time discussing with him the problem of proving the transcendence of π, whose solution five years later was to make Lindemann famous.

University teaching

For many years Henry Smith offered a cycle of three undergraduate courses on harmonic properties of figures, anharmonic properties of figures, and the geometry of three dimensions, one for each term of the University year. One can get a fair idea of the contents of these courses from *An elementary treatise on pure geometry*, written in 1893 by John Wellesley Russell, his pupil and successor as Balliol tutor; Russell acknowledges his debt to Smith in the preface.

Smith set ($m = 3$)	Cantor set

The first few stages of the construction of the Smith and Cantor sets. At each stage the Cantor set is topologically the same as the following stage of the Smith set.

Smith's lectures were later varied to include advanced plane geometry, focal properties of conic sections, reciprocal polars, pure geometry, surfaces of the second order, the theory of numbers, and a course on geodesics. After Smith had become Keeper of the University Museum and had resigned his tutorship, these topics were joined by some more analytic courses which he had previously confined to his inter-collegiate classes: integral calculus and differential equations, calculus of variations and definite integrals, partial differential equations, calculus of finite differences, and probabilities. These seem to reflect the shift in Smith's own research interests as he worked increasingly on elliptic functions.

The course on integral calculus is particularly interesting as it was first given in 1875, the year in which Smith published a remarkable paper *On the integration of discontinuous functions*. In this paper, Smith introduced the first example of what is now called a 'Cantor set'; Cantor's own example appeared eight years later and was not presented as his own discovery. Smith's example divides an interval into a number m of subintervals, and then keeps repeating this process to each remaining subinterval, except the last. Smith seems also to have been the first person to perceive the connection between measure and integral. However, his paper received less attention than it deserved, owing to an inaccurate review in the *Fortschritte der Mathematik*. In his history of integration, Hawkins has remarked that:

Probably the development of a measure-theoretic viewpoint within integration theory would have been accelerated had the contents of Smith's paper been known to mathematicians whose interest in the theory was less tangential than Smith's.

A course on continued fractions in 1879 may similarly be linked to Smith's paper *De fractionibus quibusdam continuis*, although there is an earlier *Note on continued fractions*, based on an 1875 address to the British Association.

Smith concluded his paper on discontinuous functions with a discussion of higher-dimensional analogues of his results. He remarks that in two dimensions one may obtain regions bounded by curves of infinite length. Although he gave no specific examples, it is tempting to

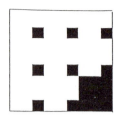
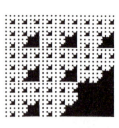

Smith's construction when $m = 3$ gives a version of the snowflake curve, a curve of infinite length.

Early stages of Smith's construction when $m = 2$ indicate the convergence to the Sierpinski gasket.

speculate that he may have applied his construction in the plane, deleting at each stage just those points whose x-coordinate and y-coordinate both lie in the interval removed from the line. For $m = 3$, this gives a region whose lower right-hand boundary, repeated elsewhere, is a version of the *snowflake curve*, first published by Helge von Koch in 1904. It is also possible to carry out Smith's construction in the plane with $m = 2$; in this case it gives the *Sierpinski gasket*.

Some of his earlier geometry lectures may also have been linked to his own research work. In 1866, the Academy of Sciences in Berlin had posed a geometrical problem on the intersection of quartic curves. Henry Smith solved this, and in 1868 was joint winner (with Hermann Kortum, a professor at Bonn) of the Academy's Steiner prize. Probably more interesting than the solution of the problem itself was one of its corollaries. It is well known that it is impossible to trisect an angle using ruler and compass constructions. The work of Smith and Kortum showed that, given a ruler, a compass, and one conic section that is neither circular nor degenerate, trisection of the angle is possible. Kortum later earned some notoriety for inducing the young analyst Alfred Pringsheim to leave Bonn, by disconcerting him with an irrelevant question in a thesis viva.

Grand Prix

We now turn to the final year of Smith's life, which was marred by events over which he had no control. In 1881, he had a bad fall and suffered from serious illness. Whilst convalescing, he chanced to notice in an old copy of the *Comptes Rendus* the announcement of a competition for the Grand Prix of the French Academy of Sciences. The problem, to complete the work of Eisenstein by finding a formula for the

Henry Smith turned to Charles Hermite for help when the French Academy announced as its 1883 Grand Prix topic a problem that Smith had already solved.

number of ways of expressing an integer as a sum of five squares, was that already solved by Smith some 15 years earlier.

At first Smith was uncertain what he should do. As he wrote to his friend James Glaisher:

In the Royal Society's *Proceedings*, vol xvi, pp. 207, 208, I have given the complete theorems not only for five, but also for seven squares: and though I have not given my demonstrations, I have (in the paper beginning at p. 197) described the general theory from which these theorems are corollaries in some fulness of detail. Ought I to do anything in the matter? My first impression is that I ought to write to Hermite, and call his attention to it. A line or two of advice would really oblige me, as I am somewhat troubled and a little annoyed.

Glaisher endorsed this course of action, and so Smith wrote to Hermite. In his reply, received a few days later, Hermite starts by asserting that none of the members of the commission which proposed the prize subject was aware of Smith's prior solution of the problem, and that he himself had only learned of it for the first time in Smith's letter. After apologizing profusely for the oversight, he suggested a way in which Smith could help to relieve the embarrassment of the Academy; he could rewrite his earlier work in French with the complete proofs, and enter it according to the rules of the competition. Then he would see that justice would be done. In any case, Hermite would make sure that the Academy was appraised of Smith's priority. Hermite was well known for his generosity towards younger mathematicians, and he may well have been trying to protect the person who made the mistake.

By now the competition had been running for well over a year and Smith had less than three months to comply with Hermite's suggestion. The fact that he managed to do so, despite illness and other commitments, is remarkable in itself. Unfortunately, Hermite's plan failed. On 9 February 1883, two months before the results of the competition were announced, Smith died of the same disease of the liver that had killed his father.

In fact, the plan was doomed, even had Smith lived. Hermite had no doubt been hoping that Smith's would be the only entry to the competition, but that was not to be the case. There were two other entries. One of these, submitted by Théophile Pépin, did not succeed in solving the problem, but the other presented a solution along exactly the lines that Smith had outlined in his paper of 1867. The prize was therefore awarded jointly to Smith and to the other entrant, the young Hermann Minkowski, a student from the University of Königsberg in East Prussia.

At that time, Minkowski was only 18 years old, and the circumstance of his sharing the prize with Henry Smith must have been unusual enough to excite general interest. It did not take the French journalists long to discover that Smith's solution to the problem had already been published many years before; Glaisher's obituary notice to

Hermann Minkowski.

the Cambridge Philosophical Society in February had already mentioned his priority. Coupled with the judge's remarks on the uncanny similarity between the two successful entries, this quickly excited the idea that Minkowski might simply have plagiarized Smith's earlier work. The Franco-Prussian War was still a recent memory, and national feelings ran high and were easily inflamed. German mathematicians saw the French accusations against Minkowski as pure chauvinism, aroused by the fact that Minkowski studied in Prussia. It took some time and an explanatory statement from the Academy defending their judgement before the affair finally simmered down. Camille Jordan's official statement on behalf of the Academy, that Smith's work was known and that the prize was specifically offered to force him to produce the details of his proofs, is contradicted by Hermite's admission in his letter that no-one had been aware of Smith's paper.

With the benefit of hindsight, knowing that Minkowski became a number theorist and mathematician of the greatest distinction, one can be confident that he was not guilty of any deliberate attempt to deceive the Academy. He may well have known of Smith's papers, particularly the earlier ones, for they are cited in the paper by Frobenius which we mentioned earlier. However, if he knew of Smith's papers, he would doubtless have assumed that the Academy knew them too, and that the competition was indeed designed to elicit the full proofs. Retrospectively, one can feel that it is almost fitting that Smith should have shared this last honour with the man who became his scientific heir.

There is another postscript to the story which is worthy of mention. In 1900, Hilbert was to give an address to the Second International Mathematical Congress in Paris. He consulted Minkowski about the form it should take, and on 5 January Minkowski replied: 'Of speeches which might interest you the only one which comes to mind is that by Henry John Stephen Smith *On the present state and prospects of some branches of pure mathematics* in the second volume of his Collected Works.' This talk was Smith's presidential address to the London Mathematical Society on 9 November 1876. Abandoning the usual pattern of such talks, he set out to indicate some of the branches and problems of pure mathematics which he felt to be neglected by English mathematicians. The title of Smith's lecture recalls Baden Powell's *On the present state and future prospects of mathematics at the University of Oxford*, but, in the 50 years between Powell's lecture and Smith's death, Oxford mathematics had progressed beyond all recognition.

Henry Smith in later life.

Further reading

Henry Smith's publications appear in the two volumes of his *Collected mathematical papers* (edited by J. W. L. Glaisher), Oxford University Press, 1894, reprinted by Chelsea Publishing Co., New York, 1965.

Popular accounts of aspects of Henry Smith's work appear in two articles by Keith Hannabuss, 'The mathematician the world forgot', *New Scientist* **97** (1983), 901–3, and 'Forgotten fractals', *The Mathematical Intelligencer* **18**, No. 3 (1996), 28–31.

Vanda Morton's *Oxford rebels, the life and friends of Nevil Story Maskelyne 1823–1911*, Alan Sutton, Gloucester, 1987, describes some of the scientific developments in the middle of the century, focusing on the circle around Smith's friend, Nevil Story Maskelyne.

CHAPTER 13

James Joseph Sylvester

John Fauvel

On the 3rd I went with the Spottiswoodes (Mr, Mrs & Cyril) to the Lyceum Theatre to hear Irving and Miss Terry in "Much Ado about Nothing." On the 8th I made a communication on Cremonian congruences to the Mathematical Society. On the 9th the news of Prof. Smith's death at Oxford reached us. He had but recently returned from Paris. It was on the 13th that I first told Admiral Luard of my intention to retire from my present appointment at the end of the Session. It was on the 17th that Spottiswoode asked me if I would become a candidate for the Savilian Professorship of Geometry at Oxford. I said no.

Thomas Archer Hirst's diary for February 1883 hints at the problem facing Oxford mathematics upon the sudden and unexpected death of the Savilian professor at the height of his powers. Henry Smith, the most distinguished mathematician to occupy the Savilian Chair since John Wallis two centuries before, had raised the status and international repute of mathematics at Oxford; it would not be easy to find a replacement. William Spottiswoode—president of the Royal Society and himself an Oxford graduate—was naturally concerned. To sound out his friend Hirst, director of studies at the Royal Naval College, Greenwich, and a former president of the London Mathematical Society, was an obvious step to take. Spottiswoode himself was to die suddenly of typhoid fever within a few months, and did not live to see the election of Smith's successor. Oxford sentiment was clear, however: 'We hope that the place left empty by the man whose loss we are mourning will be filled by another genuine Oxford man able to assert our real high mathematical standing.'

On 4 April, Arthur Cayley, Sadleirian Professor of Pure Mathematics at Cambridge, wrote to the German mathematician Felix Klein, young but already internationally renowned, about the vacancy at Oxford. Cayley had a keen interest in who would be appointed to the Savilian Chair, from the position of one who had done more than anyone else to promote pure mathematical research in England, both by example and by generous encouragement of the work of younger researchers. Yet his letter to Klein—in some respects an ideal candidate—was somewhat half-hearted: 'There are no good Oxford men for it, but I do not

The late nineteenth century saw various innovations in teaching mathematics at Oxford. In 1886 the University acquired a set of plaster models of surfaces to illustrate geometrical ideas such as the 27 lines on a cubic surface; Sylvester gave a course on surfaces as well as a public lecture on the models.

Prominent Royal Society scientists were represented in a group portrait of 1889. On the left, J. J. Sylvester is sitting behind T. H. Huxley; Arthur Cayley is on the right.

know whether or not there would be a disposition to elect a foreigner. Would the position suit you?'

Perhaps Cayley felt that, from his own experiences at Cambridge, the time was not yet ripe to create a full international research school in England, which would have been the main point of persuading Klein to come to Oxford. Or it may be that Cayley was playing a more subtle game, and was putting Klein off because he had already decided that this was the right post for his old friend James Joseph Sylvester, the 69-year-old professor of mathematics at Johns Hopkins University in Baltimore. Certainly he is likely to have known by then that Sylvester was throwing his hat into the ring, as transatlantic letters then took about two weeks to arrive. On 16 March, Sylvester had written to Cayley from the United States:

I heard with dismay and the deepest regret of Henry Smith's decease. I shall most probably offer myself as a Candidate for the succession to his professorship … Do you think I am likely to be appointed? If the chances are considerably against me—it would be impolitic to offer—but perhaps even if impolitic it would be right on my part to do so by way of testing what I consider—although you may not perhaps agree with me—an important principle.

The 'important principle' is not certain, but is likely to have been Sylvester's strong belief that the best candidate should put himself forward, irrespective of age or religion, or indeed university background.

The electors' choice in the end was indeed Sylvester, and thus Smith was succeeded by a man 12 years his senior. To add to the adverse factors, Sylvester was the first Savilian Professor of Geometry since Wallis not to have been educated at Oxford; indeed, even the formal awarding of his Cambridge degree had been comparatively recent, since although he sat the mathematical tripos at Cambridge as long ago as 1837, as a Jew he could not receive his degree until the abolition of theological tests in 1871. Sylvester commented on the fact that he was not an Oxford graduate when, after a year in post, he wrote happily to Cayley in November 1884:

Oxford is beginning to be a delightful residence to me. She is a dear good mother our University here and stretches our her arms with impartial fondness to take all her children to her bosom even those whom she had not reared at her breast.

But it was not a straightforward appointment, even though there was no more distinguished mathematician in the English-speaking world. On 26 July 1883, Sylvester wrote to Simon Newcomb, the astronomer who was to be his successor at Johns Hopkins:

I had hoped to be appointed Profr. at Oxford to succeed the late Henry Smith: the sad death of Spottiswoode I believe has frustrated at all events for the present and possibly for good and all that expectation. The election that was announced for the 7th Inst. is postponed until in December next. Possibly I may withdraw my candidature to make room for younger men—and to avoid the mortification of a refusal.

Attempts had indeed been made to persuade Sylvester to withdraw on account of his age:

Ingleby some time ago was so impressed from a conversation he had had with "a former mathematical lecturer at New College" with the impossibility of my being appointed that he urged me ... to consider whether it would not be better for me to withdraw my name.—but I never for a moment contemplated doing so—as that would have been a needless avowal of unfitness.

The election eventually took place in December 1883, with Sylvester as a candidate. In the following month he wrote to a friend that the electors had reached a unanimous decision.

It is curious that you should have been able to predict my election to the Savilian Professorship of Geometry. At one time I thought my chances more than dubious, but in the end I was elected, it seems à l'unanimité des voix. There were 7 electors (Dr. Salisbury one of them) and he is reported to have said that had his casting vote been wanted he would have given it in my favor ... most people thought he would, from his political convictions, be averse to me.

Sylvester had triumphed over not only his disadvantages of birth, education, and age, but also past quarrels of his tempestuous life: one of the electors was Spottiswoode's successor as president of the Royal Society, the biologist Thomas Henry Huxley, whose incautious views on mathematics had come under vigorous attack from Sylvester at a British Association meeting some years before. Clearly Huxley joined in recognizing the best man for the job, regardless of the past bruises he had received at Sylvester's hands.

Adventures of a shuttlecock

The path was an extraordinary one that led Sylvester from his birth within London's Jewish community, on 3 September 1814, to Oxford's Savilian Chair nearly 70 years later. Throughout his life, his proud and forthright acknowledgement of his Jewish heritage affected the development of his career. His talent for mathematics displayed itself early, at the Jewish schools he attended in Highgate and then in Islington. At the age of 14 he spent five months at the University of London (later called University College, London), where one of his teachers was Augustus De Morgan, and in 1829, at the age of 15, he entered the Royal Institution School in Liverpool, staying with his aunts.

In October 1831 Sylvester entered St John's College, Cambridge. Due to a succession of illnesses, it was not until 1837, at the age of 22, that he sat his finals, the mathematical tripos. In this fiercely competitive examination he was placed second—that is, he was 'Second Wrangler'. It was a good year: the Fourth Wrangler was the great mathematical physicist George Green, who had come to Caius College as a mature student, and the Fifth Wrangler was Duncan Gregory, scion of a distinguished Scottish mathematical family, and a notable figure in mathematical Cambridge before his untimely death.

Since Sylvester would not sign the Thirty-nine Articles of the Church of England, he could not receive his degree from Cambridge; nor could he enter for the prestigious Smith's prize, the traditional way for the best young Cambridge mathematicians to crown their university careers; nor was he eligible for a college fellowship. (At Oxford, he would not even have been able to matriculate—that is, he could not even have become a student in the first place.) Thus barred from all further progress at Cambridge, he was within a few months appointed professor of natural philosophy at University College, London, where his former teacher Augustus De Morgan was the inspiring and dynamic professor of mathematics and where there were no religious restrictions on appointments. Sylvester's first four papers, published in the *Philosophical Magazine*, were concerned with the optical properties of crystals and the motion of fluids and rigid bodies. They confirmed the promise that so many older contemporaries had already seen in him and in 1839, at the early age of 25, Sylvester was elected a Fellow of the Royal Society.

Years later, Sylvester described the trajectory of his life as having been *thrown from one side to the other* [of the Atlantic] *like a shuttlecock*. In 1841 he resigned from University College and crossed the Atlantic to take the post of professor of mathematics at the University of Virginia, the university founded by Thomas Jefferson some 20 years earlier to embody the ideals of the Enlightenment. But by the time Sylvester arrived—indeed, from its earliest years—the student body was permeated with violent disorder and with the racist, anti-semitic, and xenophobic attitudes of the Virginian gentlemen who would later fight a Civil War to defend their right to keep human beings of a different colour in slavery. A year before Sylvester arrived, the professor of law was murdered by a student. An English Jew was unlikely to make much headway. One student presciently wrote to his mother: 'I reckon our London cockney knows about as much about Virginian manners and character as a horse would about the differential calculus.' By March 1842, within four months of arrival, Sylvester had left the post under circumstances of which various conflicting accounts have been given; they apparently arose from the way in which he handled an abusive student, whom he may (or may not) have stabbed in self-defence with a sword-stick, and what he perceived as the faculty's lack of support for him. His departure was a setback to Jefferson's dream of the University of Virginia as a modern, enlightened, liberal institution which would attract cosmopolitan scholars of international repute.

A serious consequence of Sylvester's ignominious retreat from Virginia—for the world of mathematics, no less than personally—was the almost complete drying-up of his mathematical creativity. During this period he produced not a single paper, whereas he had written no fewer than five papers in the year just prior to his departure for the United States. Feeling disgraced and cut off from the world of mathematics, he returned to England in 1844 to begin yet another new life: the shuttlecock had been thrown back across the Atlantic.

In England again, Sylvester turned his mathematical skills to working as an actuary and contemplated a career in law. In 1846 he entered the Inner Temple, and in 1850 was called to the Bar. In the course of his legal studies he made the acquaintance of Arthur Cayley, a fellow refugee from mathematics into law. Cayley had been Senior Wrangler at Cambridge in 1842, as well as winner of the Smith's prize and a Fellow of Trinity College. The 50-year friendship of Cayley and Sylvester was one of the most significant relationships of British mathematical life in the nineteenth century. The striking contrast in their personalities—Sylvester proud, turbulent, and argumentative, Cayley calm, patient, and methodical—enabled each to complement and value the strengths of the other. It was thanks to Cayley that Sylvester regained his confidence and began to flourish once more as a creative mathematician, an act of support to which Sylvester paid fulsome tribute—for example, in 1851 when he referred in the course of a paper to 'Mr Cayley (to whom I am indebted for my restoration to the enjoyment of mathematical life) ...'

In 1855 Sylvester was appointed professor of mathematics at the Royal Military Academy in Woolwich. He was able to put some of the tempestuous past behind him, and devote himself to mathematical research and service to the mathematical community. During this period he wrote over 80 papers and made some of his greatest discoveries: his proof of 'Newton's rule' for the roots of equations, work on number theory, and the continuing exploration of invariant theory on which Cayley and he had embarked.

Sylvester also worked hard for the wider mathematical community. He was instrumental in the founding of a new journal, the *Quarterly Journal of Mathematics*, of which he remained an editor until 1878, and he also began to contribute a series of problems to the *Educational Times*; by the end he had contributed over three hundred of these. During this period, honours and responsibilities began to pour in from home and abroad: in 1861 Sylvester received a Royal Medal from the Royal Society; in 1863 he became Jakob Steiner's successor as mathematics correspondent to the French Academy of Sciences; from 1866 to 1868 he was president, in succession to the founding president Augustus De Morgan, of the recently formed London Mathematical Society, whose interests he actively promoted by recruiting new members; and in 1869 he presided over the mathematical and physical section of the British Association meeting at Exeter. The only group perhaps slightly short-changed in all this activity were his students at Woolwich; the teaching of elementary mathematics to bored young cadets was not a situation to bring the best out of Sylvester.

In 1870 he was superannuated—in only his mid-50s, and much against his will—because of a new government regulation for military establishments, and there was another lull in his mathematical productivity. In the same year he published a book, *The laws of verse*, in which he set out his view of the mathematical laws that govern the writing of poetry. Sylvester set great store by this book, although his enthusiasm was not always shared by others. Looking for a suitable place to carry on his work, he began to make approaches towards Oxford: in 1871 he asked Lincoln College about the possibility of a fellowship, but the college declined the opportunity. However, at the end of the decade, in 1880, Sylvester received an honorary doctorate of civil law from Oxford.

The time had come for the shuttlecock to be thrown across the Atlantic once more. A new university was being founded in Maryland, the Johns Hopkins University at Baltimore, and its president Daniel

Gilman was seeking the best scholars to give academic leadership. The position of professor of mathematics was offered to Sylvester. By this time he could virtually dictate his own terms, which were for a salary of $5000, to be paid in gold. From his arrival in 1876 until he left at the end of 1883, Sylvester worked with great vigour and considerable effectiveness to create a school of mathematical research—to build up, indeed, an American mathematical research community. Conditions were ideal: mathematically strong teaching and research assistants, the cream of America's most talented students, routine administrative tasks taken over by others, and above all a professor in his advanced prime whose enthusiasm seems to have been infectious and inspirational. As one of his students of this time observed:

Sylvester continued to revisit his poetic inspiration long after publication. In his own copy of *The laws of verse* (1870), in the Mathematical Institute in Oxford, the original text is drowned in an ocean of revisions and afterthoughts.

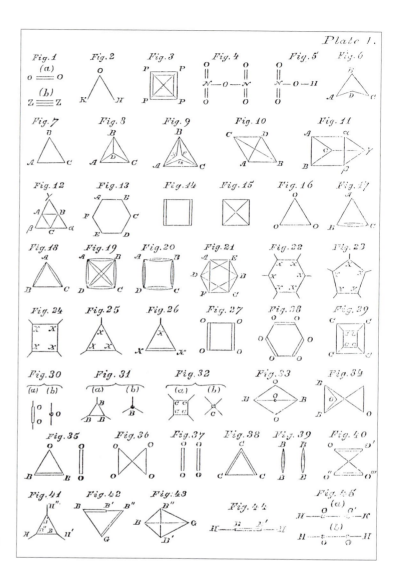

Sylvester's interest in the theory of graphs extended over 20 years. These diagrams are from a paper he contributed to the *American Journal of Mathematics* in 1878, the year that he founded the journal.

No young man of generous mind could stand before that superb grey head and hear those expositions of high and dearly-bought truths, testifying to a passionate devotion … without carrying away that which ever after must give to the pursuit of truth a new and deeper significance in his mind.

In 1878 he founded the first considerable mathematics research journal in the USA, the *American Journal of Mathematics*. This served both to promote and encourage American mathematics, and to attract international attention to the USA as a young country developing a research tradition of its own. How seriously Sylvester took his role may be seen from the explanation he gave to Felix Klein of his decision to move back to England at the end of 1883:

I did not consider that my mathematical erudition was sufficiently extensive nor the vigor of my mental constitution adequate to keep me abreast of the continually advancing tide of mathematical progress to that extent which ought to be expected from one on whom practically rests the responsibility of directing and molding the mathematical education of 55 million of one of the most intellectual races of men upon the face of the earth.

He had in fact already decided to leave Baltimore and return home, whatever the result of the Savilian election. He submitted his resignation to Johns Hopkins University in the autumn of 1883, well before the election took place, explaining to Cayley: 'If I am not appointed to Oxford I shall buy an annuity and study abroad or go into business with my small acquired capital the product of my seven years savings.' However, as it turned out, this contingency plan was unnecessary and the old shuttlecock was thrown back across the Atlantic for the last time. His vindication was complete; the stone that the builders rejected had become the head of the corner.

Only six years before, Sylvester had given a Commemoration Day address at Johns Hopkins in which he attacked the English universities as hitherto 'the monopoly of a party and the appanage of a sect', adding some pointed comments on those representatives of Oxford and Cambridge who had sought to bar non-Christians—particularly, as he bitterly remarked, those who 'professed the faith in which the founder of Christianity was educated'—from Chairs at the ancient universities. They showed

the blinding and blighting effect of early sectarian influences, one-sided culture, and narrow partisan connections, even on minds of a superior order, and on dispositions amiable by nature. There is a black drop of gall, a taint of intellectual rancour and animosity, which infects all it comes in contact with, more indelible, more difficult to wring out or efface, than that dread smear on Lady Macbeth's hand, which could "the multitudinous sea incarnadine." I doubt not that those who have taken this part, so prejudicial to their country's welfare, believe themselves to have been actuated by honest motives, just as I should not hesitate to admit that Torquemada was actuated by such and believed that he was doing a work acceptable to God when torturing heretics …

Realizing that this comparison with Lady Macbeth and the Spanish Inquisition might strike the listener as showing some personal interest in the issue, Sylvester amplified his position with one of the most telling testimonies we have to the effect of the old regulations barring Jews, Catholics, and others from full participation in the ancient universities:

If I speak with some warmth on this subject, it is because it is one that comes home to me—because I feel what irreparable loss of facilities for domestic and foreign study, for full mental development and the growth of productive power, I have suffered, what opportunities for usefulness been cut off from, under the effect of this oppressive monopoly, this baneful system of protection of such old standing and inveterate tenacity of existence.

Considering how deeply he felt the injustices and discrimination he had suffered, and how strongly he had fought such prejudice for over 60 years, Sylvester's final acclamation by the electors of the oldest mathematics Chair in England must have been sweet indeed.

The Savilian Chair

Sylvester delivered his inaugural lecture as Savilian Professor of Geometry on 12 December 1885, two years after his election. The printing of the lecture in the weekly journal *Nature* ensured wide publicity for his views on a variety of educational and other matters, as well as its ostensible subject: *On the method of reciprocants as containing an exhaustive theory of the singularities of curves*; a reciprocant is a differential invariant (an expression involving derivatives which remains unchanged if the variables are interchanged) which provides insights into the geometry of curves. The lecture was vintage Sylvester, down to the recital of a sonnet of his own composition, under the inspiration of discovering that a term involving b^4d did not appear where it might be expected to, in a table of reciprocantive protomorphs:

TO A MISSING MEMBER
Of a Family Group of Terms in an Algebraical Formula

Lone and discarded one! divorced by fate,
From thy wished-for fellows—whither art flown?
Where lingerest thou in thy bereaved estate,
Like some lost star, or buried meteor stone?
Thou mindst me much of that presumptuous one
Who loth, aught less than greatest, to be great,
From Heaven's immensity fell headlong down
To live forlorn, self-centred, desolate:
Or who, new Heraklid, hard exile bore,
Now buoyed by hope, now stretched on rack of fear,
Till throned Astræa, wafting to his ear
Works of dim portent through the Atlantic roar,
Bade him "the sanctuary of the Muse revere
And strew with flame the dust of Isis' shore."

The lecture gives a fascinating insight into Sylvester's style and approach, and into the qualities of humanity and creativity which he brought to Oxford. His words show a deep sense of the historical continuity of the mathematical community: Sylvester mentioned his predecessors Briggs, Wallis, Halley, and Smith—to whom especially warm respects were paid—as well as over two dozen other mathematicians living and dead. In a warm demonstration of collegiality, he paid tribute to at least four fellow mathematicians in the audience—to the

SAVILIAN PROFESSOR OF GEOMETRY: J. J. SYLVESTER, M.A., Hon. D.C.L.

The proposed grant from the University for the purchase of geometrical models not having yet received the necessary sanction, the previously announced course of Lectures "on Surfaces, illustrated by plaster, string, and card-board models," is postponed. The Lectures for the ensuing Term will be on Projective Reciprocants (otherwise called Differential Invariants) and their geometrical applications.

Intending members of the class are recommended to procure M. Halphen's " Thèse sur les Invariants Differentiels" (Paris: Gauthier Villars, 1878), and the Report of the Professor's Lectures on Reciprocants, delivered before the University in Hilary and Trinity Terms last, edited by Mr. James Hammond, in Part 3. Vol. 8 and the forthcoming parts of Vol. 9 of the American Journal of Mathematics (Trübner & Co., Ludgate Hill).

The days of lecture will, as previously announced, be Tuesday and Saturday, commencing Saturday, October 23, but in lieu of 11 A.M., as previously announced, at 4.30 P.M.

Professor Sylvester's lectures were too remote from the examination needs of students to attract a large undergraduate audience; from the *Oxford University Gazette*, 15 October 1886.

ELEMENTS

OF

PROJECTIVE GEOMETRY

BY

LUIGI CREMONA

LL.D. EDIN., FOR. MEMB. R. S. LOND., HON. F.R.S. EDIN.
HON. MEMB. CAMB. PHIL. SOC.
PROFESSOR OF MATHEMATICS IN THE UNIVERSITY OF ROME

TRANSLATED BY

CHARLES LEUDESDORF, M.A.

FELLOW OF PEMBROKE COLLEGE, OXFORD

Oxford

AT THE CLARENDON PRESS
1885

[*All rights reserved*]

A leading Italian text on projective geometry was translated in Oxford and published by the University Press, as part of Sylvester's promotion of modern geometry at Oxford.

point, indeed, of asking for a job for his friend and co-worker James Hammond. Sylvester also commented on his teaching strategy:

I think that I shall best discharge my duty to the University by selecting for the material of my work in the class-room any subject on which my thoughts may, for the time being, happen to be concentrated, not too alien to, or remote from, that which I am appointed to teach; and thus, by example, give lessons in the difficult art of mathematical thinking and reasoning—how to follow out familiar suggestions of analogy till they broaden and deepen into a fertilising stream of thought—how to discover errors and repair them, guided by faith in the existence and unity of that intellectual world which exists within us, and is at least as real as that with which we are environed.

Sylvester had a exalted vision of the future of mathematical studies in Oxford. He concluded with the hope that, with the help of all his new colleagues, his brother professors, and the mathematical tutors of Oxford, 'we could create such a School of Mathematics as might go some way at least to revive the old scientific renown of Oxford, and to light such a candle in England as, with God's grace, should never be put out'. It was not by accident that the first Jewish professor of mathematics at Oxford ended his inaugural lecture by recalling in these words an episode of deadly religious intolerance from Oxford's past. On 16 October 1555 Bishop Hugh Latimer, the Cambridge divine whose dying words Sylvester quoted here, was burned at the stake with Nicholas Ridley opposite Balliol College, on a spot in what is now Broad Street which many of his auditors would have passed on their way to the lecture.

Sylvester's ambitions for mathematics at Oxford were reasserted in subsequent lectures. In the following year, 1886, he gave an extended lecture course on the theory of reciprocants, during which he made very clear his view of the role of mathematics in the scheme of things. He vigorously refuted the perception, not unknown among other eminent Oxford professors, of mathematicians as mere calculators, and dismissed those who would undervalue the importance of mathematics:

Faraday, at the end of his experimental lectures, was accustomed to say—I have myself heard him do so—"We will now leave that to the calculators." So long as we are content to be regarded as mere calculators we shall be the Pariahs of the University, living here on sufferance, instead of being regarded, as is our right and privilege, as the real leaders and pioneers of thought in it.

Sylvester put considered thought and effort into how to raise the mathematical tone of the University. He carried on Henry Smith's tradition of inviting foreign mathematicians to visit Oxford, such as Leopold Kronecker in April 1884 and Henri Poincaré in June 1891. He also promoted the publication of mathematical texts in the new Clarendon Press series which Bartholomew Price, the Sedleian Professor of Natural Philosophy, had created as an academic division of Oxford University Press. As part of his plan to give modern geometrical methods more prominence in Oxford, Sylvester encouraged the translation and publication of the textbook on projective geometry by the distinguished Italian geometer Luigi Cremona, whom he had invited to Oxford in May 1884.

But there was a major stumbling block to the realization of Sylvester's vision for Oxford mathematics. It is clear that he saw his own pedagogical task as teaching how to reason and think math-

By the end of the nineteenth century, examinations were no longer popular communal events: a viva voce examination as depicted in *The Graphic* in 1882.

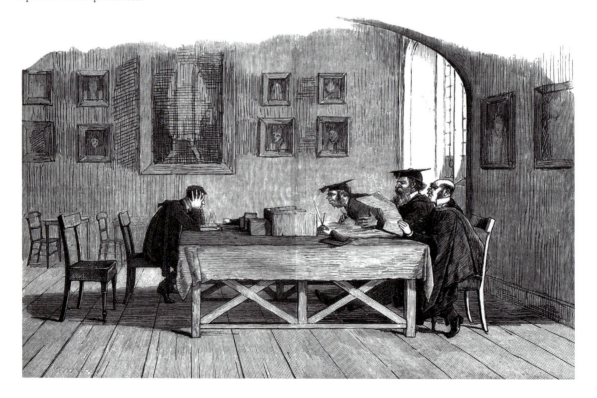

Sylvester's lecturing style

Sylvester's student A. S. Hathaway left this record of Sylvester's lectures at Johns Hopkins University in 1882–3; his Oxford students had much the same experiences a few months later:

I can see him now, with his white beard and few locks of grey hair, his forehead wrinkled o'er with thoughts, writing rapidly his figures and formulae on the board, sometimes explaining as he wrote while we, his listeners, caught the reflected sounds from the board. But stop, something is not right; he pauses, his hand goes to his forehead to help his thought; he goes over the work again, emphasizes the leading points, and finally discovers his difficulty. Perhaps it is some error in his figures, perhaps an oversight in the reasoning. Sometimes, however, the difficulty is not elucidated, and there is not much to the rest of the lecture. But at the next lecture we would hear of some new discovery that was the outcome of that difficulty, and of some article for the Journal, which he had begun. If a text-book had been taken up at the beginning, with the intention of following it, that text-book was most likely doomed to oblivion for the rest of the term, or until the class had been made listeners to every new thought and principle that had sprung from the laboratory of his mind, in consequence of that first difficulty. Other difficulties would soon appear, so that no text-book could last more than half the term.

W. P. Durfee attended Sylvester's lectures in the 1881–2 and 1882–3 sessions:

Sylvester began to lecture on the Theory of Numbers, and promised to follow Lejeune Dirichlet's book; he did so for, perhaps, six or eight lectures, when some discussion which came up led him off, and he interpolated lectures on the subject of frequency, and after some weeks interpolated something else in the midst of these. After some further interpolations he was led to the consideration of his Universal Algebra, and never finished any of the previous subjects. This finished the first year, and, although we had not received a systematic course of lectures on any subject, we had been led to take a living interest in several subjects, and, to my mind, were greatly gainers thereby.

ematically, to be achieved by leading students alongside his own mathematical creativity, on the topics he was currently researching. This strategy had worked very well at Johns Hopkins, and he may have hoped to recreate the atmosphere of Johns Hopkins by the banks of the Isis; but Oxford undergraduates, and indeed their teachers, had somewhat different aims in view from the graduate students of Baltimore. The effect was crisply described by Alexander Macfarlane, in a lecture delivered in 1902, not many years after Sylvester's death. Comparing Sylvester with his predecessor, Macfarlane noted, 'Smith had been a painstaking tutor; Sylvester could lecture only on his own researches, which was not popular in a place so wholly given over to examinations.'

Undergraduates wanting to take honours in mathematics at this time took two examinations: the First Public Examination, or Moderations, taken usually at the beginning of the second year, and the Second Public Examination, or Greats (a name also used for the classics degree), taken in the June of one's final year. The subjects for Moderations were algebra, trigonometry, plane geometry, elementary differential and integral calculus, and mechanics. For Greats, the same

Students and tutors in nineteenth-century Oxford were encouraged to stretch their bodies alongside their minds. The fencing prowess of Charles Price, mathematics tutor at Exeter College and one of the architects in the 1870s of the inter-college mathematics lectures scheme, apparently made an impression on undergraduates.

subjects were treated at a more advanced level, together with the calculus of variations, the calculus of finite differences, and the theory of chances, and in addition there was a range of topics in 'mixed mathematics'—mechanics, optics, astronomy, and some sections of Newton's *Principia*. By contrast with the Cambridge Tripos, where problem-solving was of the essence, at Oxford it was possible to pass with book-learning alone. The competitive examinations for the University mathematical scholarships, however, were closer to the Cambridge model in stressing problem-solving, and indeed in needing, for serious contestants, the assistance of a private mathematical coach.

The mathematics tutors of various colleges had collaborated since 1869 in inter-college lectures, in which tutors delivered lectures open to members of other colleges. While highly beneficial to the students, this tended to make the professors' lectures seem even less relevant. Sylvester wrote despairingly to his old friend President Gilman of Johns Hopkins University to describe how 'all the real power of influencing the studies of the place lies in the hands of the College Tutors':

It depends exclusively on the Tutors whether a Professor can get undergraduates to attend his lectures … Few of the Tutors recommend such attendance and many not merely discourage but actually prohibit it to those under their control on the ground that it will not pay in the examinations in the schools.

Sylvester found it hard to function effectively in such an atmosphere, and the fact that he found setting examinations very laborious further disenchanted him. He complained to a Cambridge friend:

Here in Oxford I am fortunate if I get an auditory of 6 persons. The Tutors Combination (carried to a great extent and now including all the principal colleges) practically reducing the Professors lectures to a Nullity … I begin to feel therefore very like the stalled Ox.

Sylvester's distress at his discovery that the examination culture of Oxford had such depressing consequences for his efforts to inspire mathematical creativity took on apocalyptic tones:

Entre nous this university except as a school of taste and elegant light literature is a magnificent sham. It seems to me that Mathematical science here is doomed and must eventually fall off like a withered branch from a Tree which derives no nutriment from its roots.

But Sylvester, nothing if not a fighter, sought ways of taking the battle into the enemy camp and encouraging the college mathematics tutors—whom he had already conceded were 'the real power of influencing the studies of the place'—to take an interest in creative mathematics. A traditional Victorian solution to a problem was to found a society. In the summer of 1888, Sylvester founded the Oxford Mathematical Society, as his old friend Arthur Cayley hastened to inform the readers of *Nature*:

Edwin Elliott on what mathematics teaching at Oxford was for.

E. B. Elliott, Fellow and mathematical tutor of The Queen's College from 1873 to 1892, first Waynflete Professor of Pure Mathematics from 1892, was more representative of the tone and spirit of Oxford mathematics in the late nineteenth century than the outsider J. J. Sylvester:

We thought, too exclusively probably, but I still hold soundly, that our business as teachers in a University was to educate, to assist young men, many of whom had no too strong a sense of responsibility, or care about the application of what mind they possessed, to make the best use of their powers. Those of mediocre capacity were as important to us, or we thought they ought to be, as the clever ones who gave us greatest intellectual satisfaction … We should have repudiated indignantly the notion, now not unpopular, that high honours in the Mathematical Schools should be reserved for those who prove inclination and the right sort of capacity to extend mathematical knowledge, and were in no hurry as teachers to encourage specialisation … The man who can turn from one problem to another, from prevailing once to grappling again, is qualifying for success in the battle of life. It is the sort of thing he will have to do if his future is to be one of business or responsible administration. If he is going to be a teacher much the same holds … If he is going to be one of the rare beings who advance knowledge, let us congratulate him on his future, and feel sure that some degree of thankfulness for firm grounding will be present in him when he occasionally grumbles at having wasted time.

Accordingly we laid great stress on elements, tried to secure adequate attention to a sufficient variety of subjects in the common ground of mathematicians, to Algebra, Geometry, the older analysis—we had not heard of the newer—the Mechanics of Solids and Fluids, Optics and Astronomy, fixing the range examined on in a subject (except perhaps in Geometry) rather low, but demanding a thoroughness of knowledge which was decidedly high. We did not encourage the introduction into papers of questions which could not be answered by those who knew well only the ordinary text-books and recognised lectures …

But how about research and original work under this famous system of yours, I can fancy someone saying. You do not seem to have promoted it much. Perhaps not! It had not yet occurred to people that systematic training for it was possible.

Wherever Dr Sylvester goes, there is sure to be mathematical activity; and the latest proof of this is the formation, during the last term at Oxford, of a Mathematical Society, which promises, we hear without surprise, to do much for the advancement of mathematical science there.

At its first meeting, on 9 June 1888, Sylvester was elected president, with two vice-presidents, William Esson, the highly regarded tutor who would eventually succeed him in the Savilian Chair, and the Sedleian professor Bartholomew Price; the secretary was Edwin Elliott. There were at first 36 members, from among mathematical graduates in Oxford, and 14 more were elected over the next five years.

The opportunities and encouragement offered by the Oxford Mathematical Society were initially and for some time of great value. It met six times a year, and provided for many years the only Oxford forum for discussion of research-level mathematics, where people with mathematical interests could come together regularly and learn of

> # THE OXFORD MATHEMATICAL SOCIETY.
> ### NOVEMBER, 1888.
> *[List of Original members].*
>
> **President.**
> J. J. SYLVESTER, M.A., F.R.S., Savilian Professor of Geometry.
>
> **Vice-Presidents.**
> REV. BARTHOLOMEW PRICE, M.A., F.R.S., Sedleian Professor of Natural Philosophy.
> WILLIAM ESSON, M.A., F.R.S., Merton College.
>
> **Secretary.**
> E. B. ELLIOTT, M.A., Queen's College.
>
> ---
>
> R. E. BAYNES, M.A., Ch. Ch.
> REV. J. BELLAMY, D.D., Vice-Chancellor, and President of St. John's.
> C. E. BICKMORE, M.A. New College.
> T. BOWMAN, M.A., Merton College.
> J. E. CAMPBELL, B.A., Hertford College.
> A. W. CAVE, M.A., Magdalen College.
> REV. R. H. CHARSLEY, M.A., St. Mary Hall.
> REV. J. CHEVALLIER, M.A., New College.
> R. B. CLIFTON, M.A., F.R.S., Professor of Experimental Philosophy.
> REV. H. DEANE, B.D., St. John's College.
> J. M. DYER, M.A., Worcester College.
> C. J FAULKNER, M.A., University College.
> H. T. GERRANS, M.A., Worcester College.
> J. GRIFFITHS, M.A., Jesus College.
> J. HAMMOND, M.A., Queen's College, Camb.
> C. E. HASELFOOT, B.A., Hertford College.
>
> REV. R. HARLEY, M.A., F.R.S.
> E. H. HAYES, M.A., New College.
> C. LEUDESDORF, M.A., Pembroke College.
> A. LODGE, M.A., St. John's College.
> D. B. MONRO, M.A., Provost of Oriel College.
> C. J. C. PRICE, M.A., Exeter College.
> REV. C. PRITCHARD, M.A., F.R.S., Savilian Professor of Astronomy.
> L. J. ROGERS, M.A., Balliol College.
> J. W. RUSSELL, M.A., Merton College.
> C. H. SAMPSON, M.A., B.N.C.
> REV. E. F. SAMPSON, M.A., Ch. Ch.
> A. L. SELBY, M.A., Merton College.
> REV. F. J. SMITH, M.A., Trinity College.
> E. J. STONE, M.A., F.R.S., Ch. Ch., Radcliffe Observer.
> REV. G. S. WARD, M.A., Hertford College.
> J. COOK WILSON, M.A., Oriel College.

The Oxford Mathematical Society, founded by J. J. Sylvester in 1888, provided the functions of a research seminar.

research developments. It had a rule that it should not publish any of the papers read to it, which both encouraged people to share work in progress with fewer inhibitions, and also enabled the writers of more developed work to submit their papers for publication to a grander society, such as the London Mathematical Society or the Royal Society. About three or four papers were read at each meeting—22 members and one visitor read 98 papers in the first five years. Sylvester topped the bill, with 17 papers. Elliott recalled one of these with wry amusement.

Sylvester once expatiated to us on a remarkable simplification of what is known as Hilbert's first proof of Gordan's theorem. He introduced the name of Cayley, and was I feel sure putting before us something that Cayley had just put before him. Anyhow, very shortly afterwards, Cayley, who was no plagiarist, published it as his own. Our young Society may be said to have had the honour of a joint paper from the two most noted of English mathematicians. But here comes the human interest. The joint paper of the two great men was all wrong, though we of course did not find it out. It was I fear no new thing for Sylvester to make hasty mistakes, but I believe that we are associated with the only occasion on record when Cayley was convicted of a serious howler.

The society's role in providing the equivalent of a research seminar became less vital as Oxford mathematics grew more institutionalized, as will be seen in the next chapter; the society eventually petered out, as an active organization, in the 1970s. So although it did not take off into national life and flourish in the way that two earlier societies, the London Mathematical Society and the Edinburgh Mathematical Society, have done, it played a valuable role in mathematical Oxford's transition towards twentieth-century professionalism. The *Oxford Magazine*, in a memorial article after Sylvester's death, commented:

Perhaps the most valuable direct outcome of his residence among us was the foundation of the Oxford Mathematical Society, a still vigorous institution of which he was originator and first president. The members of that Society, even more perhaps than the attendants at his formal lectures, have been impressed and excited to emulation as they have seen his always commanding face grow handsome with enthusiasm, and his eyes flash out irresistible fascination, while he expounded his latest discovery or brilliant anticipation.

Thus the spirit of Sylvester lived on—the overwhelming effect of his enthusiasm was attested by many contemporaries. As Thomas Archer Hirst ruefully recorded in his diary: 'Sylvester joined me at my dinner at the Club and by his excited and exciting mathematical talk caused me to pass a sleepless night.'

Sylvester's life-experiences left him keenly supportive of disadvantaged sectors of the community. In 1872, for example, his enthusiasm for extending education to the working classes led him to stand as a candidate for election to the London School Board. Like his friend Cayley, Sylvester had also for long supported the movement for women's education, and mathematical education in particular—indeed, Florence Nightingale took mathematics lessons from him in the 1840s. At Johns Hopkins University it was his intervention which in 1878 induced the authorities to admit Christine Ladd, the first woman in the United States to attend regular classes in mathematics as a graduate student. He secured her a fellowship for three years; she qualified for her Ph.D. but the University would not grant it until 1926, when she was 79 years old.

Sylvester came to Oxford in the 1880s at a period in which limited advances in the availability of mathematics education for women were being consolidated. Somerville Hall, later Somerville College—named after Mary Somerville, the distinguished translator of Laplace's *Mécanique céleste*—opened in 1879, and the first woman to study mathematics at Oxford, Georgina Nicholson, took her finals in 1883. As a woman, she could not take her degree, however, in much the same way as Sylvester had been ineligible for his Cambridge degree in 1837.

Mary Somerville, the most famous of British scientific ladies of the nineteenth century, was a keen advocate of women's mathematical and scientific education. Her *Physical geography* was an early textbook for Oxford's new School of Natural Science in the 1850s, perhaps the first book by a woman to be used in Oxford's science teaching.

MUSIC AND MATHEMATICS

YESTERDAY afternoon meeting at a friend's house a lady visitor to Oxford who was to sing that evening at one of the hebdomadal concerts in Balliol College, and the conversation happening to turn on the gifted mathematical lady Professor in the University of Stockholm, my thoughts shaped themselves, as I was walking home, into the following lines, which, if likely to interest any of your readers, I shall be happy to see appear in the world-wide-diffused columns of NATURE.

New College, November 15   J. J. SYLVESTER

SONNET

To a Young Lady about to sing at a Sunday Evening Concert in Balliol College

Fair maid ! whose voice calls Music from the skies
Weaving amidst pale glimpses of the moon
Tones with fresh hues of glowing fancy strewn
And soft as dew that falls from pitying eyes–
Let from their virgin fount those accents rise
That bid sad Philomel suspend her tune,
Thinking the lark doth chant his lay too soon–
Whose else that trill which with her own note vies !
To her whose star shines bright o'er Maelar lake
And thee who beautifi'st glad Isis' shore
Grant ! I one joint harmonious garland bind :
Thou canst with sounds our senses captive take–
She the true Muse, fond poets feigned of yore,
Strike Heaven's own lyre, Nature's o'erruling mind.

J. J. Sylvester's mathematical, musical, poetic, and Oxford interests intertwined in a sonnet, published in *Nature* in December 1886. His support for the mathematical education of women is seen in his reference to the Russian mathematician Sonya Kowalevskaya, 'her whose star shines bright o'er Maelar lake'. Comparing the printed form of Sylvester's sonnet with this manuscript text illustrates how his mind was constantly changing about the final form of his compositions.

Women were not admitted to Oxford degrees and full membership of the University until 1920; the equivalent progress in Cambridge did not happen until after another World War.

The first woman to achieve first class honours in mathematics at Oxford was not in fact an Oxford student at all, but from Cambridge, Grace Chisholm of Girton College. In May 1892, she and her friend Isabel Maddison sat their tripos examination at Cambridge, and then visited Oxford for a few days—possibly as the result of a dare—to sit the final honours School in mathematics there. Chisholm obtained the highest marks of any student at Oxford that year, whereas her position as a Wrangler was 'between 23 and 24'; this may indicate something of the relative mathematical training afforded undergraduates in the two universities: Oxford students were less intensively coached in problem-solving. Chisholm went on to take a Ph.D. under Felix Klein at Göttingen, leading to a distinguished mathematical career in partnership with her husband William Young.

Grace Chisholm was the second woman to take a Ph.D. at Göttingen. The first was Sonya Kowalevskaya in 1874, the brilliant Russian appointed in her mid-30s to a teaching position in mathematics at Stockholm University, at about the time of Sylvester's appointment to Oxford. In December 1886 Sylvester published a characteristic sonnet in *Nature* making reference to Kowalevskaya so obscurely that it is as well that his covering letter explained to whom he was referring.

Not all of Oxford's mathematicians were so supportive of women's educational aspirations. When in March 1884 Congregation debated the proposed admission of women to some of the honours Schools, one of the most vigorous speeches against the proposal was from Charles Dodgson, the former mathematical lecturer at Christ Church, on the grounds of the injury to health which this would inflict upon 'girl-undergraduates'. The response of mathematically inclined girl-undergraduates to such attitudes may be conjectured from the experiences of a French visitor to Oxford. In 1895, Jacques Bardoux was shown around Somerville College by a somewhat disdainful 'lady student'.

As we went up a delightful staircase she asked me about the most famous French geometrician. I muttered the name of M. Poincaré, but acknowledged my ignorance. My guide's disdain became more marked.

Ironically, it was the visit of the brilliant young French geometer to Oxford four years earlier, in June 1891, that seems to have provoked in

If he could read it, Felix Klein may have been surprised to receive a letter from J. J. Sylvester complaining about his handwriting: 'I find some difficulty in making out some of the words in your highly esteemed letter and would take it as a great favor if you could write a little more clearly…'

Shortly after Sylvester's death, the Royal Society instituted a medal in his memory, given every three years. Among the mathematicians associated with Oxford to have been awarded the Sylvester Medal are Augustus Love, G. H. Hardy, and Mary Cartwright.

Sylvester a sad realization that the mathematical baton was already being seized by a fresh generation. Writing to his old friend Arthur Cayley, Sylvester confessed, 'I expect Poincarre [sic] tomorrow and he will have rooms in College. I rather dread the encounter as there is so little in the way of Mathematics upon which I can hope to talk to him'.

It is true that Sylvester had always felt less comfortable with geometry than with algebra, a feeling which his appointment to the Savilian Chair of Geometry did little to dispel. But he was keenly interested in geometry education and in 1891 was president of the Association for the Improvement of Geometrical Teaching. This body, forerunner of the Mathematical Association, was working on a document critical of Oxford's pass examination in geometry. As a member of the Faculty Board, as well as the Association's president, Sylvester felt awkward and wrote to the secretary of the Association that although he agreed with the Association's stance he would be placing himself in an anomalous position by saying so, and offering to resign in favour of 'Dr Hirst or Henrici or Esson'. Sylvester did indeed resign the presidency of the Association, but as much through failing health as because of a conflict of interest in seeking to reform geometry education at Oxford.

In these last years, Sylvester's creative interest in mathematics continued with spasmodic vigour, while his body gradually weakened. A particular interest was the theory of graphs—the word itself, in this mathematical sense, is due to Sylvester and W. K. Clifford around 1878. During the years 1889–91, there was an intense correspondence between Sylvester, the Danish mathematician Julius Petersen, and the Germans David Hilbert and Felix Klein, concerning invariants and the factorization of graphs. In 1891, Petersen wrote one of the first major papers on graph theory, acknowledging his great debt to Sylvester. Klein and Hilbert were fascinated by Sylvester's use of partitions and generating functions in enumeration problems, and letters passed between the four men with great frequency—although not always trouble-free. On Christmas Day 1889, Klein wrote irritably to Hilbert: 'Sylvester is too irregular, altogether a "genius"; in matters of business there is no way of getting anywhere with him'.

Sylvester's own letters testify to the growing strain on his health:

Since that last mental effort ... the trouble in my eyes and certain causes of mental inquietude have incapacitated me for the last five months and more for all mathematical work of investigation—and had it not been for Mr Hammond's aid, I doubt whether I could even have prepared the ordinary lectures for my class. I hope gradually to be able to recover my mental tone and resume work.

James Hammond had worked with Sylvester on reciprocants in the mid-1880s, and was still helping him in secretarial and research capacities, besides developing his own career. Despite his long and close association with Sylvester—and despite the warm tribute Sylvester paid

Hammond during his inaugural lecture, when he said that apart from Cayley 'there is no one I can think of with whom I ever have conversed, from my intercourse with whom I have derived more benefit'—it was a more established Oxford figure, William Esson, who was appointed Sylvester's formal deputy when increasing eye trouble forced Sylvester's retirement from active duties in 1894, and who succeeded him after his death in 1897.

Further reading

There is much of relevance to Oxford in Karen Parshall's *James Joseph Sylvester: life and work in letters*, Clarendon Press, Oxford, 1998. Apart from this, the fullest accounts of Sylvester's life and work appear in standard sources such as John North's article in the *Dictionary of scientific biography* and H. F. Baker's 'Biographical notice' in *The collected mathematical papers of James Joseph Sylvester*, Vol. IV, Cambridge University Press, 1912, pp. xv–xxxvii.

The most measured account of Sylvester's earlier tribulations in Virginia is given in Lewis S. Feuer, 'America's first Jewish professor: James Joseph Sylvester at the University of Virginia', *American Jewish Archives* **36** (1984), 152–201. His later period in the United States is well covered in Karen Parshall and David E. Rowe, *The emergence of the American mathematical research community 1876–1900: J. J. Sylvester, Felix Klein, and E. H. Moore*, American Mathematical Society, Providence, and London Mathematical Society, London, 1994.

CHAPTER 14

The twentieth century

Margaret E. Rayner

Oxford mathematicians were slow to show an interest in twentieth-century mathematics, and the University gave them very little encouragement to change. The University was confidently proud of its undergraduate course *Literae Humaniores*, or 'Greats', a course in classical languages and literature, together with philosophy, and was sufficiently convinced of the importance of the training given by a study of classical languages to insist that all undergraduates, in every discipline, should pass examinations in Latin and Greek. It was left to Cambridge to excel in mathematics; the challenge and the high standard of the tripos examination were world famous, and the most able young British mathematicians were attracted to Cambridge to compete in the race for success. Many of the cleverest then stayed on to teach the following generations. It was a formula for excellence. Oxford appears to have started the century believing what many sixth-formers still say, that it is Oxford for arts and Cambridge for science.

In 1900, there were three established Chairs of mathematics in Oxford: the Savilian Chair of Geometry, the Sedleian Chair of Natural Philosophy, and the recently founded Waynflete Chair of Pure Mathematics. The professors were required to keep residence in Oxford for prescribed periods and to give a certain number of lectures and classes, but there was no formal requirement to do research. Professors were appointed for life, and it was unusual for them to retire. The appointments were made by the University, but each professor was attached to a college. Stipends came from benefactions administered by the University and augmented by college funds. No office or accommodation was provided by the University, and married professors were known to emerge from their homes in North Oxford only to give their lectures and classes. Colleges varied in what rooms and accommodation they gave to professorial Fellows.

At the turn of the century, two of the three professors, William Esson and Edwin Elliott, were Oxford men and the other, Augustus Love, came from Cambridge. All had been fairly recently appointed.

William Esson held the Savilian Chair of Geometry from 1897, after a career as mathematics tutor at Merton College and three years as

G. H. Hardy, Savilian Professor of Geometry from 1920 to 1931, in his rooms at New College.

William Esson succeeded J. J. Sylvester as Savilian Professor of Geometry in 1897.

Edwin Elliott, first Waynflete Professor of Pure Mathematics.

deputy professor when failing eyesight prevented Sylvester (then the holder of the Chair) from discharging his duties. As a young tutor, he had collaborated with an experimental chemist, Vernon Harcourt, to establish a law of chemical reaction; after that work, Esson turned to University business and the reform of teaching methods. He had taken the lead in setting up the Mathematical Association of College Lecturers, the forerunner of the formal faculty structure, and also replaced the expensive private coaching for some undergraduates by a system of tutorials for all by college Fellows, a pattern which still continues. But by the time William Esson became a professor, his interests had largely moved to finance, both in the University and in college. Although his Chair came with a fellowship at New College, he remained the Estates Bursar at his old college, Merton. He held the Chair until his death in 1916 at the age of 77.

Edwin Bailey Elliott had been elected to the Waynflete Chair in 1892. He was very much an Oxford man, having been a schoolboy at Magdalen College School before going on to a very successful undergraduate career at Magdalen College. In 1873, he obtained a first in finals, and immediately went as Fellow and mathematical tutor at The Queen's College. Nineteen years later, he was elected to the professorship and held the Chair for 29 years, retiring in 1921 at the age of 70. His mathematical interests were largely in algebra, and he wrote a textbook, well known in its time, on quantics. He had no sympathy with 'foreign modern' symbolic methods, and little patience with unsupported speculation; it is not surprising that he did not encourage research by young mathematicians. He also made an unsympathetic impression on undergraduates, one of whom, William Ferrar, wrote that:

Elliott, the great man on Algebra & all its developments, a man who has written books which have put the works of his rivals on the bookshelves, is the worst lecturer who ever picked up chalk. I have spent hours on him last term & not got a scrap of good from him.

Elliott was deeply involved in University business, and his advice was frequently sought on financial matters. His colleague Theodore Chaundy of Christ Church wrote that in such matters Elliott's strength lay in exactness of detail rather than in boldness of conception. It is possible that this description also applied to his mathematical work.

The most significant research in mathematics was being done by the third professor, Augustus Edward Hough Love. In 1899, he had come from Cambridge to take up the Sedleian Chair of Natural Philosophy. He remained professor for 41 years, until he died in 1940 in his late 70s. He was a geophysicist as well as a mathematician and his interests were in mechanics, continuum mechanics, and electrodynamics. His work was extremely well known outside Oxford, and his book on elasticity has remained a standard work of reference. One of his colleagues

The moustache of Augustus Love, Sedleian Professor of Natural Philosophy from 1899 to 1940, was widely admired. In later years a student (C. K. Thornhill) sketched it during a lecture, shortly before the Second World War when Love was in his mid-70s.

described him as a typical representative of the characteristic English school of applied mathematics. His lectures were extremely popular with students for their clarity and intelligibility and, in his early years in Oxford, he attracted good research students to work with him. His whimsicality of manner and appearance endeared him to many friends. His magnificent moustache was described by a friend as 'charmingly reminiscent of a frozen waterfall'.

College tutors

At the beginning of the century, there were seven tutorial Fellows in mathematics in addition to the three professors. These Fellows held permanent appointments and had rooms in college, even if they lived out. They carried the responsibility for the individual instruction of undergraduates and assisted the professors in setting and marking examinations. All the tutors at that time were Oxford men; Cambridge graduates were not tempted here. There was a firmly established tradition that did not fade until after the Second World War that only an Oxford graduate *could* teach the Oxford syllabus.

Part of a tutor's responsibility was to select the mathematical scholars for his college, usually by means of an examination he had helped to set. Colleges worked together in small groups for this exercise. Candidates came up to Oxford to write the papers and to be interviewed. The commoners (students who did not gain awards) were admitted by the head of the college, usually without reference to any tutor; even so, it is said, some commoners did better than some scholars in finals.

When a college needed a new mathematics tutor, other colleges would be asked to suggest the name of a very bright undergraduate who had won one of the University mathematical scholarships and was clearly going to get a first in finals. Research potential, or a willingness to pursue research, were not highly rated: an ability to do the examination questions and to fit into college life were much more important. The predominance of Oxford graduates amongst tutors, and the lack of emphasis on research, without doubt held back syllabus changes and other innovations.

There were also college lecturers who came and went, depending on the mathematical teaching needs of the college. It was not unusual for a tutorial Fellow at one college to be a lecturer for a few years at another. There was probably also a certain amount of informally arranged private teaching by people who held no appointment in the University or the colleges. Colleges frequently admitted students in mathematics without having either tutor or lecturer within the college. The mathematical activity was thus scattered throughout the colleges and the town. There was no central department in which professors could give their lectures and classes, and they had to persuade colleges to lend lecture rooms and dining halls.

Tutors also lectured in their own colleges, and could open these lectures to undergraduates in other colleges. Tutors—holding no University appointments in those days—were under no obligation to give these inter-collegiate lectures, but they had, some years earlier, come together to form the Mathematical Association of College Lecturers in order to provide a comprehensive programme of lectures for all undergraduates. The Association met once a term, had lunch, and got down to business in the afternoon. It was much later, around 1915, that these informal arrangements were replaced by a formal faculty structure. The tradition of a termly lunch continued after that, but was later abandoned.

There were other social occasions. The mathematicians met on one or two evenings a term as the Oxford Mathematical and Physical Society, founded in 1888 by James Joseph Sylvester (see Chapter 13). On these occasions, visitors were sometimes invited from Cambridge and a speaker, often from Oxford, gave a paper. These meetings provided rare opportunities for professors, tutors, and research students to discuss advances in mathematics. However, there were very few graduate students at this time. The introduction of the D.Phil. degree in the 1920s had attracted some students, and a doctorate gradually became an essential prerequisite for an academic career, but many older mathematicians by-passed it.

For the first few years of the century, the number of students taking the final honours School in mathematics did not exceed 20: far more students studied mathematical Moderations as the first part of a degree course, before going on to science or some other subject for finals. However, it was already fairly uncommon for a student to take finals in mathematics after taking another subject in the First Public Examination. College tutors were responsible for teaching all the mathematics needed by the college, some of it at a very low level for the simplified form of mathematical Moderations available for intending scientists.

Women had studied mathematics in rather small numbers from the late nineteenth century; one of these, Isabel Duncan, who matriculated at Lady Margaret Hall in 1892, was to be the mother of the future Oxford topologist Henry Whitehead. Throughout this time there were very few women students resident in Oxford, since women taking the Oxford finals examinations were permitted to live elsewhere; in particular, a significant group of them were entered by Royal Holloway College.

It was not until 1923 that a woman who had been trained in Oxford, Mary Cartwright of St Hugh's College, graduated with first class honours. After a spell of schoolteaching, she returned to Oxford in

G. H. Hardy and Mary Cartwright at a seminar in the 1930s.

(below) Mary Cartwright in later life.

1927 to carry out research in mathematical analysis under the supervision of G. H. Hardy, gaining her D.Phil. in 1930. In 1934 she became Fellow and lecturer in mathematics at Girton College, Cambridge, where she remained for the rest of her career, later becoming Mistress of Girton. One of the most distinguished analysts of recent times, Dame Mary was elected to a fellowship of the Royal Society in 1947 (until 1994 the only woman mathematician to be so honoured) and was awarded the Society's prestigious Sylvester Medal in 1964.

The first woman to be awarded a D.Sc. at Oxford (in any subject) was Dorothy Wrinch in 1929, a Cambridge-trained mathematician who was married to John Nicholson, Fellow and director of studies in mathematics at Balliol. Wrinch lectured in mathematics at Lady Margaret Hall and other women's colleges throughout much of the inter-war period, but never achieved a permanent college appointment. Latterly her interests moved increasingly towards mathematical biology, and in 1939 she left Oxford for the United States, following a controversy about protein structures with a rising star of Oxford science, the crystallographer Dorothy Crowfoot Hodgkin.

Planning for change

When Edwin Elliott retired from the Waynflete Chair in 1921, he was succeeded by Arthur Lee Dixon, also an Oxford graduate, who had held a tutorship at Merton College. Dixon made very few mathematical innovations, but he was remembered by one undergraduate of his time as 'knowing some very pretty stuff' in nineteenth-century geometry. He was the last mathematician to be elected with life tenure to an Oxford Chair; in fact, he stayed for 22 years and retired in 1945.

If Arthur Dixon's election did not produce mathematical leadership, that of Godfrey Harold Hardy certainly did. Oxford succeeded in attracting Hardy from Cambridge to the Savilian Chair of Geometry, vacant since the death of William Esson. This brought to Oxford one of the outstanding mathematicians of the time—some would say *the* outstanding mathematician. Hardy had initially been inspired by his Cambridge teacher Augustus Love (by whom, he said, 'my eyes were first opened') and his work in analysis had a profound influence on the subsequent development of mathematics, particularly in the United Kingdom; indeed, he made the theory of functions (created by the great European mathematicians) into one of the chief mathematical studies in English universities.

Hardy arrived in Oxford in early 1920 at the top of his creative powers. During the 11 happy and productive years which followed, he was to have an enormous influence on every aspect of Oxford's mathematical life. In addition to his superb mathematical ability, Hardy had great charm and generosity which helped him to generate in others a

Increased international contacts were a feature of the Hardy years. The Hungarian mathematician George Pólya came to Oxford in 1924, on Hardy's recommendation, as the first international Rockefeller Fellow; their collaboration led to a text on *Inequalities* (1934).

INEQUALITIES

By

G. H. HARDY
J. E. LITTLEWOOD
G. PÓLYA

CAMBRIDGE
AT THE UNIVERSITY PRESS
1934

In 1928–9 the Princeton mathematician Oswald Veblen spent a year in Oxford, in an exchange with G. H. Hardy. A product of this visit was a collaboration between Veblen and an Oxford research student, Henry Whitehead, resulting in *The foundations of differential geometry* (1932).

Cambridge Tracts in Mathematics
and Mathematical Physics

General Editors
P. HALL, F.R.S., AND F. SMITHIES, Ph.D.

No. 29

THE FOUNDATIONS OF
DIFFERENTIAL GEOMETRY

BY
OSWALD VEBLEN
AND
J. H. C. WHITEHEAD

CAMBRIDGE UNIVERSITY PRESS

lively enthusiasm for mathematics. He encouraged college tutors to collaborate in their research where they had previously worked in isolation, and he became the centre of a group of committed and clever young mathematicians, including E. C. Titchmarsh, Mary Cartwright, L. S. Bosanquet, Gertrude Stanley, Alexander Oppenheim, and Edward Wright. Hardy was engagingly eccentric, never used a watch or a fountain pen, avoided using a telephone whenever possible, and usually corresponded by prepaid telegrams and postcards. He had a passionate interest in cricket, an enthusiasm he bequeathed to some of his successors at Oxford (see Chapter 15).

G. H. Hardy's influence went well beyond mathematics itself. He was convinced that the reputation and study of mathematics at Oxford would remain at a very low ebb until colleges were persuaded to appoint more mathematics Fellows, and thereby create more career opportunities for the most promising students. In 1930 he wrote 'Mathematicians are reasonably cheap, but they cannot be had for nothing.' Hardy also made a strongly argued case to the University for its providing an institute—a meeting place for professors, other faculty members, and graduate students and a place in which lectures could be given. This was a development taking place in a number of universities in Europe and America, and of which Hardy had some experience. Additional support for the idea was given by Oswald Veblen of Princeton University, who had exchanged places with Hardy for a year, in 1928–9.

The University was persuaded by the sub-faculty to accept the *idea* of an institute, but the size, location, and shape of what emerged were disappointing. At that time, an extension was being planned for the Radcliffe Science Library, and it was decided that the proposed institute should occupy a part of the new library wing, adjacent to the section containing the mathematical collection. The Institute consisted of a room for each professor, one for a secretary, and one for lectures. It was a good deal less than the mathematicians had asked for and, in any case, the rooms would no longer be available when the space was needed for books by the Radcliffe Science Library.

If Hardy's plea in 1930 for an institute was partly successful, his plea for more college Fellows was less so, and the number crept up only from eight to ten between 1930 and 1939. The number of male undergraduates then dropped very sharply during the Second World War, and by the end of the War more women than men were taking finals.

Hardy emphasized the purity of pure mathematics, and proclaimed on several occasions, not least during his inaugural lecture on number theory in 1920, that his kind of mathematics had no practical consequences for the wider community. To this extent his presence in Oxford encouraged a polarization, and with Love no longer giving a clear direction there were few voices to promote the applied side of mathematics.

G. H. Hardy in 1930, the year in which he made a public plea for the formation of a mathematical institute at Oxford.

Edward Arthur Milne, first holder of the Rouse Ball Chair of Applied Mathematics, did much to strengthen the applied side of Oxford mathematics.

In the mid-1920s an opportunity presented itself to restore the balance. The Cambridge mathematician and historian of mathematics W. W. Rouse Ball left £25 000 to Oxford (besides a rather larger legacy to Cambridge) for the establishment of a Chair of mathematics which he hoped would not neglect its historical and philosophical aspects. After some discussion within the University it was decided to advertise it as a Chair in mathematical physics, to which Edward Arthur Milne was appointed. A former student of G. H. Hardy at Trinity College, Cambridge, Milne had been tipped by Hardy as an exceptionally promising pure mathematician, but his experience with the Ministry of Munitions in the First World War had diverted him to mathematical physics and the theory of astrophysics. At Oxford Milne inspired a group of brilliant young mathematicians interested in his own research areas of stellar structure and kinematic relativity, besides continuing the style of international contacts that Hardy had fostered.

In 1931 Hardy returned to Cambridge to assume the Sadleirian Chair of Pure Mathematics, vacated on the resignation of E. W. Hobson; in Cambridge, he said, he could live in Trinity College for the rest of his life. His successor as Savilian professor was his former pupil Edward Titchmarsh, a quieter personality who did not attempt to sustain Hardy's public flair or internationalism. A distinguished Oxford analyst in the Hardy tradition, Titchmarsh was particularly known for his texts on the theory of functions and the Riemann zeta-function. He continued in the Savilian Chair for over 30 years, being succeeded by Michael Atiyah after his death in 1963.

One effect of Hardy's return to Cambridge was that there was no-one of influence in Oxford to speak for refugee mathematicians from Germany during the 1930s. While several other universities saw the opportunity to assist refugees while bolstering their own academic profile, in which Hardy was a moving force, nothing happened at Oxford beyond abortive negotiations for the great algebraist Emmy Noether to come to Somerville College on her dismissal from Göttingen; she went to the USA instead.

Another Oxford product, the brilliant topologist Henry Whitehead read mathematics at Balliol College, and later returned to continue research before going to study with Oswald Veblen at Princeton University. In 1933, he succeeded his former tutor at Balliol; his tutor, John Nicholson, had been a student of Whitehead's uncle, the mathematician–philosopher Alfred North Whitehead. Henry Whitehead was a brilliant, but not a patient, man. There is a vivid account of his tutorials by one of his students:

At our first meeting, he told me that there were three ways of learning: his tutorials, lectures and the work I did on my own (including use of libraries and discussion with others). He added that the value of these was in the reverse order. I had an hour with him every week ... We

The Oxford University Invariant Society

Henry Whitehead introduced many students to a new world of mathematics by founding the Invariant Society with two Balliol students, Graham Higman and Jack de Wet; the word *Invariant* was chosen at random from the first book that came to hand from Whitehead's bookshelf. To this society, which sometimes met in his own college room in Balliol, he invited distinguished mathematicians to give talks. At the first meeting, in Hilary Term 1936, G. H. Hardy came over from Cambridge to talk on 'Round numbers'. The Invariant Society still flourishes, and meets several times each term to engage in a variety of mathematical activities, amongst which is the original objective of listening to outstanding contemporary mathematicians.

OXFORD UNIVERSITY

INVARIANT SOCIETY

———

HILARY TERM, 1941.

President : J. H. C. WHITEHEAD, Esq.
Chairman : D. C. PACK (New College).
Secretary : E. J. F. PRIMROSE (St. John's).
Treasurer : A. W. BABISTER (New College).
Women's Secretary : A. P. COBBE (Somerville).

In the event of an air-raid warning meetings
will be held as arranged.

Oxford University **Invariant Society**

Presents

Professor Benoit Mandelbrot

Negative Dimension, Lacunarity and Other New
Developments in Fractal Geometry.

Maths Institute, Tuesday 3rd week. 8:15 pm.

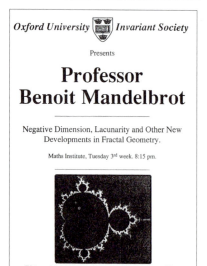

Tickets are available for non-members at £2.50 each, from Juliette White (Worcester College).
Cheques should be made payable to the Oxford University Invariant Society.
Some tickets may possibly be available on the door.

INVARIANTS OF
QUADRATIC DIFFERENTIAL
FORMS

BY

OSWALD VEBLEN

Henry B. Fine Professor of Mathematics,
Princeton University

CAMBRIDGE
AT THE UNIVERSITY PRESS
1933

Hardy manuscript on round numbers.
Title-page of Veblen's book on invariants.

Invariant Society card from 1941.

Poster advertising a lecture by Benoit
Mandelbrot in 1994.

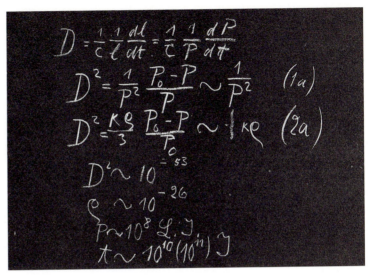

Albert Einstein visited Oxford in 1931 to receive an honorary degree. He delivered three lectures on relativity at Rhodes House; a blackboard from the second lecture, given on 16 May, is preserved at the Museum of the History of Science.

usually sat side by side on a sofa, with the floor as depository for a growing heap of paper. I soon learnt that he was not keen to solve problems which were giving me difficulty. He either talked in a general way about the issues involved, or generalized the problem, or—most often—discussed his own research in algebra.

Change is achieved

The outbreak of the Second World War posed severe problems for Oxford mathematics teaching, even more than for other subjects, since skilled mathematicians were especially needed for war service. Two people, Ida Busbridge of St Hugh's College and William Ferrar at Hertford, sustained the mathematical work of the University through the war by carrying heavy loads of lecturing, teaching, and examining. Ferrar was well known for his textbooks, and it was said that whenever a new topic (such as convergence) was introduced into the syllabus, Ferrar was told to produce a book on it. Ida Busbridge, looking after the tutorial arrangements for all of the women's colleges, had an especially heavy load, as women formed a much higher proportion of the undergraduate population over this period, although she was afforded some assistance by Kathleen Sarginson, who after the war was appointed Somerville's first Fellow and tutor in mathematics.

In the immediate post-war years the number of undergraduates quickly rose and soon exceeded the pre-war number. The quality of these students was very high; for example, 19 out of the 46 finalists in 1947 obtained firsts. With much larger numbers reading mathematical Moderations, it was no longer possible for lectures to be held in colleges. They were frequently given in the Examination Schools (with

Ida Busbridge (left) was the first woman mathematics Fellow of an Oxford College (St Hugh's). In her later years she developed an interest in the history of Oxford mathematics. In this 1936 photograph, the physicist and astronomer Madge Adam is on the right.

Sydney Chapman was Love's successor as Sedleian Professor of Natural Philosophy from 1946 to 1953.

postage-stamp-sized blackboards) or in lecture theatres in laboratories at highly unpopular times of day. There were still only ten mathematics Fellows, who now included one woman, Ida Busbridge. The applied mathematics provision was strengthened with the election in 1946 of the geophysicist Sydney Chapman to the Sedleian Chair, vacant since Love's death in the early war years, although in the event he spent half of his time at the University of Alaska.

By this time, the mathematics sub-faculty was again looking at the need for a new mathematical institute. The matter was urgent, because the curators of the Radcliffe Science Library needed back all the space temporarily lent to the mathematicians. In any case, the existing Institute was by now totally inadequate to accommodate the rapidly increasing new mathematical activity—much of which was generated by Henry Whitehead, who had been elected to the Waynflete Chair in 1947.

The enormous importance of Henry Whitehead's work in topology, and his ability as a leader of a research group, attracted round him a large number of extremely able graduate students and academic visitors. The old Mathematical Institute was much too small for this University-based mathematical activity, but there was still no prospect of a building designed especially for mathematicians. Instead, they were moved across the road in 1952 into 10 Parks Road, a large red brick house of the standard North Oxford type, which contained some small lecture rooms and a common room, in addition to rooms for professors. It was a big step forward, however, in that at least some of the graduate students could have desks there. By this time, the number of graduate students was rapidly growing as grants became available for doctoral research.

By the early 1960s, the Mathematical Institute was overflowing with seminars, graduate students, and academic visitors, and there was a great need for a new building. By some quick thinking, Jack Thompson of Wadham College had put in an early claim for a site in St Giles, thereby taking the building of the Institute out of a long waiting list. In 1966, 36 years after Hardy had asked for an institute, the mathematicians moved into a new purpose-designed building.

The work of planning and overseeing the move was led by Charles Coulson. Coulson had been elected to the Rouse Ball Chair in 1951 from a Chair of theoretical physics at King's College, London, and became well known, both inside and outside Oxford, for his research and textbooks on theoretical chemistry and mathematics, and also for his religious publications and talks. Coulson's own research in quantum theory and theoretical chemistry drew to Oxford many research workers for whom space was urgently needed in the new building.

The expansion in the number of college tutors went in parallel with the growing reputation of Oxford mathematics. As a consequence of

Charles Coulson lecturing to first-year undergraduates in 1972, in the Mathematical Institute that he created.

The search for a Mathematical Institute

(top right) The Radcliffe Science Library (before 1952).

(top left) 10 Parks Road (1952–66).

(middle) St Giles, before the building of the Mathematical Institute.

(bottom) The Mathematical Institute, opened in 1966.

Edward Titchmarsh, Savilian Professor of Geometry from 1932 to 1963, wrote several influential books.

THE THEORY OF THE
RIEMANN
ZETA-FUNCTION

BY

E. C. TITCHMARSH

F.R.S.

SAVILIAN PROFESSOR OF GEOMETRY IN THE
UNIVERSITY OF OXFORD

OXFORD
AT THE CLARENDON PRESS
1951

the 1963 Robbins Report on university expansion, the University was invited to increase the number of undergraduate places, and to provide graduate courses that would train academic staff for the new institutions of higher education. With these increases came—at last—the funds to appoint new college Fellows. Between 1954 and 1961, the number of Fellows had almost doubled from ten to 19, and in the next four years, there was an additional 50 per cent increase. Apart from the sudden rise in undergraduate numbers, the growing population of graduate students required supervisors in an ever-expanding range of mathematical fields.

Since 1960, the mathematics syllabus has been in a state of constant change. It is said that before the war a syllabus lasted for five years. Today, a syllabus is barely in print before new proposals come forward. Some of the post-war changes are the result of changes in sixth-form syllabuses, while others were the consequence of the wish to include topics of particular interest to new faculty members, more of whom were not themselves Oxford graduates. Another reason for extensive syllabus revision was the greatly increased career opportunities for those who studied mathematics, particularly in the areas of statistics, operations research, and computation.

All these newer applications put pressure on the faculty to include specialized elements in the syllabus and, in particular, led to the establishment of the Joint School of Mathematics and Computation and to a School of Computation. A move for *less* specialized courses came from a national report encouraging inter-disciplinary degrees, and this led to the introduction of the Joint School of Mathematics and Philosophy.

The total number of undergraduates in the four courses has now risen to about 250 each year, and many colleges now have two or more tutors. Compared with the three named Chairs in 1900, there are now seven, as well as several *ad hominem* professorships. Nevertheless, it could be argued that mathematics is less well provided for now than in 1900, in view of the massive increase in undergraduate and graduate numbers. Certainly numbers have increased, but it is in the area of research, both in output and distinction, that Oxford mathematics has changed most. The University can now claim that it has a mathematics faculty that attracts scholars from all over the world to take appointments; but even now the Institute cannot provide enough space for all the academic visitors who wish to come.

It would be unwise to claim that the change in the reputation of Oxford mathematics had been brought about solely by the efforts of G. H. Hardy in 1930. There have been many other factors outside the University—political, financial, and social. But there is no doubt that Hardy put forward a plan that enabled Oxford University and its mathematics faculty to take the greatest advantage of the opportunities which arose after the Second World War.

Computing in Oxford

The Computing Laboratory was established in 1957, under the directorship of the numerical analyst, Leslie Fox, to assist scientific research in the University with the assistance of a 'digital computer', and to provide postgraduate supervision in numerical analysis and the theory of computation, as well as undergraduate teaching.

The theory of computation was reinforced by the creation less than ten years later of the Programming Research Unit, with Christopher Strachey at its head. However, the Unit was externally funded, its staff were untenured, and the University would make no commitment to adopt the Unit when its external funding expired. As Strachey was firmly against an undergraduate degree in computer science, the Unit concentrated on research, attracting postgraduate students in order, in his words, 'to introduce a degree of scientific and intellectual discipline into the subject of computing'.

Strachey died in 1975. With the arrival of Tony Hoare in 1977 as his successor, the teaching of computer science took on a higher profile, and by 1994 the University had added the single honours School in Computation to the previously established Joint Schools of Mathematics and Computation and of Engineering and Computing Science. Earlier concerns about combining the academic and industrial viewpoints, so important to Strachey, have led to postgraduate programmes for industrial practitioners, to build on the continued interest at Oxford in the theory of computation.

Statistics in Oxford

As late as 1978 Sir Maurice Kendall said that Oxford 'has always been rather stony ground for the development of statistics ever since Florence Nightingale tried to found a Chair there a century ago'. From 1891 to 1922 Oxford possessed, in the person of the economist Francis Edgeworth, a professor of high reputation as a statistician; but statistics did not begin to take root in Oxford until well into the twentieth century, and even then developments were rather spasmodic and fragmented.

An Institute of Statistics, financed initially by the Rockefeller Foundation, was founded in 1935 under the supervision of the Social Studies Research Committee. The Institute's first director, Jakob Marschak, reader in statistics, was required to 'conduct independent statistical research and give informal assistance to members of the University engaged in statistical research'.

After the Second World War, various posts of a statistical nature were created, including a lectureship in the design and analysis of scientific experiment, to which David Finney was appointed in 1947. In the following year, David Champernowne became Oxford's first professor of statistics, in the area of social studies. Twenty years later, a Chair in applied statistics was created in the Department of Biomathematics, to which Maurice Bartlett was appointed.

It was not until 1988 that a Department of Statistics was established, within the Faculty of Mathematics, by bringing together members of the Department of Biomathematics with statistically inclined members of the Mathematical Institute. In 1989 David Hinkley was appointed professor of statistical science and in the following year Brian Ripley became professor of applied statistics. The prominence of statistics in Oxford continues to increase and was marked by the establishment in 1997 of the annual Florence Nightingale lecture—over a century after the fruitless attempts of Nightingale and Jowett to establish the study of statistics in Oxford, as told in Chapter 11.

Graham Higman, one of the founders of the Invariant Society and Waynflete Professor of Pure Mathematics from 1960 to 1985. This painting hangs in the 'Higman Room' at the Mathematical Institute.

Further reading

Further information about Hardy, Love, Titchmarsh, and Whitehead appears in the *Dictionary of scientific biography*.

Information about Oxford University in the first half of the twentieth century can be found in *The history of Oxford University*, Vol. VIII, *The twentieth century* (edited by B. H. Harrison), Clarendon Press, Oxford, 1994. The development of the scientific research tradition at Oxford during the inter-war period is described by Jack Morrell in *Science at Oxford 1914–1939*, Clarendon Press, Oxford, 1997. The study and teaching of astronomy at Oxford over this period is surveyed in M. G. Adam's article 'The changing face of astronomy in Oxford 1920–60', *Quarterly Journal of the Royal Astronomical Society* **37** (1996), 153–79.

CHAPTER 15

Some personal reminiscences

Sir Michael Atiyah

I came to Oxford in 1961 and left in 1990, with a few years at Princeton in between. During my almost 30 years at the Mathematical Institute, Oxford mathematics was transformed out of all recognition. It was a period of unique expansion and interesting developments, in which the Oxford mathematical scene grew from a small nucleus of people to a very large organization, with many ramifications and a much larger presence on the national and international scene.

I should like to say a little about my own personal and mathematical involvement in Oxford. My background was in algebraic geometry, although tending slightly in the direction of topology, so when I arrived in Oxford, which had been much influenced by Henry Whitehead, I found it easy to fit in. Although I did not regard myself in any sense as a professional topologist—I could not untie knots, or anything like that—I was probably more competent at geometry than some of the previous Savilian professors had been.

In an understandable oversight, William Hodge, formerly Lowndean professor at the University of Cambridge and my mentor there in the 1950s, once confused the Savilian Chair of Geometry with the Savilian Chair of Astronomy. This was entirely understandable, because the Chair I held had been occupied by Halley whom most people think of as an astronomer, whereas the Chair of astronomy had been held by Christopher Wren. However, the corresponding position in Cambridge is called the Lowndean Chair of Astronomy and Geometry, so there they still go together.

When G. H. Hardy was appointed Savilian Professor of Geometry, his geometry credentials were not very strong. It is claimed that the only theorem of geometry that Hardy ever proved is the following: that if a rectangular hyperbola is also a parabola, then it is an equiangular spiral! I am not enough of a scholar to attest to either the historical veracity or the truth of this claim, but it is such a marvellous statement about Hardy's view of geometry that I think it deserves to be correct.

Hardy was not like H. F. Baker, the Lowndean professor at Cambridge before Hodge. Baker had been appointed to the Chair of

Michael Atiyah at his desk in the Mathematical Institute, 1974.

G. H. Hardy on geometry

During his tenure of the Savilian Chair of Geometry, G. H. Hardy gave his presidential address to the Mathematical Association (1925), under the title '*What is geometry?*,' in the course of which he said:

You might object ... that geometry is, after all, the business of geometers, and that I know, and you know, and I know that you know, that I am not one; and that it is useless for me to try to tell you what geometry is, because I simply do not know. And here I am afraid that we are confronted with a regrettable but quite definite cleavage of opinion. I do not claim to know any geometry, but I do claim to understand quite clearly what geometry is.

He had, nevertheless, contributed to the geometrical literature with the following note, published in the *Mathematical Gazette* in 1907:

> **224.** [M^1. **8. g.**] *A curious imaginary curve.*
>
> The curve $\qquad\qquad (x + iy)^2 = \lambda\,(x - iy)$
>
> is (i) a parabola, (ii) a rectangular hyperbola, and (iii) an equiangular spiral. The first two statements are evidently true. The polar equation is
>
> $$r = \lambda e^{-3i\theta},$$
>
> the equation of an equiangular spiral. The intrinsic equation is easily found to be $\rho = 3is$.
>
> It is instructive (i) to show that the equation of any curve which is both a parabola and a rectangular hyperbola can be put in the form given above, or in the form
>
> $$(x + iy)^2 = x \text{ (or } y),$$
>
> and (ii) to determine the intrinsic equation directly from one of the latter forms of the Cartesian equation. $\qquad\qquad$ G. H. HARDY.

astronomy and geometry, despite the fact that he was an analyst whose work had been on transcendental functions and elliptic functions. But he took his job seriously, and immediately converted himself into a real geometer, producing enormous tracts on pure projective geometry. It was an extreme case of dedication. Since neither Hardy nor Titchmarsh had followed that route, I was the first respectable geometer to hold the Savilian Chair for some time, and I was happy to integrate with the Oxford scene. Of course, it was growing very fast, with new people continually coming in, and I was able to interact with people in other fields—functional analysts, algebraists, and subsequently theoretical physicists and people working on differential equations. We were a large happy group, interacting in different ways.

When I first arrived in Oxford, there were several other professors of whom I have vivid personal memories. Of these, three are no longer with us, so I can feel free to recount my personal memories of them. The personality of somebody has a major effect on their influence, as can be seen when one contrasts the different people that have been at Oxford. Perhaps historians looking back on the past should ask themselves not only what a person contributed intellectually in terms of written work, but what his personality was like, so that they can assess what influence he had on his subject in a way that is harder to measure than by original work.

Henry Whitehead

First there was Henry Whitehead, a very genial figure. His life was entirely devoted to three activities, in various proportions—mathematics, drinking in the pub, and cricket—and he was equally good at all of them. In later life, he added a fourth, looking after pigs at the farm. Henry had a remarkable personality. Most people who met him had no idea that he was an intellectual: they thought he was either a barman or a pub-keeper.

Henry was a man of many talents. His natural exuberance and enjoyment of life had a big effect on the mathematics students in Oxford at the time. He naturally attracted a large school of students, and built up a substantial group of people in algebraic topology, many of whom (or their disciples) are still there. Henry Whitehead had a big impact, and left a lasting legacy on the mathematical scene in Oxford.

There are a couple of stories I like that indicate his seriousness. Although Henry had done many fine things of quality, he had never proved any single major theorem that stood out. He once told me that Hodge had done the opposite, having proved one big theorem about harmonic integrals that made him famous. Henry said that he would have sold his soul to the devil to have proved a theorem like that.

Henry had many friends working in topology all over the world. He had great friends in America and elsewhere, who made a kind of community, and he felt this was very important for his life. On one occa-

Henry Whitehead and friend at his home, Manor Farm, Noke.

Henry Whitehead with two of the younger topologists that he encouraged, Ioan James (later Savilian Professor of Geometry) and Peter Hilton.

sion, he said to me, 'Wouldn't it be terrible if one day I woke up and had a brilliant idea in functional analysis?' He really meant it quite seriously. Obviously he meant that if he *had* this brilliant idea then he would be duty bound to drop all his friends, follow up this new line, and meet an entirely new bunch of people; it was something that really worried him. Actually, there is a happy ending to this story because, in more recent times, many of us have had bright ideas in functional analysis. Instead of dropping our friends, we just made a bigger crowd of friends, and the functional analysts and topologists are now part of a larger club. So you do not have to drop your friends when you have a new idea.

Edward Charles Titchmarsh

Titchmarsh was rather at the opposite extreme. He was a scholarly man who sat in his room and wrote beautiful books—impeccable, effectively written textbooks, from which many students have learnt their complex analysis. But he was a man of very few words; his influence was not due to personal contact, but through his writings.

When I first came to Oxford, he was the curator of the Mathematical Institute, so as a newly arrived member I had to go to him to get the key to my office. I was duly ushered into his big room, where he was sitting at his desk. I sat down and he handed over the key, and I then expected a speech of welcome or some words of advice, but we just sat in silence. After five minutes, I left.

Titchmarsh had something in common with Henry Whitehead, and that was an interest in cricket—not quite a fundamental passion, but more an obligatory interest. He held the Savilian Chair of Geometry,

G. H. Hardy leads a cricket team of Oxford mathematicians during a British Association meeting in 1926: H. K. Salvesen, W. L. Ferrar, W. J. Langford, E. C. Titchmarsh, E. H. Neville, E. H. Linfoot, and L. S. Bosanquet. Hardy dubbed this photograph 'Mathematicians v. The Rest of the World'.

which had been occupied by Hardy, an enormous enthusiast for cricket and many other games. When Hardy came over from Cambridge, he established what became an important annual event on the Oxford cricket scene, the annual match between the Fellows of New College and the boys of New College School. After that, the Savilian Professor of Geometry was always expected to captain the team, and Titchmarsh duly did so most conscientiously for 30 years. I am afraid that I was not quite so conscientious.

Charles Coulson

The third memorable Oxford professor at that time was Charles Coulson. A leading churchman, chairman of Oxfam, and very much a public figure, he was well known both in the University and in the country at large.

Coulson was the driving force behind the Mathematical Institute, the man who decided that Oxford needed a centre for mathematics; so he persuaded the University that we had to have an institute. Hardy had tried to do that many years before, but Coulson was the one who really got it going. He had enormous drive and personality, and essentially ran the whole thing. Although, as a theoretical chemist, he was in some sense on the fringes of the mathematical scene, he was very much at the centre of things by virtue of his personality, and he held the Mathematical Institute together. It was a great achievement to prevent it from fragmenting into pure mathematicians on the one hand, and applied mathematicians on the other. He had a vision of the unity of mathematics, applied mathematics, and physics, which served it well in subsequent years.

As chairman of Oxfam (The Oxford Committee for Famine Relief, founded in 1942) Charles Coulson travelled widely. This photograph was taken in Uganda in the late 1960s.

One advantage of moving from one place to another is that some of the books at the bottom of your pile in one place appear at the top in the other. While tidying up, I had occasion to look at *The spirit of applied mathematics*, an inaugural lecture delivered before the University of Oxford on 28 October 1952 by C. A. Coulson, FRS, which I read with great interest. It is a fine peroration on the nature of applied mathematics, with no formulas—he was a very literate man. Professors of mathematics nowadays do not go in for giving inaugural lectures so much, but in those days it was still common. Let me give you one or two extracts from it.

Coulson was skilful at producing quotations from other people. He was very methodical about this, and had card indexes full of brilliant quotations. There were two in particular that I liked. One of them was from J. L. Synge, the distinguished theoretical physicist, who said:

... physics will become more mathematical. This does not mean that physics will use more mathematical symbols (it is near saturation at present), but rather that physicists will learn more widely what mathematics is, in the modern sense.

That was written in 1951, and those who followed what happened in the 1980s and 1990s will recognize it as a very far-reaching, far-seeing statement—it is essentially what happened over those 20 or so years.

The other quotation, also from Professor Synge (obviously Synge was one of Charles Coulson's favourite authors), said:

The use of applied mathematics in its relation to a physical problem involves three steps. First, a dive from the world of reality into the world of mathematics; two, a swim in the world of mathematics; three, a climb from the world of mathematics back into the world of reality, carrying the prediction in our teeth.

In Coulson's vision, the Mathematical Institute enabled mathematicians of different disciplines to socialize as well as work together. Here a table tennis match at Coulson's house in 1955 is watched by, from the right, Coulson (Rouse Ball Professor of Mathematics), E. C. Titchmarsh (Savilian Professor of Geometry), and George Temple (Sedleian Professor of Natural Philosophy).

Coulson had a firm grip on the unity of mathematics and the need to keep pure and applied mathematicians together.

It was through Charles Coulson that the Mathematical Institute eventually moved from its original rather dingy quarters to the magnificent buildings it now occupies, even though they are too small for its present purposes. I should mention in passing that other people were also involved in that move; Jack Thompson, a very powerful figure on University scenes and a chairman of almost every committee (in particular, the important committee dealing with buildings), saw to it that, when the right moment came, the interests of mathematics were not neglected. So that is how the Mathematical Institute came into being. It is useful to have on your side people who hold important positions in the University.

Visitors to the Institute

Over the years, the Mathematical Institute has had very many visitors. One in particular stays in my memory—Professor Izrail Gelfand from Russia, to whom we gave an honorary degree in 1973.

The history of that is as follows. Gelfand was a very famous Russian mathematician, but he had never been allowed out of Russia; in those days it was less easy than it is now. Many people unsuccessfully tried to invite him to various places or to conferences. He had in the past worked on atomic physics, and the authorities said that it was too secret and that he could not be allowed out. I was then told by André Weil that the only way to get him out was to persuade Oxford University to offer him an honorary degree, so I spoke to my University friends, who thought it was a good idea. We offered him an honorary degree, which remarkably had the desired effect—the prestige of an honorary degree at Oxford was sufficient to unlock the doors of the KGB. The letter that went out was quite impressive. It said that, amongst the people who have received honorary degrees from Oxford University there were six Russians. The first was Tchaikovsky, whom they had heard of, so it did the trick.

Mind you, it did the trick only in terms of getting him to accept the degree. We then had to see whether he was going to turn up, and for a long time we heard nothing at all. Even the day before the honorary degree ceremony was due to take place, we had heard nothing, although we had pulled every conceivable string. We had even gone to the highest possible authorities. When Harold Macmillan was Prime Minister, he visited Moscow—there are pictures of him wearing his big Russian hat—and one of the things he did there was to attend the famous Gelfand seminar, a major scene of mathematical talent in Russia, to learn what happened in intellectual circles in Moscow. Since Macmillan was the Chancellor of Oxford University at that time, we thought it proper to use this channel to try to ensure that Gelfand came.

On the day of the actual ceremony, we had given up hope that he would arrive, and I was playing cricket for the Mathematical Institute's annual match between the staff and the students. I remember being on the field and being summoned by someone, rather like Drake playing bowls. A telegram had arrived, announcing that Gelfand was arriving at Heathrow in two hours' time, so we got into our cars and dashed down there to pick him up. We brought him back to Oxford, but arrived just too late for the ceremony. He came to dinner at Christ Church, and on the next day there happened to be a minor ceremony, so they slotted him in and he got his degree. I still have a photograph on my shelf showing Gelfand in his honorary degree gown.

There is a nice little further touch to add. He liked his Oxford gown very much, and he thought that when universities gave you an honorary degree, they also gave you the gown. So as not to disappoint him, I gave it to him as a present, and he returned to Moscow with it. A few years later, Harvard University offered him an honorary degree; by this time, once you had received one honorary degree, you could go out again. So he went to Harvard and took his Oxford gown with him. When he arrived for the magnificent ceremony in Harvard Yard, he asked them whether he could wear it. Nobody had ever asked this before, but the upshot was that he was allowed to wear his Oxford robes when he received his Harvard degree.

We had many other visitors during my time in Oxford. One year, I invited Peter Lax from New York University, and Mark Kac, at that time at Cornell and Rockefeller, to spend a term in Oxford together to strengthen the area of analysis. They gave lectures on scattering theory and energy levels in quantum mechanics, much to everybody's enlightenment.

Mark Kac was a Polish mathematician who had emigrated early on to the United States. He was a man of considerable character, charm, and wit, and a very likeable fellow all round. Everybody enjoyed his lectures, and I remember him telling me a story about the time he was teaching complex variable theory to a rather dim student who obviously was not understanding the central point. So Kac said to him: 'this function, $1/z$, what kind of singularity does it have at the origin?'—a basic question in order to get the ideas across—but the student still could not understand. Finally, in desperation, Kac said, 'Look at me, what am I?' So the student rising up said, 'Ah, a simple Pole.' In fact, Mark Kac was anything but simple...

The Mathematical Institute

I will now return to the development of the Mathematical Institute over its first 20 years or so. Its main feature was that it did not split into

Izrail Gelfand visited Oxford in 1973 to receive an honorary degree.

Roger Penrose, Coulson's successor as Rouse Ball Professor of Mathematics from 1973 to 1998, worked to popularize mathematics as well as to establish the Mathematical Institute as a world-renowned centre for relativity theory.

A recent named Chair at the Mathematical Institute hearkens back to a former age of Oxford mathematics: the second Wallis Professor of Mathematics was Simon Donaldson, shown here lecturing to a group of students.

separate departments of pure and applied mathematics. At that time in the 1960s and 1970s, when there was a lot of expansion, many British universities found it convenient, or perhaps desirable, to separate mathematics into two or more departments; this certainly happened in Cambridge, and in several other places as well. In Oxford it did not, and for a department of that size it was rather singular in that respect. As I mentioned, that was mainly due to the influence of Coulson in holding it together, but it was helped by having a new building for people to meet in. Perhaps it was just an accident of timing. Cambridge, to which I returned, still had two departments, partly because they were in the act earlier. At a certain stage, their interests diverged, and they split apart. Oxford came on to the scene later, and had not had long enough to quarrel!

Later, new people came in to replace those who had retired. We had the appointments of Roger Penrose and Brooke Benjamin, who further strengthened the bridges between pure and applied mathematics, much of this centring around a common interest in partial differential equations. As time moved on, people working in geometry moved more in the direction of differential geometry, and then of analysis, which was also connected with partial differential equations. So the general interaction between the different parts of pure and applied mathematics produced some kind of natural cohesion which held things together. By the 1980s it could be said that Oxford was the world centre in many of these fields.

The future

The purpose of history is to predict the future, and maybe it is worth thinking not just about the future direction of mathematics at Oxford University, but about the whole future of higher education. The Royal Society has advocated increasing the number of people who go on to full-time education or training after 16, which, despite the rapid expansion of this area in recent years, is still behind what happens in other countries. In order to make this possible, it has proved necessary to provide more suitable courses than at present, since the present pattern of education has been much too rigid and specialized, and unsuitable for many students. The Royal Society has proposed a framework which would unify all kinds of courses, from the most vocational to the most academic, under some larger umbrella, thereby making it possible for people to study various combinations of topics at levels suitable to their abilities.

Such proposals would have an impact for the whole of higher education. Already, changes in schools are beginning to have an effect, and students will soon arrive at university with a broader-based background and less specialized knowledge. New types of vocational courses may well open up higher education to new types of students who have not benefited from higher education in the past. Accordingly, university courses will have to adapt, and courses will need to go more slowly to start with. In the past, universities decided what was good for people, and everybody else had to follow suit. But pressure is now coming up from the bottom, and universities will have to adapt their courses to

Sir Michael Atiyah, pictured at the Mathematical Institute in 1989.

deal with the new intake. After three years, students cannot be expected to reach the degree of specialization which they reach at present—that may not be a bad thing—and for some a fourth year may be needed to bring them up to the level necessary for further work. In that scenario, there may need to be four-year courses for some students and three-year courses for others. Indeed, this pattern is already in place for subjects such as physics and mathematics.

Education is one of those areas that constantly involve discussions, especially in democratic institutions like Oxford University. As one enters into more difficult times, all universities will have to adapt in various ways, and the Mathematical Institute will have to adapt accordingly.

Oxford's mathematical Chairs

In this appendix we list the holders of the Savilian, Sedleian, Waynflete, Rouse Ball, and Wallis Chairs since their foundations, with the date of appointment or accession to the Chair.

Savilian professors of geometry and astronomy

Savilian professors of geometry

1619	**Henry Briggs** (1561–1631)
1631	**Peter Turner** (1586–1652)
1649	**John Wallis** (1616–1703)
1704	**Edmond Halley** (1656–1742)
1742	**Nathaniel Bliss** (1700–64)
1765	**Joseph Betts** (1718–66)
1766	**John Smith** (*c.* 1721–97)
1797	**Abraham Robertson** (1751–1826)
1810	**Stephen Peter Rigaud** (1774–1839)
1827	**Baden Powell** (1796–1860)
1861	**Henry John Stephen Smith** (1826–83)
1883	**James Joseph Sylvester** (1814–97)
1897	**William Esson** (1839–1916)
1920	**Godfrey Harold Hardy** (1877–1947)
1932	**Edward Charles Titchmarsh** (1899–1963)
1963	**Michael Francis Atiyah** (b. 1929)
1969	**Ioan MacKenzie James** (b. 1928)
1995	**Richard Lawrence Taylor** (b. 1962)
1997	**Nigel James Hitchin** (b. 1946)

Savilian professors of astronomy

1619	**John Bainbridge** (1582–1643)
1643	**John Greaves** (1602–52)
1649	**Seth Ward** (1617–89)
1661	**Christopher Wren** (1632–1723)
1673	**Edward Bernard** (1638–96)
1691	**David Gregory** (1659–1708)
1708	**John Caswell** (1656–1712)
1712	**John Keill** (1671–1721)
1721	**James Bradley** (1693–1762)
1763	**Thomas Hornsby** (1733–1810)
1810	**Abraham Robertson** (1751–1826)
1827	**Stephen Peter Rigaud** (1774–1839)
1839	**George Henry Sacheverell Johnson** (1808–81)

1842 **William Fishburn Donkin** (1814–69)

1870 **Charles Pritchard** (1808–93)

1893 **Herbert Hall Turner** (1861–1930)

1932 **Harry Hemley Plaskett** (1893–1980)

1960 **Donald Eustace Blackwell** (b. 1921)

1988 **George Petros Efstathiou** (b. 1955)

1999 **Joseph Ivor Silk** (b. 1942)

Sedleian professors of natural philosophy

1621 **Edward Lapworth** (1574–1636)

1636 **John Edwardes** (b. 1600)

1648 **Joshua Crosse** (*c.* 1615–76)

1660 **Thomas Willis** (1621–75)

1675 **Thomas Millington** (1628–1704)

1704 **James Fayrer** (*c.* 1655–1720)

1720 **Charles Bertie** (*c.* 1695–1744)

1741 **Joseph Browne** (1700–67)

1767 **Benjamin Wheeler** (*c.* 1733–83)

1782 **Thomas Hornsby** (1733–1810)

1810 **George Leigh Cooke** (1779–1853)

1853 **Bartholomew Price** (1818–98)

1899 **Augustus Edward Hough Love** (1863–1940)

1946 **Sydney Chapman** (1888–1970)

1953 **George Frederick James Temple** (1901–92)

1968 **Albert Edward Green** (1912–99)

1979 **Thomas Brooke Benjamin** (1929–96)

1996 **John MacLeod Ball** (b. 1948)

Waynflete professors of pure mathematics

1892 **Edwin Bailey Elliott** (1851–1937)

1922 **Arthur Lee Dixon** (1867–1955)

1947 **John Henry Constantine Whitehead** (1904–60)

1960 **Graham Higman** (b. 1917)

1985 **Daniel Gray Quillen** (b. 1940)

Rouse Ball professors of mathematics

1928 **Edward Arthur Milne** (1896–1950)

1952 **Charles Alfred Coulson** (1910–74)

1973 **Roger Penrose** (b. 1931)

1999 **Philip Candelas** (b. 1951)

Wallis professors of mathematics

1969 **John Frank Charles Kingman** (b. 1939)

1985 **Simon Kirwan Donaldson** (b. 1957)

Sir Michael Atiyah was Savilian Professor of Geometry (1963–69) and Royal Society Research Professor (1973–90) at Oxford, before becoming Master of Trinity College, Cambridge (1990–97). He has worked in many fields of mathematics, including geometry, topology, and differential equations, and was awarded the Fields Medal in 1966. Elected a Fellow of the Royal Society in 1962, he was awarded its Royal Medal and Copley Medal and was president from 1990 to 1995. He became the first director of the Isaac Newton Institute for Mathematical Sciences, Cambridge, in 1990. Knighted in 1983, he was made a member of the Order of Merit in 1992. His *Collected mathematical papers* have been published in five volumes by Oxford University Press.

Allan Chapman teaches and researches in the history of science in the Faculty of Modern History, University of Oxford, where he is attached to Wadham College. His interests lie in the history of astronomy, with particular emphasis on the development of astronomical instruments, observatories, and astronomical biography. He has edited John Flamsteed's *Historia coelestis* (1982), and written *Dividing the circle* (1990, 1995), *Astronomical instruments and their uses* (1996), and *The Victorian amateur astronomer* (1998).

John Fauvel is Senior Lecturer in Mathematics at the Open University, and former President of the British Society for the History of Mathematics. He is currently involved in an international study of the relationships between the history and pedagogy of mathematics. He has been an editor and co-editor of several books, including *Darwin to Einstein: historical studies on science and belief* (1980), *Conceptions of inquiry* (1981), *The history of mathematics: a reader* (1987), *Let Newton be!* (1988), and *Möbius and his band* (1993).

Raymond Flood is University Lecturer in Computing and Mathematics at the Department for Continuing Education, Oxford University, and Senior Tutor of Kellogg College. His main research interests lie in statistics and the history of mathematics. He is a co-editor of *The nature of time* (1986), *Let Newton be!* (1988), and *Möbius and his band* (1993).

Robert Goulding is Elizabeth Wordsworth Junior Research Fellow at St Hugh's College, Oxford. His current research is on Henry Savile's mathematical papers, and other interests include the history of trigonometry and parallactic methods.

Willem Hackmann is the Senior Assistant Keeper at the Museum of the History of Science and Reader in the History of Science at Oxford University. He is editor of the *Bulletin of the Scientific Instrument Society*, and his books include *Electricity from glass: the history of the frictional electricity machine* (1978), *Apples to atoms: great scientists portrayed from Newton to Rutherford* (1986), and *Learning, language and invention* (1994).

Keith Hannabuss is Billmeir Fellow and Tutor of Mathematics at Balliol College, Oxford. His main research areas are in representation theory and quantum field theory, but he has also had a long-standing interest in the history of mathematics and theoretical physics. He is the author of *An introduction to quantum theory* (1997).

John North is Chairman of the History of Philosophy Department in the Philosophical Institute at the University of Groningen, The Netherlands. His books include *The measure of the universe* (a history of modern cosmology, 1965, 1990), a three-volume edition of the writings of Richard of Wallingford (1976), *Chaucer's universe* (1988), *The Fontana history of astronomy and cosmology* (1994), and *Stonehenge* (1996).

Margaret E. Rayner is Emeritus Fellow of St Hilda's College, Oxford, having previously been Tutor in Mathematics. Her main interest in mathematics was isoperimetric inequalities, but in recent years she has worked on the nineteenth- and twentieth-century history of Oxford.

Robin Wilson is Senior Lecturer in Mathematics at the Open University and Fellow of Keble College, Oxford. He has written and edited a number of books on graph theory and combinatorics, including *Introduction to graph theory* (1972), *Graph connections* (1997), and *An atlas of graphs* (1998). He is involved with the popularization of mathematics and with the history of mathematics, and is a co-editor of *Let Newton be!* (1988) and *Möbius and his band* (1993).

REFERENCES

CHAPTER 1

1 'In the year of our Lord 886'
William Camden, *Britannia*, 1586, in *The Oxford book of Oxford* (edited by Jan Morris), Oxford, 1978, p. 4.

1 'that catarrhal point of the English Midlands'
The Oxford book of Oxford (edited by Jan Morris), Oxford, 1978, p. 3.

2 'The usefulness of considering lines'
A. C. Crombie, *Robert Grosseteste and the origins of experimental science*, Oxford University Press, 1953, p. 110.

2 'He who knows not mathematics cannot know the other sciences'
cited in A. G. Molland, 'The geometrical background to the "Merton School"', *British Journal for the History of Science* 4 (1968), 110.

2 'it is mathematics which reveals every genuine truth'
cited in J. A. Weisheipl, 'The place of John Dumbleton in the Merton School', *Isis* **50** (1959), 446.

5 'the foundations of that learning' and 'to manage our reason'
cited in Mordechai Feingold, 'The humanities', *The history of the University of Oxford*, Vol. IV, *Seventeenth-century Oxford* (edited by Nicholas Tyacke), Clarendon Press, Oxford, 1997, pp. 211–357, especially p. 282.

6 'Logical Precepts are more useful'
Anon, 'An essay on the usefulness of mathematical learning', Oxford, 1701, p. 6.

6 'Logic will, most probably, be untried ground'
Algernon M. M. Stedman, *Oxford: its social and intellectual life*, Trubner and Co, London, 1878, pp. 177–8.

8 'Dyvers Englyshe menne have written right well'
Robert Record, *The castle of knowledge*, 1556, p. 99.

14 'one of the most important bodies of unpublished English scientific manuscripts'
A. C. Crombie *et al.*, 'Thomas Harriot (1560–1621): an original practitioner in the scientific art', *Times Literary Supplement*, 23 October 1969, pp. 1237–8.

15 'the architect and founder of the modern Press'
Peter Sutcliffe, *The Oxford University Press: an informal history*, Oxford, 1978, p. 39.

16 'the recent renascence of mathematical studies in Oxford'
Peter Sutcliffe, *The Oxford University Press: an informal history*, Oxford, 1978, p. 229.

17 'be apt to look upon it as Ominous'
Edmond Halley, *A description of the passage of the shadow of the moon over England*, 1715.

18 'the advantages attending'
Baden Powell, *An historical view of the progress of the physical and mathematical sciences*, Longman, London, 1834, p. 390.

18 'would not neglect its historical and philosophical aspects'
Jack Morrell, *Science at Oxford 1914–1939*, Clarendon Press, Oxford, 1997, p. 315.

19 'His Almageste and bokes grete and smale'
Geoffrey Chaucer, *The Miller's tale*, lines 22–5.

20 'It is related of him that one night'
The Oxford book of Oxford (edited by Jan Morris), Oxford, 1978, p. 28.

20 'Of studie took he most cure'
Geoffrey Chaucer, *The prologue* to *The Canterbury tales*, lines 303–8.

22 'But how about research and original work'
Edwin Elliott, *Why and how the Society began and kept going*, lecture to Oxford Mathematical and Physical Society, 16 May 1925, p. 9.

23 'will always be the first moving principle'
Baden Powell, *On the present state and future prospects of mathematics at the University of Oxford*, Oxford, 1832.

23 'a farce in my time' and ' "What is the Hebrew for the place of a skull?" '
Horace Twiss, *The public and private life of Lord Chancellor Eldon*, Vol. i, London, 1844, p. 57.

23 'this horrible incubus'
letter from J. J. Sylvester to Arthur Cayley, 1 February 1886, from Karen Hunger Parshall, *James Joseph Sylvester: life and work in letters*, Clarendon Press, Oxford, 1998, p. 262.

23 'In 1926 or so I began my chore of examining'
William Ferrar, *An account of my career*, ms 1976, The Queen's College, p. 167.

24 'Together, in tandem or in counterpoint'
The history of the University of Oxford, Vol. II, *Late medieval Oxford* (edited by J. I. Catto and Ralph Evans), Clarendon Press, Oxford, 1992, p. vi.

24 'At Redbrick they treat mathematics'
cited in I. W. Busbridge, *Oxford mathematics and mathematicians*, Oxford Mathematical Society, 1974, p. 33.

25 'it is known to be the intention'
The Times, 21 July 1825, p. 2.

26 'it is undeniable that the average candidate'
Arthur Joliffe, 'Examinations for mathematical scholarships at Oxford', *Special reports on educational subjects* **27**, HMSO, 1912, p. 263.

27 'We are justly proud of the names'
Baden Powell, *The present state and future prospects of mathematical and physical studies in the University of Oxford*, Oxford, 1832.

27 'G. H. Hardy's inaugural lecture, thirty-five years on'
G. H. Hardy, *Some famous problems of the theory of numbers*, Clarendon Press, Oxford, 1920.

CHAPTER 2

33 'one of the first comprehensive trigonometrical texts in Europe'
for more information, see J. D. North, *The universal frame*, Hambledon Press, London, 1989, Chapter 1 (Coordinates and categories).

33 'using such a difficult medium of expression'
for an example, see *Quadripartitum 1.8* Vol. 1 (edited by J. D. North), with its modern equivalent on p. 50 of Vol. 2.

36 'having worked constantly with tables of differences'
for more details, see J. D. North, *Stars, minds and fate*, Hambledon Press, London, 1989, pp. 343–6.

CHAPTER 3

42 'The Student freshly come'
M. Feingold, *The mathematicians' apprenticeship*, Cambridge 1984, p. 24.

43 'my Memory reacheth the time'
M. Feingold, *The mathematicians' apprenticeship*, Cambridge 1984, p. 30.

43 'The mathematical sciences are particularly disciplinary'
Foster Watson, *Vives: on education*, Cambridge 1913, p. 202.

44 'causa quia non sine maxima incommodo'
J. W. Shirley, *Thomas Harriot*, Oxford, 1983, pp. 54–5.

46 'The which two lectures are not to be redd'
M. Feingold, *The mathematicians' apprenticeship*, Cambridge, 1984, p. 39.

47 'No man can worthely praise Ptolemye'
Robert Record, *The castle of knowledge*, 1556, pp. 126–7.

47 ''Tis said that while he was of All Souls coll.'
Anthony à Wood, *Athenae Oxonienses* (edited by P. Bliss), Vol. i, 1813, col. 255.

48 'an eminent mathematician called Whytehead'
Anthony à Wood, *Athenae Oxonienses* (edited by P. Bliss) Vol. i, 1813, col. 761.

48 'When Whytehead died' and 'what I have said relating to him'
Anthony à Wood, *Athenae Oxonienses* (edited by P. Bliss), Vol. i, 1813, col. 762.

49 'For, the Honour, and Estimation of the Vniuersities' and 'Well may all men coniecture'
John Dee, *Mathematical praeface*, 1570, p. A.iiij.

50 'The great Dudley, Earle of Leicester'
Aubrey's Brief lives (edited by O. L. Dick), Penguin 1962, p. 114.

51 'My old cosen Parson Whitney'
Aubrey's Brief lives (edited by O. L. Dick), Penguin 1962, p. 21.

51 'Sure I am that such books'
Anthony à Wood, *History and antiquities of the colleges and halls in the university of Oxford*, Vol. ii, (edited by J. Gotch, 1796), p. 108.

52 'He could not abide Witts'
Aubrey's Brief lives (edited by O. L. Dick), Penguin 1962, p. 328.

52 'famous for his learning, especially for the Greek tongue'
Anthony à Wood, *Athenae Oxonienses* (edited by P. Bliss) 1813, Vol. ii, col. 310.

55 'Can military and literary studies thrive'
Register, ii.1, 232.

55 'alone, or almost alone', 'carefully thought out, very ingenious hypotheses' and '*ad usum professorum meorum*'
H. Savile, *Praelectiones tresdecim in principium Elementorum Euclidis*, Oxford, 1621, p. 41.

56 'At length upon the hearing'
Anthony à Wood, *Athenae Oxonienses* (edited by P. Bliss) 1813, Vol. ii, col. 622.

57 'in my publike lectures'
Richard Hakluyt, *Principall navigations*, 1589, dedication.

58 'Harriot in fact possessed'
D. T. Whiteside, 'In search of Thomas Harriot', *History of Science* **13** (1975), 61–2.

60 '*ad imitationem Ptolemaei et Copernici*'
S. Gibson, *Statuta antiqua universitatis oxoniensis*, Oxford, 1931, p. 529, l. 35.

CHAPTER 4

63 'What more pleasing studies can there be' and 'what so intricate or pleasing'
Robert Burton, *Anatomy*, Oxford, 1621, p. 352, and Oxford, 1628, p. 264; quoted by N. Tyacke, 'Science and religion at Oxford before the Civil War', *Puritans*

and revolutionaries (edited by P. Pennington and K. Thomas), Clarendon Press, Oxford, 1978, p. 79.

63 'Some years later, that scion'
R. E. W. Maddison, *The life of the Honourable Boyle*, Taylor and Francis, London, 1969, pp. 17–18.

63 'industry [which] is a lodestone'
Robert Burton, *Anatomy*, Oxford, 1621, pp. 51–8 and T. Birch, *The works of the Honourable Robert Boyle*, new edn., Vol. III, London, 1722, pp. 396–9, 415, 442; for a discussion on this point, see W. D. Hackmann, 'Attitudes to natural philosophy instruments at the time of Halley and Newton', *Polhem* **6** (1988), 143–58.

64 'I Greatly rejoyce to see'
To the Reader, in T. Blundeville, *His exercises, containing six treatises,the titles whereof are set down in the next printed page*, 7th edn, London, 1636.

64 'Both were interested in the practical aspects'
E. G. R. Taylor, *The mathematical practitioners of Tudor and Stuart England*, Cambridge, 1954, p. 173, entry 27 on Blundeville; Blagrave's *Mathematical jewell* was revived in 1658 by John Moxon in a paper version, see J. Palmer, *The catholique planisphere which Mr Blagrave calleth the mathematical jewell*, London, 1658, which accompanied this paper instrument.

64 'The complexity of the origins of natural philosophy'
see, for instance, R. Harré, 'Knowledge', pp. 11–54, and S. Schaffer, 'Natural philosophy', pp. 55–81, in *The ferment of knowledge. Studies in historiography of eighteenth-century science* (edited by S. G. Rousseau and R. Porter), Cambridge University Press, 1980.

64 'to use Eduard Dijksterhuis's evocative phrase'
E. J. Dijksterhuis, *The mechanization of the world picture*, Oxford, 1961, especially pp. 241–7; also cited by J. A. Bennett, 'The mechanic's philosophy and the mechanical philosophy', *History of Science* **24** (1986), 1–28.

64 'In his otherwise admirable paper'
T. S. Kuhn, 'Mathematical versus experimental tradition in the development of physical science', *Journal of Interdisciplinary History* **7** (1976), 1–31, reprinted in his *The essential tension*, Chicago University Press, 1977, pp. 31–65, especially p. 53.

64 'A more fruitful way forward'
J. A. Bennett, 'The mechanic's philosophy and the mechanical philosophy', *History of science* **24** (1986), 1–28, especially p. 6.

65 'In England, the early development was centred on Gresham College'
I. R. Anderson, 'The administration of Gresham College and its fluctuating fortunes as a scientific institution in the seventeenth century', *History of Education* **9** (1980), 13–25; this paper reflects poorly on Gresham College as an institution for popular education, but reinforces its reputation as a centre for research in mathematical sciences, see J. A. Bennett, 'The mechanic's philosophy and the mechanical philosophy', *History of science* **24** (1986), 1–28, pp. 20–3, and also M. Feingold, *The mathematicians' apprenticeships: science, universities and society in England 1560–1640*, Cambridge, 1984, p. 188. See also Allan Chapman, 'Gresham College: scientific instruments and the advancement of useful knowledge in seventeenth-century England', *Bulletin of the Scientific Instrument Society* **56** (1998), 6–13. The relation of Gresham College to the universities had previously been discussed in C. Hill, *Intellectual origins of the English revolution*, Oxford, 1965, pp. 18–84, 301–14, and in *The intellectual revolution of the seventeenth century* (edited by C. Webster), London, 1974, pp. 243–83.

65 'Dee, a mathematical genius and founder fellow'
P. J. French, *John Dee. The world of an Elizabethan magus*, London, 1972, pp. 28–9.

66 'While in the Low Countries, Dee worked with Gemma Frisius's *Johannis, confratis & monachi, Glastoniensis, chronica sive historia de rebus Glastoniensibus* (edited by T. Hearne), Vol. II, *The compendius rehersal of John Dee his dutiful declaration, and proofe of the course of his studious life ...*, Oxford, 1726, p. 501; also J. A. Bennett, 'The Mechanic's philosophy and the mechanical philosophy', *History of science* **24** (1986), 1–28, p. 10.

66 'The cross-staff brought back by Dee from Louvain'
J. J. Roche, 'The radius astronomicus in England', *Annals of Science* **38** (1981), 1–32, especially 16–19.

66 'An example of this type of staff, made in 1571'
F. A. B. Ward, *A catalogue of European scientific instruments in the Department of Medieval and Latin Antiques of the British Museum*, London, 1981, p. 98, no. 288, made of sheet-brass-covered wooden rod, signed 'Nepos Gemmae Prisy Louanÿ ano 1571'.

66 'our country hath no man (that I ever yet could here of)'
John Dee to Sir William Ceceyl, 16 February 1562–3, *Philosophical Society biographical and historical miscellanies*, Vol. I, London, 1854, pp. 6–7, quoted by J. A. Bennett, 'The mechanic's philosophy and the mechanical philosophy', *History of science* **24** (1986), 1–28, p. 10.

66 'the French educational reformer Pierre de la Ramée'
The elements of geometrie of ... Euclide (translated by H. Billingsley), London, 1570, *With a very fruitful Preface by M. I. Dee ...* ; see P. J. French, *John Dee. The world of an Elizabethan magus*, London, 1972, p. 167.

66   'if you thinke it a blessed thing, to compasse the worlde'
T. Hood, *A copie of the speache made by the mathematical lecturer ...* , London, 1588, sig. B and Biii; quoted by J. A. Bennett, 'The mechanic's philosophy and the mechanical philosophy', *History of of science* **24** (1986), 1–28, pp. 10–11.

66   'Accounts were done using rule boards or cloth'
J. M. Pullen, *The history of the abacus,* London, 1969, pp. 43–56, and F. P. Barnard, *The casting-counter and the counting board. A chapter in the history of numismatics and early arithmetic,* Oxford, 1917, reprinted Castle Cary, Somerset, 1981.

66   'the book that opened the way to self-education'
A. W. Richeson, 'The first arithmetic printed in English', *Isis* **37** (1947), 47–56; see also J. B. Easton, 'The early editions of Robert Recorde's *Ground of artes*', *Isis* **58** (1967), 515–32, and her biography of Robert Recorde in the *Dictionary of scientific biography.*

66   'A generation of craftsmen and instrument makers'
E. G. R. Taylor, *The mathematical practitioners of Tudor and Stuart England*, Cambridge, 1954, p. 167, entry 5.

67   'Indeed, throughout the sixteenth century'
E. G. R. Taylor, *The mathematical practitioners of Tudor and Stuart England*, Cambridge, 1954, p. 313, and G. L'E. Turner, 'Mathematical instrument-making in London in the sixteenth century', *English map-making 1500–1650* (edited by S. Tyacke), London, The British Library, 1983, pp. 93–106; see also G. L'E. Turner, 'The instrument makers of Elizabethan England', *Sartoniana* [Ghent] **8** (1995), 19–31, and S. Johnston, 'Mathematical practitioners and instruments in Elizabethan England', *Annals of Science* **48** (1991), 319–44.

67   'that early group of brilliant men which included Geoffrey Chaucer'
J. D. North, 'Kalenderes enlumymed ben they. Some astronomical themes in Chaucer', *Review of English Studies*, new series, **20** (1969), 129–54, 257–83, 418–44, and J. D. North, *Richard of Wallingford. An edition of his writing with introductions, English translation and commentary*, Vol II, Clarendon Press, Oxford, 1976, pp. 309–70; see also his 'Monasticism and the first mechanical clocks', *The study of time* (edited by J. T. Fraser and N. Lawrence), Vol. II, New York, 1975, pp. 381–98.

68   'he constructed two famous large polyhedral sundials'
J. D. North, 'Nicolaus Kratzer—The king's astronomer', *Studia Copernica XVI. Science and history. Studies in honor of Edward Rosen*, Studia Copernica, 1978, pp. 205–34, and P. Pattenden, *Sundials at an Oxford college*, Roman Books, Oxford, 1979, pp. 12–24; Kratzer's Oxford sojourn is based on Anthony ˆ Wood, *History and antiquities of the colleges and halls in the university of Oxford* (edited by J. Gutch, 1796), Vol. II, part 2, p. 834.   For the Kratzer portrait with instruments by Holbein, see P. Ganz, *The paintings of Hans Holbein*, London, 1950, p. 233. For a recent summary and description of contemporary instruments, see W. D. Hackmann, 'Nicolaus Kratzer: the king's astronomer and renaissance instrument-maker', *Henry VIII: a European court in England* (edited by D. Starkey), Collins and Brown, in association with the National Maritime Museum, Greenwich, pp. 70–3, and pp. 152–3. On the sundial, see L. Evans, 'On a portable sundial of gilt brass made for Cardinal Wolsey', *Archaeologia* **62** (1901), 1–2.

68   'This task Harriot tackled both theoretically and practically'
D. W. Waters, *The art of navigation in England in Elizabethan and early Stuart times*, Hollis and Carter, London, 1958, pp. 320, 341, 373, and in particular Appendix no. 30, pp. 584–91.

68   'old collegiat, good friend and late fellow student'
N. Tyacke, 'Science and religion at Oxford before the Civil War', *Puritans and revolutionaries* (edited by P. Pennington and K. Thomas), Clarendon Press, Oxford, 1978, p. 75; J. Ward, *The lives of the professors of Gresham College*, London, 1740, pp. 77–81; J.V. Pepper's article on Edmund Gunter in the *Dictionary of scientific biography*, Vol. V, pp. 593–4; Burton's copy of Gunter's book is in Christ Church Library.

68   'Henry Briggs's table of logarithms'
Briggs's work was preceded by John Napier's *Merifici logarithmorum canonis descriptio*, 1614, translated by Edward Wright in 1616.

68   'According to Aubrey, Henry Savile sent for Gunter' and 'and brought with him his sector'
Aubrey's 'Brief lives', (edited by Andrew Clark), Vol. II, Oxford, 1898, p. 215; see also G. L'E. Turner, 'The physical sciences', *The history of the University of Oxford*, Vol. V, *The eighteenth century* (edited by L. S. Sutherland and L. G. Mitchell), Oxford University Press, 1986, pp. 659–81, especially pp. 669–70.

69   'Gemini may have imported his first rough-cast instruments'
A. J. Turner, *Early scientific instruments Europe 1400–1800*, Sotheby's, London, 1987, p. 48, and his end-notes, pp. 84–7.

69   'Both instruments have on the back a horizontal projection'
J. A. Bennett, *The divided circle. A history of instruments*

for astronomy, navigation and surveying, Phaidon, Oxford, and Christie's, 1987, p. 45; G. L'E. Turner, 'Mathematical instrument-making in London in the sixteenth century', *English map-making 1500–1650* (edited by S. Tyacke), London, The British Library, 1983, pp. 93–106, especially p. 98; R. T. Gunther, 'The great astrolabe and other scientific instruments of Humphrey Cole', *Archaeologia* **76** (1927), 273–317; Joyce Brown, *Mathematical instruments in the Grocers' Company*, London, 1979, p. 58.

70 'His son and grandson, also called Benjamin' information supplied by A. V. Simcock, librarian of the Museum of the History of Science, Oxford; he has written a brief biography on Elias Allen for the *Dictionary of national biography*, supplementary volume from the beginnings to 1985; see also E. A. Loftus, *A history of the descendants of Maximilian Cole*, London, 1938, and E. G. R. Taylor, *The mathematical practitioners of Tudor and Stuart England*, Cambridge, 1954, p. 291, entry 492.

70 'the accommodator, not the inventor' this is a deliberate misquote by David Bryden to William Robinson; see D. J. Bryden, 'A patchery and confusion of disjointed stuffe: Richard Delamain's *Grammelogia* of 1631/3', *Transactions of the Cambridge Bibliographical Society* **6** (1974), 158–66, and especially endnotes 26 and 27, and A. J. Turner, 'William Oughtred, Richard Delamain and the horizontal instrument in the seventeenth century', *Annali dell'Istituto e Museo di Storia della Scienza di Firenze* **6** (1981), 99–125. On Oughtred, see J. F. Scott's brief biography in the *Dictionary of scientific biography*, F. Cajori, *William Oughtred, a great seventeenth century teacher of mathematics*, Chicago and London, 1916, and the many references in M. Feingold, *The mathematicians' apprenticeship. Science, universities and society in England, 1560–1640*, Cambridge University Press, 1984.

70 'It is believed to have belonged to the Savilian Professor of Astronomy' there is no mention of an instrument conforming to this description in the 1697 list of Savilian instruments, reprinted in R. T. Gunther, *Early science in Oxford*, Vol. II, Oxford, 1923, p. 79; see also his article, 'The first observatory instruments of the Savilian professors at Oxford', *The Observatory* **60** (1937), 190–7.

70 'Robert Davenport, the earliest recorded instrument maker in Scotland' D. J. Bryden, 'Scotland's earliest surviving calculating device: Robert Davenport's circle of proportion of *c.* 1650', *The Scottish Historical Review* **55** (1976), 54–60.

71 'the Oxford Museum of the History of Science purchased an equinoctial ring dial' information supplied by A. V. Simcock; another equinoctial ring dial by Greatorex, but with a slightly different signature, and a horizontal dial are known by his hand. See A. J. Turner, 'William Oughtred, Richard Delamain and the horizontal instrument in the seventeenth century', *Annali dell'Istituto e Museo di Storia della Scienza di Firenze* **6** (1981), 99–125, especially p. 123, item 9; see also D. J. Bryden, *Sundials and related instruments*, Catalogue 6, Whipple Museum of the History of Science, Cambridge, 1988.

71 'He may have been attracted to establish a workshop in Oxford' J. A. Bennett, *The mathematical sciences of Christopher Wren*, Cambridge University Press, 1982, pp. 16–17, 40; this book also cites other sources.

71 'The named instruments are: *Holland's* universal quadrant …' Richard Holland, *An explanation of Mr Gunter's quadrant as it is enlarged with an analemma*, Oxford, 1676, which states on the title page, 'This Book, with all Mathematical Instruments, are Made and Sold by *John Prujean*, Mathematick-Instrument-maker, near *New College*, in Oxon; A catalogue of instruments, made and sold by John Prujean, near New-College, in Oxford. With notes of the use of them', printed on the final page of Richard Holland, *Globe notes*, 4th edn, 1701, p. 40.

71 'The names have recently been identified' D. J. Bryden, 'Made in Oxford: John Prujean's 1701 catalogue of mathematical instruments', *Oxoniensia* (Oxfordshire Architectural and Historical Society) **LVIII** (1993), 263–85. This catalogue was previously reprinted in R. T. Gunther, *Early science in Oxford*, Vol. I, Oxford, 1922, p. 180; see also D. J. Bryden, 'Evidence from advertising for mathematical instrument making in London 1556–1714', *Annals of Sciences* **49** (1992), 329–30.

71 'John Napier was the Scottish mathematician' D. J. Bryden, *Napier's bones. A history and instruction manual*, Harriet Wynter, London, 1992.

72 'the Bodleian Library holds an unmounted sheet of an astrolabe' Bodleian Library, engraving of latitude plate 51° 45', signed *Joha. Prujean Fexit Oxo:*, 175 mm diam, inv. no. Savile Mm.d 141*.

72 'the Widow of one Mr John Prujean' *Remarks and Collections of Thomas Hearne*, Vol. III (edited by C. E. Doble *et al.*), Oxford Historical Society, **xiii** (1888), p. 347.

72 'he still remains to be wrested from obscurity'
R. V. and P. J. Wallis, *Bibliography of British mathematics and its applications, Part II 1701–1769*, Project for Historical Bibliography, Epsilon Press, Newcastle upon Tyne, 1986, p. 68. My attention was drawn by A. V. Simcock to Cookcow and its variant spellings; he may have been a local person related to William Cuckoo, a cook and innkeeper in Oxford, who worked for some time in Brasenose College.

73 'The remains of the Orrery collection have been'
A. J. Turner, *Early scientific instruments Europe 1400–1800*, Sotheby's, London, 1987, pp. 231–54, and his article, 'Mathematical instruments and the education of gentlemen', *Annals of science* **30** (1973), 51–88.

CHAPTER 5

79 'Bainbridge assembled an impressive set of instruments'
R. T. Gunther, *Early science in Oxford*, Vol. II, Oxford, 1923, pp. 79–80.

82 'Naper, lord of Markinston'
John Ward, *Lives of the professors of Gresham College*, London 1740, p. 121.

82 'Great Briggs who raced with stars'
Greek poem by Henry Jacob; this translation is taken from D. M. Hallowes, 'Henry Briggs, mathematician', *Transactions of the Halifax Antiquarian Society* (1961–2), 79–92, at 86.

82 'a most exact latinist and Grecian'
Anthony à Wood, *Athenae Oxonienses* (edited by P. Bliss), 1813.

83 'Dr. Bambridge Came from Cambridge'
Anthony à Wood, *Athenae Oxonienses* (edited by P. Bliss), 1813, Vol. ii, p. 68.

85 'in his *Mathematical magick* of 1648'
John Wilkins, *Mathematical magick*, 1648, reprinted in *The mathematical and philosophical works of the Right Rev. John Wilkins*, Vol. II, London, 1802, pp. 194–211, and Allan Chapman, 'A world in the Moon; John Wilkins and his lunar voyage of 1640', *Quarterly Journal of the Royal Astronomical Society* **32** (1991), 121–32.

86 'Although the condemnation of Galileo in 1632'
Allan Chapman, 'A year of gravity', *Quarterly Journal of the Royal Astronomical Society* **34** (1993), 33–51, especially 35–8.

86 'Francis Godwin, whose speculations about a voyage to the Moon'
Francis Godwin, *The man in the moone*, 1638, reprinted in F. K. Pizor and T. A. Camp, *The man in the moon*, Sedgwick and Jackson, London, 1971, pp. 3–40.

86 'Boyle came to reside in Oxford'
letter from John Wilkins to Robert Boyle, 6 September 1653, cited from Boyle's *Works*, Vol. V, p. 629, in R. E. W. Maddison, *The life of the honorable Robert Boyle F.R.S.*, Taylor and Francis, London, 1969, pp. 81, 85; Boyle was invited to Oxford, but did not permanently reside there until 1656.

87 'Hooke realized that new *quantitative* knowledge'
Robert Hooke, *Micrographia*, London, 1665, 'The preface'.

88 'This was a *zenith sector*'
Robert Hooke, *An attempt to prove the motion of the Earth*, London, 1674, pp. 16–28.

88 'one of the first persons to recognize and publicize'
Robert Hooke, *Animadversions on the first part of the Machina Coelestis of … Johannes Hevelius*, London, 1674, pp. 46–77, and Robert Hooke, *A description of helioscopes*, London, 1676, pp. 1–32.

88 'it was Hooke who developed the tangent-screw'
Robert Hooke, 'A description of an instrument for dividing a foot into many thousand parts …', *Philosophical Transactions* **2** (1667), 541–4.

88 'Realizing that the energy given out by a spring'
Robert Hooke, Lecture *'De potentia restitutiva' or of Spring …* , London, 1678, p. 1.

90 'His "weather clock" of 1663'
see 'A weather clock. Begun by Sir Chr. Wren', in Nehemiah Crew, *Catalogue & description of the natural & artificial rarities belonging to the Royal Society*, Pt. IV, London, 1681, p. 357, and Robert Hooke, 'Method for making a history of the weather', in Thomas Sprat, *A history of the Royal Society*, London, 1667, pp. 173–9.

90 'Wren wrote four tracts on the cycloid'
John Wallis, *Opera mathematica*, 3 vols., Oxford, 1693–5. Vol. I deals with the geometry of the cycloid, pp. 496–541; for Wren's tracts, see pp. 532–41.

91 'Wren was one of the first persons to administer an intravenous drug'
Parentalia: or Memoirs of the family of the Wrens (edited by Sir Christopher Wren, junior), London, 1750, p. 230.

93 'These were devised in accordance with Hooke's own theories'
Derek Howse, *Greenwich Observatory, III, Buildings and instruments*, Taylor and Francis, London, 1975, pp. 1–6, and Derek Howse, *Francis Place and the early history of the Greenwich Observatory*, Science History Publications, London, 1975, pp. 48–61.

CHAPTER 6

97 'In the year 1649 I removed to *Oxford*'
Christoph Scriba, 'The autobiography of John Wallis,
F.R.S.', *Notes and Records of the Royal Society* **25** (1970),
40.

98 'but that some private spirits'
John Webster, *Academiarum examen*, 1654, p. 41,
reprinted in *Science and education in the seventeenth
century* (edited by Allen G. Debus), London, 1970,
p. 123.

98 '*Arithmetick* and *Geometry* are sincerely & profoundly
taught'
Seth Ward, *Vindiciae academiarum*, 1654, pp. 28, 30,
reprinted in *Science and education in the seventeenth
century* (edited by Allen G. Debus), London, 1970,
pp. 222, 224.

99 'of easy access to the studious'
cited in M. Feingold, 'Mathematical sciences and new
philosophies', in *The Oxford history of the University of
Oxford*, Vol. IV, *Seventeenth-century Oxford* (edited by
Nicholas Tyacke), 1997, p. 386.

99 'did draw his Geometricall Schemes'
Aubrey's *Brief lives* (edited by O. L. Dick), Penguin,
1962, p. 18.

99 'Mr Oughtred Clavis being one of ye best'
Mathematical papers of Isaac Newton, Vol. 1 (edited by
D. T. Whiteside), Cambridge, 1967, p. 16.

100 'harsh and difficult, so that few would venture to
approach them'
J. F. Scott, *The mathematical work of John Wallis*, Taylor
and Francis, 1938, p. 23.

100 'At the beginning of my mathematical studies'
Isaac Newton, 'Epistola posterior', 24 October 1676,
in *The correspondence of Isaac Newton*, Vol. ii, *1676–1687*
(edited by H. W. Turnbull), Cambridge, 1960,
p. 130.

100 'Essentially Wallis in his interpolation approach'
D. T. Whiteside, 'Patterns of mathematical thought in
the later seventeenth century', *Archive for History of
Exact Sciences* **1** (1960), 178–388; especially 236.

101 'why I doubt whether it be advisable'
letter from Wallis to the Earl of Nottingham, 20
February 1690, reproduced in David Eugene Smith,
'John Wallis as a cryptographer', *Bulletin of the
American Mathematical Society* **24** (1917), 82–96;
especially 89–90.

103 'He was 40 years old before he looked on
Geometry'
Aubrey's *Brief Lives* (edited by O.L. Dick), Penguin,
1962, p. 230.

103 'the propriety of the separate authority'
Simon Schaffer, 'Wallifaction: Thomas Hobbes on
school divinity and experimental pneumatics', *Studies
in History and Philosophy of Science* **19** (1988), 275–98;
especially 288.

104 'he told me he loved not to be expostulated with'
Anthony à Wood, *Athenae Oxonienses* (edited by
P. Bliss), Vol. I, London 1813, p. xcix.

105 ' covered over with the scab of symbols'
Thomas Hobbes, *Six lessons to the professors of
mathematics*, 1656.

105 'Something was offered about a design'
from the first *Journal Book of the Royal Society*, cited in
The Royal Society: its origins and founders (edited by
H. Hartley), The Royal Society, 1960, p. 31.

106 'a semi-official mathematical sub-committee'
The Royal Society: its origins and founders (edited by
H. Hartley), The Royal Society, 1960, p. 150.

109 'And (herein as in other Studies)'
Christoph Scriba, 'The autobiography of John Wallis,
F.R.S.', *Notes and records of the Royal Society* **25** (1970),
pp. 40–1.

109 'ancient Mathematicians Greek and latin'
cited in Peter Sutcliffe, *The Oxford University Press: an
informal history*, Clarendon Press, Oxford, 1978, p. xxi.

110 'Books and experiments do well together'
letter from Edward Bernard to John Collins, 3 April
1671, in *Correspondence of scientific men of the
seventeenth century* (edited by S. Rigaud), Vol. i, Oxford,
1841, p. 158.

110 'I can by no means admit your excuse'
The correspondence of Isaac Newton, Vol. iv: *1694–1709*
(edited by J. F. Scott), Cambridge, 1967, pp. 116–17.

110 'The Art of printing was first brought into England'
cited in Peter Sutcliffe, *The Oxford University Press: an
informal history*, Clarendon Press, Oxford, 1978, p. xiii.

111 'I think Oxford the most convenient place for it'
letter from Wallis to Newton, 30 May 1695, *The
correspondence of Isaac Newton*, Vol. iv: *1694–1709*
(edited by J. F. Scott), Cambridge, 1967, p. 129.

111 'hath taught (in a manner)'
John Wallis, *A treatise of algebra*, 1685, Preface.

113 'I do account him prudent sober industrious'
letter from Newton to Arthur Charlett, 27 July 1691, in
The correspondence of Isaac Newton, Vol. iii: *1688–1694*
(edited by H. W. Turnbull), Cambridge, 1961,
pp. 154–5.

115 'He says 83 is an incurable distemper'
Arthur Charlett to Samuel Pepys, 17 May 1699, *Private*

correspondence and miscellaneous papers of Samuel Pepys 1679–1703, vol. i, (edited by J. R. Tanner), London 1926, p. 175.

115 'the perfection wherewith the Very Mathematicks'
Private correspondence and miscellaneous papers of Samuel Pepys 1679–1703, vol. ii (edited by J. R. Tanner), London, 1926.

CHAPTER 7

118 'Edmond Halley, a talented young man of Oxford'
J. Flamsteed, *Philosophical Transactions* **10** (1675), 368–70, original in Latin, translated here by Alan Cook.

122 'corrupt the youth of the University'
C. A. Ronan, *Edmond Halley, genius in eclipse*, MacDonald, London, 1970, p. 124; this quotation is given by Ronan without an exact source, and comes from one of the three unpublished drafts of the same letter from Flamsteed to Newton, 20–2 February 1691, Roy. Obs. Ms. 42, p. 129; for details, see H. W. Turnbull, *The correspondence of Sir Isaac Newton*, Vol. III, Cambridge University Press, 1961, pp. 199–205, although the version of the letter published (the final draft of 24 February 1691) does not contain the quotation.

123 'Flamsteed frequently spoke of Halley as "Newton's Creature"'
Flamsteed castigated Halley in a suppressed section of his *Historia coelestis Britannica*, the original manuscript of which is in the Royal Society Library; see *The 'Historia coelestis Britannica' of John Flamsteed, 1725*, edited by Allan Chapman, National Maritime Museum Monograph **52** (1982), 160–80.

125 'was the same as the first 1680 comet'
D. K. Yeomans, *Comets*, John Wiley, New York, 1991, p. 99.

127 'When the German scholar Conrad von Uffenbach visited England'
Zacharias C. von Uffenbach, *London in 1710* (translated by W. H. Quarrell and M. Mare), London, 1934, p. 142.

127 'recording astronomical events such as that from Palmyra'
Edmond Halley, 'Some remarks of the ancient state of the city of Palmyra', *Philosophical Transactions* **19** (1695), 160–75.

128 'It may justly be reckoned one of the principal uses'
Edmond Halley, 'An account of the causes of the late remarkable appearance of the planet Venus', *Philosophical Transactions* **29** (1716), 466–8.

128 'But by now, Halley's astronomical and scientific reputation'

A. H. Cook, 'The election of Edmond Halley to the Savilian Professorship of Geometry', *Journal for the History of Astronomy* **15** (1984), 34–6.

129 'Mr Hally made his Inaugural Speech'
Letter of Thomas Hearne, 8 June 1704, in Bodleian Library; cited in Alan Cook, *Edmund Halley: Charting the heavens and the seas*, Clarendon Press, Oxford, 1997, p. 322.

129 'a reconstruction of the fragmentary *Sectio spatii*'
Apollonii Pergaei conicorum libro octo (edited by Edmond Halley), Oxford, 1710.

130 'But his real contributions to astronomy during 1704–20'
H. E. Bell, 'The Savilian professors' houses and Halley's observatory at Oxford', *Notes and Records of the Royal Society* **16**, no. 2 (November 1961), 179–86, and R. T. Gunther, *Early science in Oxford*, Vol. II, Oxford, 1923, pp. 82–5.

131 'In 1715–16, Halley published a study of six variable stars'
Edmond Halley, 'A short history of the several new-stars that have appeared within these last 150 years', *Philosophical Transactions* **29** (1715), 354–6.

131 'perhaps no less than our whole solar system'
Edmond Halley, 'An account of some nebulae', *Philosophical Transactions* **29** (1716), 390–2.

134 'His critics in the 1720s always accused him'
letter from Joseph Crossthwaite to Abraham Sharp, 29 August 1730, *An account of the Revd. John Flamsteed* (edited by Francis Baily), London, 1832, p. 363.

134 'The transit, clock, and mural quadrant'
Derek Howse, *Greenwich Observatory, III, The buildings and instruments*, Taylor & Francis, London, 1975, pp. 21, 32, 126.

CHAPTER 8

138 'was the first who publickly taught Natural Philosophy'
J. T. Desaguliers, preface to *A course of experimental philosophy*, London, 1734.

139 'their promotion of the Newtonian philosophy'
R. T. Gunther, *Early science in Oxford*, Vol. II, Oxford, 1923. For a recent article on Whiteside, see A. V. Simcock's entry in the *Dictionary of national biography: Missing persons*, Oxford University Press, 1993, pp. 712–13.

141 'Bradley seized upon an invitation from Halley'
Stephen Peter Rigaud, *Miscellaneous works and correspondence of the Rev. James Bradley*, Oxford, 1832, p. iii.

142 'With this, he was able to observe a wider range of zenith stars'
James Bradley, 'A letter to Dr Edmond Halley, Astronom. Reg. & Co., giving an Account of a new-discovered Motion of the Fixed Stars', *Philosophical Transactions* **35** (1718), 637; reprinted in Stephen Peter Rigaud, *Miscellaneous works and correspondence of the Rev. James Bradley* (edited by P. Rigaud), Oxford, 1832, pp. 1–16.

143 'He named the new motion "nutation"'
James Bradley, 'A letter to the Rt. Hon. George, Earl of Macclesfield, concerning an apparent Motion observed in some of the Fixed Stars', *Philosophical Transactions* **45** (1748), 1; reprinted in Stephen Peter Rigaud, *Miscellaneous works and correspondence of the Rev. James Bradley* (edited by P. Rigaud), Oxford, 1832, pp. 17–41.

145 'When one adds together his various salaries and fees'
R. T. Gunther, *Early science in Oxford*, Vol. XI, Oxford, 1937, pp. 65–70, and A. Chapman, 'Pure research and practical teaching; the astronomical career of James Bradley, 1693–1762', *Notes and Records of the Royal Society* **47** (2) (1993), 205–12.

147 'his Greenwich observations were not published until 1805'
James Bradley, *Astronomical Observations, made … from … mdccl to mdcclxii …* (edited by Thomas Hornsby), Vol. II, Oxford, 1805, pp. 345–420.

148 'It is unfortunate that his untimely death'
R. T. Gunther, *Early science in Oxford*, Vol. XI, Oxford, 1937, pp. 275–7.

CHAPTER 9

151 'If any number of schollars'
David Gregory, *Mercurius Oxoniensis*, 1707.

152 'You know, I believe, that poor Dr Gregory is dead'
letter from Jonathan Swift to Robert Hunter, 12 January 1709, *Correspondence of Jonathan Swift*, Vol. i (edited by Harold Williams), Oxford, 1963, p. 121.

154 'proud, Fanaticall Whigg'
Remarks and collections of Thomas Hearne, Vol. iii, *1710–12* (edited by D. W. Rannie), Oxford Historical Society **xiii** (1888), p. 439.

154 'was turn'd out upon the D. of Brunswick's coming to the Crown'
Remarks and collections of Thomas Hearne, Vol. iv, *1712–14* (edited by D. W. Rannie), Oxford Historical Society **xxxiv** (1898), p. 69.

154 'One Mr Sterling, a Non-juror of Bal. Coll.'
Remarks and collections of Thomas Hearne, Vol. v, *1714–16*

(edited by C. E. Doble *et al.*), Oxford Historical Society **xlii** (1901), p. 268.

154 'the turning-point in contemporary appreciation' and 'produced the first proficient exposition'
D. T. Whiteside, *The mathematical papers of Isaac Newton*, Vol. vii: *1691–1695*, Cambridge University Press, 1976, pp. 574–5.

155 'Mathematics are studied, Your Grace knows very well'
letter from George Carter to Archbishop Wake, 7 September 1721, Wake MS 16, f. 88, cited in *The Oxford History of Oxford University*, Vol. V, *The eighteenth century* (edited by L.S. Sutherland and L. G. Mitchell), Oxford University Press, 1986, p. 481.

155 'I have known a profligate *Debauchee*'
Nicholas Amhurst, *Terrae filius*, 1721, no. 21.

156 'In the university of Oxford, the greater part'
Adam Smith, *An inquiry into the nature and causes of the wealth of nations* (1776), Vol. 2 (edited by R. H. Campbell and A. S. Skinner), Oxford University Press, 1976, p. 761.

156 'it will be his own fault if anyone should endanger'
letter from Adam Smith to William Smith, 24 August 1740, from Adam Smith, *An inquiry into the nature and causes of the wealth of nations* (1776), Vol. 2 (edited by R. H. Campbell and A. S. Skinner), Oxford University Press, 1976.

157 'small and complex society, set in an equally small and beautiful … town'
Dame Lucy Sutherland, *The University of Oxford in the eighteenth century: a reconsideration*, Somerville College, 1973, p. 12.

157 'I read here much, and like vastly'
letter from C. J. Fox to Sir George Macartney, 13 February 1765, Lord John Russell, *Memorials and correspondence of Charles James Fox*, Vol. i, London, 1853, p. 19.

157 'had a wonderful capacity for calculation'
Lord John Russell, *Memorials and correspondence of Charles James Fox*, Vol. i, London, 1853, p. 19.

157 'I expect you will return with much keenness for Greek'
Lord John Russell, *Memorials and correspondence of Charles James Fox*, Vol. i, London, 1853, p. 20.

157 'The two years of my life I look back to'
Diaries and correspondence of James Harris, first earl of Malmesbury, Vol. I (edited by Lord Malmesbury), London, 1844, p. ix.

158 'I rise about nine, get to breakfast by ten'
The Oxford sausage, 1764, quoted in Albert Mansbridge,

The older universities of England, Longman, London, 1923, p. 113.

158 'We have gone through the Science of Mechanics'
letter from Jeremy Bentham to Jeremiah Bentham, 1767, *The correspondence of Jeremy Bentham*, Vol I: *1752–76* (edited by T. L. S. Sprigge), Athlone Press, London, 1968, p. 67.

158 'the Course will last no longer than a week'
letter from Jeremy Bentham to Jeremiah Bentham, 15 March 1763, *The correspondence of Jeremy Bentham*, Vol I: *1752–76* (edited by T. L. S. Sprigge), Athlone Press, London, 1968, p. 108.

158 'only classical and historical knowledge could make able statesmen'
The Countess of Minto, *Life and letters of Sir Gilbert Elliot*, Vol. i, London, 1874, p. 38.

158 'the collections books from 1699 onwards have been preserved'
P. Quarrie, The Christ Church collections books, *The history of the University of Oxford, Vol. V: The eighteenth century* (edited by L. S. Sutherland and L. G. Mitchell), Oxford, 1986, pp. 493–511.

159 ' "the Execution of the Criminal interfered with another lecture" '
S. Love, *I am at a loss*, Bodleian Library, shelfmark Gough Oxf. 90 (10).

159 'Williamson was a real disciple of Euclid'
Augustus De Morgan, *Arithmetical books*, 1847, p. 78.

160 'As to his manuscripts lately discovered'
Charles Hutton, *Mathematical and philosophical dictionary*, 1797, Vol. i, p. 586.

161 'undoubtedly went far to prepare a way'
Robert T. Gunther and the Old Ashmolean (edited by A. V. Simcock), Museum of the History of Science, Oxford, 1985, p. 14.

161 'Whilst sitting apart, the junior examiner'
Josiah Bateman, *The life of the Right Rev. Daniel Wilson*, Vol. i, John Murray, London, 1860; pp. 66–7.

162 'except as regards conic sections and matters of that kind'
Norman Gash, *Mr Secretary Peel: the life of Sir Robert Peel to 1830*, London, 1961, p. 56.

162 'I should like a little previous consideration before I move'
letter from Robert Peel to John Wilson Croker, 8 March 1823, cited in Anthony Hyman, *Charles Babbage: pioneer of the computer*, Oxford University Press, 1982, p. 52.

162 'all the undergraduates at Christ Church read the "Principia" '
The diary of the Rt. Hon William Wyndham 1784–810 (edited by H. Baring), London, 1866, p. 244.

162 'I had but 38 hours between my first and second examinations'
Vanda Morton, *Oxford rebels: the life and friends of Nevil Story Maskelyne 1823–1911*, Alan Sutton, Gloucester, 1987, p. 31.

164 ' "They are very dull people here", Shelley said to me'
Thomas Jefferson Hogg, *The life of Percy Bysshe Shelley*, Vol. I, London, 1858, p. 70.

164 'A half-burnt match, an ivory block'
Percy Bysshe Shelley, letter to Maria Gisborne.

164 'He gained a Double First in 1826'
William Tuckwell, *Reminiscences of Oxford*, London, 1907, p. 204.

165 'December 9. In the Schools six hours'
W. E. Gladstone, cited in *The Oxford book of Oxford* (edited by Jan Morris), Oxford University Press, 1978, p. 303.

166 'where the dictates of Aristotle are still listened to'
John Playfair, *Edinburgh Review* **xi** (January 1808), 283.

166 'What are the mere elements of geometry?'
Edward Copleston, *A reply to the calumnies of the Edinburgh Review against Oxford*, Oxford, 1810, p. 18.

Chapter 10

169 'Hornsby was already acquiring his own kit of instruments'
A. D. Thackeray, *The Radcliffe Observatory, 1772–1972*, Radcliffe Trust, 1972, especially Appendix A, p. 33, for Hornsby's 'Petition' of 1768 outlining the need for an observatory.

169 'an observatory at Blenheim Palace'
Ivor Guest, *Dr John Radcliffe and his Trust*, Radcliffe Trust, 1991, p. 242. This twelve foot Gregorian Reflector was given to the Radcliffe Observatory by the Duke of Marlborough, but remained in its packing case for a century, by which time it was obsolete; it is now on display in the Museum of the History of Science, Oxford.

170 'Hornsby, moreover, was to be provided with a spacious house'
Ivor Guest, *Dr John Radcliffe and his Trust*, Radcliffe Trust, 1991, p. 224 onwards for building details.

172 'For the sum of £1260'
John Bird (and his executors after 1776) received a total of £1392 16s for his instruments; see A. D. Thackeray, *The Radcliffe Observatory, 1772–1972*, Radcliffe Trust, 1972, p. 4.

172 'The entire set of instruments was estimated at £2500' the instruments supplied are listed in R. T. Gunther, *Early science in Oxford*, Vol. II, Oxford, 1923, under the various classes supplied; Dolland's bill for £140 for the supply of a three-foot and a $10\frac{1}{2}$-foot achromatic refractor and their fittings is given on p. 396.

172 'no doubt the best in Europe' Thomas Bugge's 'Travel diary' to England, 13 October 1777, Royal Library, Copenhagen, Ms. Ny kgl Saml, 377e; I wish to thank Professor Gerard L'E. Turner for permission to quote from a manuscript translation.

173 'the Astro. Roy. will be before you' Bird to Hornsby, 29 January 1771, Royal Astronomical Society Library, Ms. A. 1. 13.

173 'The lessons in precision that he learned' John Bird, *The method of constructing mural quadrants*, London, 1768, p. 7, and Derek Howse, *Greenwich Observatory III, The buildings and instruments*, Taylor and Francis, London, 1975, pp. 21–6.

173 'Hornsby's quadrants were regarded' Arthur A. Rambaut, 'Note on the unpublished observations made at the Radcliffe Observatory, Oxford, between 1774 and 1838 with some results for the year 1774', *Monthly Notices of the Royal Astronomical Society* **1x** (4 January 1900), 265.

173 'scale of equal parts' John Bird, *Method of dividing astronomical instruments*, London, 1767, pp. 2–3.

174 'Apart from monitoring the aberration and nutation' Thomas Bugge left a detailed description of the zenith sector in his 'Travel diary' to England, pp. 60 recto and verso; Royal Library, Copenhagen, Ms. Ny kgl Saml, 377e; see also H. Knox-Shaw, J. Jackson, and W. H. Robinson, *The observations of the Reverend Thomas Hornsby, D.D.*, Oxford University Press, 1932, p 8.

175 'In October 1777, the equatorial sector was examined' Thomas Bugge's 'Travel diary' to England, p. 66v. Royal Library, Copenhagen, Ms. Ny kgl Saml, 377e.

176 'The priorities of the Radcliffe Trustees' the person who foresaw the need for an assistant at the Radcliffe Observatory was John Bird himself, and in 1776 he left £1000 to provide an endowment for an assistant's salary, although nothing was paid from it until 1847; see A. D. Thackeray, *The Radcliffe Observatory, 1772–1972*, Radcliffe Trust, 1972, p. 7. Part of the problem with donations to the observatory came from the fact that although Hornsby was Savilian professor at Oxford University, the observatory was a separate body owned and administered by the Radcliffe Trustees and not by the Congregation of Oxford University.

176 'Thomas Hornsby himself never expected to be paid' Ivor Guest, *Dr John Radcliffe and his Trust*, Radcliffe Trust, 1991, pp. 247 (Sedleian professor) and 497 (Radcliffe librarian).

176 'Returning to the quadrant ... he could then read off the scales' H. Knox-Shaw, J. Jackson, and W. H. Robinson, *The observations of the Reverend Thomas Hornsby, D.D.*, Oxford University Press, 1932, p. 9; on p. 70, the authors discuss another variant on this procedure used by Hornsby.

177 'The first thing that any eighteenth-century astronomer' A. Chapman, *Dividing the circle: the development of critical angular measurement in astronomy 1500–1850*, Chichester and New York, 1990, pp. 83–7; originally, Hornsby used his 32-inch quadrant to obtain a slightly less precise latitude, before fixing it exactly with the eight-foot Bird quadrant.

177 'Thomas Bugge describes this modern marker' Thomas Bugge's 'Travel diary' to England, p. 61v. Royal Library, Copenhagen, Ms. Ny kgl Saml, 377e.

177 'Hornsby made full atmospheric corrections' the Radcliffe Observatory can also boast of having maintained one of the longest and most complete meteorological records in the history of the science; see C. G. Smith, *The Radcliffe meteorological station, Oxford*, **xxii**, 1 May, 1968, pp. 77–89. In spite of its removal to Pretoria in 1935, the Radcliffe Meteorological Observer still operates from the University Department of Geography.

178 'Harold Knox-Shaw and his colleagues' H. Knox-Shaw, J. Jackson, and W. H. Robinson, *The observations of the Reverend Thomas Hornsby, D.D*; Oxford University Press, 1932; for Hornsby's values, see Arthur A. Rambaut, 'Note on the unpublished observations made at the Radcliffe Observatory, Oxford', between 1774 and 1838 with some results for the year 1774', *Monthly notices of the Royal Astronomical Society* **1x** (4 January 1900), pp. 277–8.

179 'read a course of lectures in Practical Astronomy' A. D. Thackeray, *The Radcliffe Observatory, 1772–1972*, Radcliffe Trust, 1972, especially Appendix A, p. 33, for Hornsby's 'Petition' of 1768.

179 'In 1790 the eminent instrument maker Edward Nairne' Edward Nairne's unpublished valuation is in Bodleian Ms. Top. c. 236, fols. 3–13.

180 'As soon as I have breakfasted I must go to the Museum'
letter from Thomas Hornsby to the Duke of Marlborough, November 1781.

181 'Arthur Rambaut, the Radcliffe Observer'
A. D. Thackeray, *The Radcliffe Observatory, 1772–1972*, Radcliffe Trust, 1972, p. 6, and Arthur A. Rambaut, 'Note on the unpublished observations made at the Radcliffe Observatory, Oxford', between 1774 and 1838 with some results for the year 1774', *Monthly notices of the Royal Astronomical Society* **lx** (4 January 1900), pp. 277–8.

181 'It was not until 1932 that Harold Knox-Shaw obtained funding'
H. Knox-Shaw, J. Jackson, and W. H. Robinson, *The observations of the Reverend Thomas Hornsby, D.D.*, Oxford University Press, 1932, p. 8 for table of reduced observations. It is unfortunate for Hornsby that none of his immediate Radcliffe successors devoted themselves to the publication of his work as he had done for Bradley's; indeed, S. P. Rigaud, Radcliffe Observer from 1826 to 1839, spent much of his time writing his massive *Life of Bradley*.

181 'Knox-Shaw concluded that, while Hornsby's declinations'
H. Knox-Shaw, J. Jackson, and W. H. Robinson, *The observations of the Reverend Thomas Hornsby, D.D.*, Oxford University Press, 1932, p. 90.

181 'Jesse Ramsden and John Troughton … in London were experimenting'
for the astronomical circle, see A. Chapman, *Dividing the circle: the development of critical angular measurement in astronomy 1500–1850*, Chichester and New York, 1990 and 1995, Chapter 7.

182 'Robertson's income was augmented to £300 per annum'
Ivor Guest, *Dr John Radcliffe and his Trust*, Radcliffe Trust, 1991, pp. 251–3.

182 'When Robertson died in 1826, he was followed as Radcliffe Observer'
for Hornsby's successors, see A. D. Thackeray, *The Radcliffe Observatory, 1772–1972*, Radcliffe Trust, 1972, pp. 8–12.

182 'Newman spent his last night in Oxford'
John Henry Newman, *Apologia pro vita sua*, part VI (end), 1864; Scott edition, 1913, Vol. 2, p. 97.

182 'In 1843, William Simms was also engaged to build'
for examination of the Jones circle, see Manuel Johnson, *Astronomical observations made at the Radcliffe Observatory for the year 1840*, Vol. 1, Oxford, 1842,

pp. xiii–xxii, and R. T. Gunther, *Early science in Oxford*, Vol. II, Oxford, 1923, p. 326.

182 'Although an eight-inch reflector of ten feet focal length'
Ivor Guest, *Dr John Radcliffe and his Trust*, Radcliffe Trust, 1991, p. 253, states that the Radcliffe Observatory obtained a ten-foot achromatic telescope from Sir William Herschel in 1813. This is not quite correct, for 'achromatic' is generally taken to imply a refracting telescope, and Herschel never made refractors; this instrument is the ten-foot-focus Newtonian reflector in Ackerman's 1814 print of the interior of the observatory.

183 'the magnificent 7½-inch Merz-Repsold heliometer from Hamburg'
for a description of the heliometer, see Manuel Johnson, *Astron. observ. at the Radcliffe Observatory*, Vol. XI onwards, Oxford, 1852, and H. C. King, *History of the telescope*, Griffin, London, 1955, p. 242.

183 'not until 1840, when Lucas was appointed'
for assistants' salaries, see Ivor Guest, *Dr John Radcliffe and his Trust*, Radcliffe Trust, 1991, p. 263; see also the first reference for p. 10.9.

183 'he secured money for building a new observatory'
for an account of the new Savilian Observatory and its moving spirit, Professor Pritchard, see *Charles Pritchard, Memoirs of his life* (edited by Ada Pritchard), London, 1897, pp. 260–316.

184 'the Radcliffe Observatory was finally moved to Pretoria'
A. D. Thackeray, *The Radcliffe Observatory, 1772–1972*, Radcliffe Trust, 1972, p. 21 onwards, and Ivor Guest, *Dr John Radcliffe and his Trust*, Radcliffe Trust, 1991, Chapter VII (The Radcliffe Observatory in Pretoria).

CHAPTER 11

189 'Hamilton's work on dispersion'
T. L. Hankins, *Sir William Rowan Hamilton*, Johns Hopkins University, Baltimore, 1980, p. 158.

189 'According to Oxford gossip, Charles Babbage'
W. Tuckwell, *Pre-Tractarian Oxford*, Cassell, London, 1909, p. 166.

189 'this is not borne out by Babbage's correspondence'
A. Hyman, *Charles Babbage*, Oxford University Press, Oxford, 1982, p. 63.

190 'though a certain portion had "got up" the four books of Euclid'
Baden Powell, *On the present state and future prospects of mathematics at the University of Oxford*, Oxford, 1832, p. 40.

190 'His complaint is corroborated'
J. Pycroft, *Oxford memories,* Bentley, London, 1886, p. 82.

190 'Early holders of these scholarships included'
I. Anderson, 'Cyclic designs in the 1850s; the work of Rev. R. R. Anstice, *Bulletin of the Institute of Combinatorics and its Applications* **15** (1995), 41–6.

190 'The visits of the BAAS gave science a higher profile'
J. Morrell and A. Thackray, *Gentlemen of science,* Clarendon Press, Oxford, 1981, pp. 386–96.

192 'The relationship with the BAAS turned sour'
W. Odling *et al.,* letter to Captain Douglas Galton, 20 June 1882, Bodleian ms, Oxon 8¡, p. 1079 (13).

192 'Charles Darwin later acknowledged'
C. Darwin, *The origin of species,* John Murray, London, 1894, p. xx.

192 'a hut in the grounds would meet his needs perfectly'
V. Morton, *Oxford Rebels, the life and friends of Nevil Story Maskelyne,* Alan Sutton, Gloucester, 1987, p. 71.

193 'His early papers on the least squares method'
T. M. Porter, *The rise of statistical thinking,* Princeton University Press, Princeton, 1986, p. 87.

193 'Miss Nightingale asked Jowett to come'
G. Faber, *Jowett,* Faber and Faber, London, 1957, p. 310.

193 'The only fitting memorial to Q.'
M. Diamond and M. Stone, 'Nightingale on Quetelet', *J. Royal Statistical Soc.* **A144** (1981), 66–79, 176–213, 332–51.

193 'overloaded ... with mathematics'
E. Abbott and L. Campbell, *The life and letters of Benjamin Jowett,* John Murray, London, 1897, p. 378.

194 'only end in endowing some bacillus or microbe'
M. Diamond and M. Stone, 'Nightingale on Quetelet'. *J. Royal Statistical Soc.* **A144** (1981) 66–79, 176–213, 332–51.

194 'give or bequeath 2000£ for the endowment'
E. V. Quinn and J. M. Prest, *Dear Miss Nightingale,* Clarendon Press, Oxford, 1987, p. 314.

194 'as I fear that there is no possibility'
E. Abbott and L. Campbell, *The life and letters of Benjamin Jowett,* John Murray, London, 1897, p. 478.

194 'It is often said that modern mathematical statistics was founded'
S. M. Stigler, *The history of statistics,* Harvard University Press, Cambridge, Mass., 1986, p. 266.

194 'stimulated his interest in the 'moral sciences' '
J. M. Keynes, *Francis Ysidro Edgeworth,* in the *Dictionary of national biography*; see also *Essays in biography* (edited by J. M. Keynes), Macmillan, London, 1933.

195 'his contributions to mathematical economics'
J. Creedy, *Edgeworth and the development of neoclassical economics,* Blackwell, Oxford, 1986.

195 'Edgeworth and Graves at one time'
R. Graves, *Goodbye to all that,* Penguin, London, 1960, p. 246.

195 'continued to publish a steady stream of papers'
Anonymous, 'Scientific worthies: William Spottiswoode', *Nature* **27** (1883), 597–601.

195 'In an 1861 paper, he attempted to analyse'
W. Spottiswoode, 'On typical mountain ranges: an application of the calculus of probabilities to physical geography', *J. Royal Geographical Soc.* **31** (1861), 149–54.

195 'Yet what are all these gaieties to me'
Lewis Carroll, *The collected works of Lewis Carroll,* Nonesuch, London, 1939, p. 802.

196 'The new method of evaluation, as applied to π'
Lewis Carroll, *The collected works of Lewis Carroll,* Nonesuch, London, 1939, p. 1011.

196 'Who shall say what germ of romance'
Lewis Carroll, *The collected works of Lewis Carroll,* Nonesuch, London, 1939, p. 1016.

197 'THE OFFER OF THE CLARENDON TRUSTEES'
Lewis Carroll, *The collected works of Lewis Carroll,* Nonesuch, London, 1939, pp. 1009–11.

198 'Some of this work on elections and tournaments'
F. Abeles, 'The mathematico-political papers of C. L. Dodgson', *Lewis Carroll, a celebration: essays on the occasion of the 150th anniversary of the birth of Charles Lutwidge Dodgson* (edited by E. Guiliano), C. N. Potter, New York, 1982, and N. T. Gridgeman, 'Charles Lutwidge Dodgson', *Dictionary of scientific biography*; see also D. Black, *The theory of committees and elections,* Cambridge University Press, 1958, which contains Dodgson's work as an appendix.

198 'A similar fate overtook his work in logic'
W. W. Bartley III, 'Lewis Carroll's lost notebooks on logic', *Scientific American* (July 1972), 38–46, and *'Lewis Carroll's symbolic logic* (edited by W. W. Bartley III), Potter, New York, 1977.

198 'his more advanced *Treatise on determinants*'
C. L. Dodgson, *Treatise on determinants,* Macmillan, London, 1867.

198 'it has recently excited the attention of combinatorialists'
D. P. Robbins, '1, 2, 7, 42, 429, 7436, ...', *Mathematical Intelligencer* **13** (1) (spring 1991), 12–19.

198 'many reflect his interest in mathematical education'
D. B. Esperson, 'Lewis Carroll, mathematician',

Mathematical Gazette **17** (1933), 92–100, and W. Weaver, 'Lewis Carroll', *Scientific American* (April 1956), 116–28.

198 'much of his teaching was not for the final honours School'
T. B. Strong, 'Lewis Carroll', *Cornhill Magazine* **4** (1898), 303–10.

198 'a new algorithm for long division'
C. L. Dodgson, letter in *Nature* **57** (1898), 269–71.

198 'There one finds a description of a Möbius band'
Lewis Carroll, *The collected works of Lewis Carroll*, Nonesuch, London, 1939, pp. 521–3.

198 'how to make a model of a projective plane'
M. Cohen and R. L. Green, *The Letters of Lewis Carroll*, Macmillan, London, 1973, p. 969.

198 'a tea party in a freely falling room'
Lewis Carroll, *The collected works of Lewis Carroll*, Nonesuch, London, 1939, pp. 312–13.

199 'Price's most enduring contribution to the University'
Peter Sutcliffe, *The Oxford University Press* (Part II, Chapters 2 and 4), Oxford University Press, 1978.

199 'influenced Edwin Abbott's more famous *Flatland*'
C. H. Hinton, *Speculations on the fourth dimension* (edited by R. v. B. Rücker), Dover, New York, 1980, p. viii.

200 'modern compactified Kaluza–Klein theories'
M. Gardner, *The ambidextrous universe*, Penguin, Harmondsworth, 1967, p. 226.

200 'Hinton married Mary Boole'
D. MacHale, *George Boole*, Boole Press, Dublin, 1985, p. 259.

200 'a mathematician of great talent'
G. H. Hardy, *Ramanujan*, 3rd edn, Chelsea, New York, 1978, p. 91.

200 'A generalization of the above identity'
G. H. Hardy and E. M. Wright, *The theory of numbers*, 5th edn, Oxford University Press, Oxford, 1979, p. 296.

200 'remarkable among our fellows'
A. L. Dixon, 'Leonard James Rogers', *Obituary Notices of Fellows of the Royal Society* **3** (1934), 299–301.

200 'the mathematician who discovered the gap'
N. L. Biggs, E. K. Lloyd, and R. J. Wilson, *Graph theory 1736–1936*, Clarendon Press, Oxford, 1998, p. 105.

201 'the first English monograph on Lie groups'
J. E. Campbell, *Continuous groups*, Oxford University Press, Oxford, 1903.

CHAPTER 12

204 'in the course of one particularly tedious meeting'
E. G. Sandford, *Memoirs of Archbishop Temple*, Macmillan, London, 1906, appendix to Chapter V.

204 'they prefer having work ill done'
C. H. Pearson, *Biographical sketches and recollections (with early letters) of Henry John Stephen Smith*, Oxford University Press, Oxford, 1894, p. 39.

205 'Smith ran the first college science laboratory'
T. Smith, 'The Balliol–Trinity laboratories', in *Balliol studies* (edited by J. M. Prest), Leopard's Head Press, London, 1982, p. 190.

205 'His very first pupil'
K. J. Laidler, 'Chemical kinetics and the Oxford college laboratories', *Archive for the History of the Exact Sciences* **38** (1988), 197–283, and M. King, 'The life and times of Vernon Harcourt (1834-1919)', *Ambix* **31** (1984), 17–31.

205 'worked with Brodie on his eclectic atomic theory'
W. H. Brock, *The atomic debates*, Leicester University Press, Leicester, 1967, p. 113, and A. Vernon Harcourt, 'The Oxford Museum and its founders', *Cornhill Magazine* **28** (1910), 350–63.

205 'with Maskelyne on crystallography'
H. A. Miers, 'Prof. N. Story Maskelyne', *Nature* **86** (1910), 452–3.

206 'Minkowski cited the publication of Smith's report'
H. Minkowski, *Briefe an David Hilbert*, Springer, Berlin, 1973, p. 78.

206 'Jowett's essay "the *best*... the worst"'
G. C. Smith, *The Boole–De Morgan correspondence 1842–1864*, Oxford University Press, Oxford, 1982, p. 83.

206 'the impression that his application was not serious'
D. MacHale, *George Boole*, Boole Press, Dublin, 1985, p. 168.

207 'the electors for the chair'
H. J. S. Smith, *Collected mathematical papers*,, Vol. I (edited by J. W. L. Glaisher), Oxford University Press, Oxford, 1894, p. xxv.

209 'which appeared in 1861'
H. J. S. Smith, 'On systems of linear indeterminate equations and congruences', *Philosophical Transactions* **151** (1861), 293–326.

210 'practically impossible to test the results'
Proceedings of the Royal Society **16** (1867), 197–208.

210 'His genius and accomplishments'
E. Abbott and L. Campbell, *The life and letters of Benjamin Jowett*, John Murray, London, 1897, p. 238.

210 'he sometimes forgets that he is only the editor'
H. J. S. Smith, *Collected mathematical papers*, Vol. I (edited by J. W. L. Glaisher), Oxford University Press, Oxford, 1894, p. xliv.

210 'It is the peculiar beauty of this method'
H. J. S. Smith, *Collected mathematical papers*, Vol. I (edited by J. W. L. Glaisher), Oxford University Press, Oxford, 1894, p. xxxiii.

211 'I do not know what Henry Smith may be'
H. J. S. Smith, *Collected mathematical papers*, Vol. I (edited by J. W. L. Glaisher), Oxford University Press, Oxford, 1894, p. xix.

211 'It is probable that of the thousands'
Obituary notice, *The Times*, 10 February 1883.

211 'the honour … and in order to be near the croquet ground'
E. Smith, Letter to Charles Pearson, 29 October 1874, Bodleian Library, Pearson Collection, Eng. lett. d. l91.

212 'Helmholtz mentioned learning of the plan'
N. Kurti, 'Opportunity lost in 1865?', *Nature* **308** (1984), 313–14.

212 'In 1876, Smith entertained … Ferdinand Lindemann'
H. J. S. Smith, *Collected mathematical papers*, Vol. I (edited by J. W. L. Glaisher), Oxford University Press, Oxford, 1894, p. xci.

213 'Probably the development of a measure-theoretic viewpoint'
T. Hawkins, *Lebesgue's theory of integration*, Chelsea, New York, 1970, p. 40.

214 'Kortum later earned some notoriety'
G. Kowalewski, *Bestand und Wandel*, Oldenbourg, Munich, 1949, p. 187.

214 'he chanced to notice in an old copy of the *Comptes Rendus*'
Comptes Rendus **XCIV** (6 February 1882), 330–1.

215 'In the Royal Society *Proceedings*' and 'In his reply, received a few days later'
H. J. S. Smith, *Collected mathematical papers*, Vol. I (edited by J. W. L. Glaisher), Oxford University Press, Oxford, 1894, p. lxvi.

215 'The prize was therefore awarded jointly
Comptes Rendus **XCVI** (2 April 1883), 879–83.

216 'Of speeches which might interest you'
H. Minkowski, *Briefe an David Hilbert*, Springer, Berlin, 1973, p. 120; Smith's talk appears in the *Journal of the London Mathematical Society* **8** (1876/7), 6–29 = H. J. S. Smith, *Collected mathematical papers*, Vol. II (edited by J. W. L. Glaisher), Oxford University Press, Oxford, 1894, pp. 166–190.

CHAPTER 13

219 'On the 3rd I went with the Spottiswoode's'
The journals of Thomas Archer Hirst FRS, p. 2128 [1883], Royal Institution, London; edited by W. H. Brock and R. M. MacLeod and published in microfiche by Mansell, London, 1980.

219 'We hope that the place left empty'
Oxford Magazine, 27 February 1883, p. 107.

219 There are no good Oxford men for it'
letter from Arthur Cayley to Felix Klein, 4 April 1883, Cod. Ms. Klein 8374, University of Göttingen.

220 'I heard with dismay and the deepest regret'
letter from J. J. Sylvester to Arthur Cayley, 16 March 1883, Sylvester Papers, St John's College, Cambridge, Box 11.

221 'Oxford is beginning to be a delightful residence to me'
letter from J. J. Sylvester to Arthur Cayley, November 1884, cited in R. C. Archibald, 'Material concerning James Joseph Sylvester', *Studies and essays offered to George Sarton*, New York, 1947, pp. 213–14.

221 'I had hoped to be appointed Profr.'
R. C. Archibald, 'Unpublished letters of James Joseph Sylvester and other new information concerning his life and work', *Osiris* **1** (1936), 85–154, at 150.

221 'Ingleby some time ago'
letter from J. J. Sylvester to Arthur Cayley, 13 December 1883, Sylvester Papers, St John's College, Cambridge, Box 11.

221 'It is curious that you should have been able'
letter from J. J. Sylvester to Barbara Bodichon, January 1884, cited in Barbara McCrimmon, 'Johns Hopkins University's second professor', *Manuscripts* **33** (1981), 173–84, at 183.

223 'thrown from one side to the other'
J. J. Sylvester, address at his farewell reception, 20 December 1883, Daniel Coit Gilmin Paper Ms. 1, Coll #1 Corresp., Box 48, Johns Hopkins University, Baltimore.

223 'I reckon our London cockney'
letter from R. L. Dabney to his mother Elizabeth, in Dabney Collection, University of Virginia, cited in Lewis S. Feuer, 'America's first Jewish professor: James Joseph Sylvester at the University of Virginia', *American Jewish Archives* **36** (1984), 152–201, at 158.

224 'Mr Cayley (to whom I am indebted …)'
J. J. Sylvester, 'On the relation between the minor determinants of linearly equivalent quadratic functions', *Philosophical Magazine* (1851), 295–305 = *The collected mathematical papers of James Joseph*

Sylvester, Vol. 1 (edited by H. F. Baker), Cambridge University Press, 1904, p. 246.

221 'in 1871 he asked Lincoln College'
V. H. H. Green *Oxford common room: a study of Lincoln College and Mark Pattison,* Edward Arnold, London, 1957, p. 272.

226 'No young man of generous mind'
Memorial address by Fabian Franklin, cited in Alexander Macfarlane *Lectures on ten British mathematicians of the nineteenth century,* John Wiley and Sons, New York, 1916, p. 117.

227 'I did not consider that my mathematical erudition'
letter from J. J. Sylvester to Felix Klein, 17 January 1884, cited in Karen H. Parshall and David E. Rowe, *The emergence of the American mathematical research community 1876–1900, J. J. Sylvester, Felix Klein, and E. H. Moore,* American Mathematical Society, Providence, and London Mathematical Society, London, 1994, p. 142.

227 'If I am not appointed to Oxford'
letter from J. J. Sylvester to Arthur Cayley, 5 September 1883, cited in Karen H. Parshall and David E. Rowe, *The emergence of the American mathematical research community 1876–1900, J. J. Sylvester, Felix Klein, and E. H. Moore,* American Mathematical Society, Providence, and London Mathematical Society, London, 1994, p. 138.

227 'the monopoly of a party', 'professed the faith', 'the blinding and blighting effect' and 'If I speak with some warmth'
J. J. Sylvester, address at Baltimore, 22 February 1877, *The collected mathematical papers of James Joseph Sylvester,* Vol. III (edited by H. F. Baker), Cambridge University Press, 1909, pp. 72–87.

229 'I think that I shall best discharge', and 'we could create such a School'
J. J. Sylvester, inaugural lecture, *Nature* 33 (1886), 222–231 = *The collected mathematical papers of James Joseph Sylvester,* Vol. IV (edited by H. F. Baker), Cambridge University Press, 1912, p. 298.

230 'Faraday, at the end of his experimental lectures'
J. J. Sylvester, 'Lectures on the theory of reciprocants', *American Journal of mathematics* 8 (1886), 196–260 = *The collected mathematical papers of James Joseph Sylvester,* Vol. IV (edited by H. F. Baker), Cambridge University Press, 1912, p. 329.

231 '*I can see him now*'
Florian Cajori, *The teaching and history of mathematics in the United States,* Washington, 1890, p. 266.

231 '*Sylvester began to lecture*'
Florian Cajori, *The teaching and history of mathematics in the United States,* Washington, 1890, p. 267.

231 'Smith had been a painstaking tutor'
Alexander Macfarlane, *Lectures on ten British mathematicians of the nineteenth century,* John Wiley & Sons, New York, 1916, p. 120.

232 'all the real power of influencing the studies' and 'It depends exclusively on the Tutors'
letter from J. J. Sylvester to Daniel Coit Gilman, 11 March 1887, Karen Hunger Parshall, *James Joseph Sylvester: life and work in letters,* Clarendon Press, Oxford, p. 263.

232 'Here in Oxford I am fortunate'
letter from J. J. Sylvester to R. F. Scott, 22 October 1888, Sylvester Papers, St John's College, Cambridge, Box 1.

232 'Entre nous this university'
letter from J. J. Sylvester to Daniel Coit Gilman, 11 March 1887, Karen Hunger Parshall, *James Joseph Sylvester: life and work in letters,* Clarendon Press, Oxford, 1998, p. 264.

233 '*We thought, too exclusively probably*'
E. B. Elliott, *Why and how the Society began and kept going,* lecture to Oxford Mathematical and Physical Society, 16 May 1925, pp. 8–9.

233 'Wherever Dr Sylvester goes'
A. Cayley, 'Scientific worthies. xxv.—James Joseph Sylvester', *Nature* **39**, 3 January 1889, p. 217.

234 'Sylvester once expatiated to us'
E. B. Elliott, *Why and how the Society began and kept going,* lecture to Oxford Mathematical and Physical Society, 16 May 1925, pp. 14–15.

235 'Perhaps the most valuable direct outcome'
Oxford Magazine, 5 May 1897, 293–4.

235 'Sylvester joined me at my dinner'
The Journals of Thomas Archer Hirst FRS, p. 2203 [5 July 1885]. Royal Institution, London; edited by W. H. Brock and R. M. MacLeod and published in microfiche by Mansell, London, 1980.

237 'As we went up a delightful staircase'
Pauline Adams, *Somerville for women: an Oxford College 1879–1993,* Oxford University Press, 1996, p. 112.

238 'I expect Poincarre [sic] tomorrow'
Karen Hunger Parshall, *James Joseph Sylvester: life and work in letters,* Clarendon Press, Oxford, 1998, p. 280.

238 'Dr Hirst or Henrici or Esson'
letter from J. J. Sylvester to E. M. Langley, 29 Sept. 1891, R. C. Archibald, 'Unpublished letters of James Joseph Sylvester and other new information

concerning his life and work', *Osiris* **1** (1936), 85–154, at 152–3.

238 'Sylvester is too irregular'
letter from Felix Klein to David Hilbert, 25 December 1889, in Gert Sabidussi, '*Correspondence between Sylvester, Peterson, Hilbert and Klein on invariants and the factorisation of graphs 1889–1891*', *Discrete mathematics* **100** (1992), 99–155.

238 'Since that last mental effort'
letter from J. J. Sylvester to Julius Petersen, 15 June 1890, in Gert Sabidussi, '*Correspondence between Sylvester, Peterson, Hilbert and Klein on invariants and the factorisation of graphs 1889-1891*, Odense, 1990, p. 40. *Discrete mathematics* **100** (1992), 99–155.

239 'there is no one I can think of'
J. J. Sylvester, inaugural lecture, *Nature* **33** (1886), 222–31 = *The collected mathematical papers of James Joseph Sylvester*, Vol. IV (edited by H. F. Baker), Cambridge University Press, 1912, p. 300.

CHAPTER 14

242 'Elliott, the great man on Algebra'
letter from William Ferrar to his Uncle Fred, 1913.

242 'Elliott's strength lay in exactness of detail'
Dictionary of national biography, 1931–1940, Oxford University Press, 1949, pp. 257–8.

243 'His whimsicality of manner and appearance'
Dictionary of national biography, 1931–1940, Oxford University Press, 1949, pp. 245–6.

245 'knowing some very pretty stuff'
verbal recollection by E. C. Thompson.

245 'my eyes were first opened'
G. H. Hardy, *A mathematician's apology*, Cambridge University Press, 1940, p. 147.

246 'Mathematicians are reasonably cheap'
G. H. Hardy, *The Oxford Magazine*, 5 June 1930, pp. 819–21.

247 'At our first meeting, he told me'
private communication from K. C. Bowen.

CHAPTER 15

258 'You might object ... that geometry is'
'What is geometry?' (presidential address to the Mathematical Association, 1925), *Mathematical Gazette* **12**, no. 3 (1925), 309–16; reprinted in *Collected papers of G. H. Hardy*, Vol. 7 (edited by I. W. Busbridge and R. A. Rankin), Clarendon Press, Oxford, 1979, pp. 519–26.

258 '*A curious imaginary curve.*'
Mathematical Gazette **4**, no. 9 (1907), 14; reprinted in *Mathematical Gazette* **55** (1971), 221; reprinted in *Collected papers of G. H. Hardy*, Vol. VII (edited by I. W. Busbridge and R. A. Rankin), Clarendon Press, Oxford, 1979, p. 483.

262 '... physics will become more mathematical'
J. L. Synge, *Science: sense and nonsense*, Jonathan Cape, London, 1951, p. 80.

262 'The use of applied mathematics'
J. L. Synge, *Science: sense and nonsense*, Jonathan Cape, London, 1951, p. 98.

A large number of people have helped in the preparation of this book, in a variety of ways. The authors and editors are most grateful for the support, help and advice of Vernon Armitage, Simon Bailey (Archivist, University of Oxford), Penny Bulloch, Tony Crilly, Caroline Dalton (Archivist, New College, Oxford), the late Peter Hinchliff, Peter Hingley (Librarian, Royal Astronomical Society), H. J. R. Wing (Librarian, Christ Church College, Oxford), Ioan James, John Jones, George Molland, Vanda Morton, Peter Neumann, Karen Hunger Parshall, John Prest, Elizabeth Quarmby (Special Collections, St John's College, Cambridge), the late Vincent Quinn, David E. Rowe, Tony Simcock (Librarian, Museum of the History of Science), the late Euan Squires, Jackie Stedall, Alan Tadiello, Gerard L'E. Turner, and Rachel Weiss.

Picture credits

Our thanks are due to the following sources for images: Aberdeen University; the Adler Planetarium, Chicago; All Souls College, Oxford; the Ashmolean Museum, Oxford; Badischen Landesbibliothek, Karlsruhe; Balliol College, Oxford; Birkhäuser Publications; the Bodleian Library, Oxford [1393d.23: 183 h.13: 249 y.69–71: Adams 2.68.2: B 1.10 Art Seld: Buss B.103: Douce L.subt.27: GA Oxon 109(b): GA Oxon 414 4: Godw 8 394: Ms Minn 202/6: Savile B.13: Savile.MS 011]; the British Library; the British Museum; by permission of the Syndics of Cambridge University Library; courtesy of Professor Sir Michael Atiyah, Mrs E. Coulson, Mr John Fauvel, Dr Raymond Flood, Dr Keith Hannabuss, Professor I. M. James, Professor A. Kosinski, Dr Vanda Morton, Dr Roger Norris, Professor J. D. North, Professor D. C. Pack, Professor Sir Roger Penrose, Professor Robert Rankin, Dr Arthur Stonebridge, Dr Robin Wilson; the Curators of the Examination Schools, Oxford; the Department of Astrophysics, Oxford University; Dover Publications, New York; Green College, Oxford; Guildhall Library, Corporation of London; Houghton Library, Harvard University; *The Illustrated London News*; Institut Mittag-Leffler; the Institute of Mechanical Engineers; the Oxford University Invariant Society; the London Mathematical Society; the Master and Fellows of Trinity College, Cambridge; the Mathematical Institute, University of Oxford; the Museum of the History of Science, University of Oxford; the National Maritime Museum, Greenwich; the Oxford Historical Society; the Oxford University Press; Peterhouse College, Cambridge; the President and Fellows of Corpus Christi College, Oxford; the Royal

Geographic Society; the Royal Society, London; the Science Museum, London; Somerville College, Oxford; St Hugh's College, Oxford; St Lawrence's Church, Reading; Thames and Hudson Publishers, London; The Queen's College, Oxford; Thomas Photos, Oxford; Trinity College, Oxford; Universitätsbibliothek, Göttingen; the Verger, Tingewick Church; the Warden and Fellows of Merton College, Oxford; the Warden and Fellows of Wadham College, Oxford. Every effort has been made to trace all copyright holders, but if any has been inadvertently overlooked, the publishers will be pleased to make the necessary arrangements at the earliest opportunity.